IUTAM Bookseries

Volume 39

The IUTAM Bookseries publishes the refereed proceedings of symposia organised by the International Union of Theoretical and Applied Mechanics (IUTAM).

Every two years the IUTAM General Assembly decides on the list of IUTAM Symposia. The Assembly calls upon the advice of the Symposia panels. Proposals for Symposia are made through the Assembly members, the Adhering Organizations, and the Affiliated Organizations, and are submitted online when a call is launched on the IUTAM website.

The IUTAM Symposia are reserved to invited participants. Those wishing to participate in an IUTAM Symposium are therefore advised to contact the Chairman of the Scientific Committee in due time in advance of the meeting. From 1996 to 2010, Kluwer Academic Publishers, now Springer, was the preferred publisher of the refereed proceedings of the IUTAM Symposia. Proceedings have also been published as special issues of appropriate journals. From 2018, this bookseries is again recommended by IUTAM for publication of Symposia proceedings.

Indexed in Ei Compendex and Scopus.

More information about this series at https://link.springer.com/bookseries/7695

Jukka Tuhkuri · Arttu Polojärvi
Editors

IUTAM Symposium on Physics and Mechanics of Sea Ice

Proceedings of the IUTAM Symposium held at Aalto University, Espoo, Finland, 3–9 June 2019

Springer

Editors
Jukka Tuhkuri
Department of Mechanical Engineering
Aalto University
Espoo, Finland

Arttu Polojärvi
Department of Mechanical Engineering
Aalto University
Espoo, Finland

ISSN 1875-3507 ISSN 1875-3493 (electronic)
IUTAM Bookseries
ISBN 978-3-030-80438-1 ISBN 978-3-030-80439-8 (eBook)
https://doi.org/10.1007/978-3-030-80439-8

This Springer imprint is published by the registered company Springer Nature Switzerland AG
The registered company address is: Gewerbestrasse 11, 6330 Cham, Switzerland

Preface

The *IUTAM Symposium on Physics and Mechanics of Sea Ice* was held on June 3–7, 2019 at Aalto University in Espoo, Finland. The symposium was the fourth *IUTAM Symposium on Physics and Mechanics of Sea Ice*. The previous ones have been held in 1979 (Denmark), 1989 (Canada) and 2000 (USA). With global warming dramatically changing the sea ice environment, and with an increase in activities in ice covered seas, it was very timely to have an IUTAM symposium on sea ice again.

The objective of the symposium was to bring together scientist who have made significant contributions in the study of sea ice to discuss the recent achievements and ideas for future work. The symposium focused on the following five topics: (1) fracture of ice, (2) thermodynamics of sea ice ridges, (3) global and local ice loads on ships and marine structures, (4) computational ice engineering and ice mechanics, and (5) physical and engineering problems related to ice and waves. About five presentations were given on each topic and the presentations reflected the different perspectives in the research on mechanics of sea ice. This book includes 17 papers based on the talks. All the papers have been reviewed, mostly by the symposium participants.

Fracture of ice has been studied for decades but remains a challenge. We are still struggling to understand the effects of temperature, loading rate, and scale. However, advancements have been achieved in understanding the time dependent fracture of ice, small-scale crushing, strength of ice under cyclic loading, and ice-ice friction in different scales. Still, high quality experiments both in laboratories and in the field are needed to give us answers in these fundamental questions.

Both the mechanical and thermal processes related to sea ice ridges are complicated. Ridges act as energy sinks in large-scale sea ice models and impose large loads on Arctic ships and offshore structures. After forming, ridges consolidate - during spring and summer, ridges decay. The recent work includes experimental studies on the temporal development of ridge properties, including porosity, heat and salt fluxes in ridges, freeze bonds, and numerical modelling of ridge loads on cylindrical marine structures.

Sea ice research is important for naval architects and marine engineers, as the loads from sea ice are large and often dominate the structural design. Although a large number of measurements have been conducted, and we know that the ice-structure contact is concentrated on a small area, we still do not know how the ship velocity and the ice thickness affect the ice load on a ship; we do not know well enough the link between ice loads and ice conditions. Numerical simulations are getting more and more detailed, interesting models are under development and are used both in research and in engineering work. The recent advancements include multiscale modelling of ice strength, fast 3D simulations using physically based modelling, inclusion of hydrodynamics in the models (still a major challenge), and the use of simulations to create data for statistical ice load studies.

One of the main goals of the symposium was to foster discussions on wave-ice interaction. There is a wealth of literature on analytic methods in wave-ice interaction from the oceanographic perspective and, with global warming, the importance of the topic increasing. However, our knowledge on coupled wave-ice effects on ships and offshore structures is much more limited. It is believed that a close collaboration between disciplines could provide research results that are significant both for oceanography and engineering. In general, in order to understand wave-ice dynamics, we need to understand what the ice cover does to the waves and what the waves do to the ice cover. In more detail, the current research includes studies on floe-floe collision induced wave attenuation, rheologies for marginal ice zone, and viscous effects in wave-ice dynamics. It was discussed if it is worth trying to reach a perfect model, to study the fundamental physics, or to admit that different models are needed to answer different questions. The question remains open.

The symposium program was planned, and the presentations selected, by the Scientific Committee of the symposium nominated by IUTAM. The members of the Scientific Committee were: Jukka Tuhkuri, Aalto University, Finland (chair); John Dempsey, Clarkson University, USA; Robert Gagnon, National Research Council, Canada; Sveinung Løset, Norwegian University of Science and Technology, Norway; Peter Sammonds, University College London, UK; Vernon Squire, University of Otago, New Zealand; Hajime Yamaguchi, University of Tokyo, Japan; and Robert McMeeking, University of California Santa Barbara (IUTAM representative). The practical arrangements of the symposium were taken care by the Local Organising Committee: Hanyang Gong, Ida Lemström, Arttu Polojärvi and Jukka Tuhkuri.

The symposium had in total 42 participants from 13 countries. The speakers represented Canada, Finland, Japan, The Netherlands, Norway, Poland, Russia, UK, USA; the other participants represented, in addition, China, Lebanon, Pakistan, and Sri Lanka.

The symposium was financially supported by IUTAM, the Federation of Finnish Learned Societies, City of Espoo, and Aalto University. The financial support is gratefully acknowledged.

Espoo, Finland
January 2021

Jukka Tuhkuri
Arttu Polojärvi

Photo: Adolfo Vera

Contents

Part I
Ice Mechanics and Ice Fracture

Chapter 1
Mechanisms of Cyclic Strengthening and Recovery of Polycrystalline Ice

Erland M. Schulson, Andrii Murdza, and Carl E. Renshaw

Abstract Strengthening and subsequent recovery of relatively warm polycrystalline ice cyclically loaded is attributed to a dynamic competition between the buildup of an internal back stress that originates either from dislocation pileups or grain boundary sliding and a process of creep-driven stress relaxation.

Keywords Flexural strength · Strengthening upon cyclic loading · Dynamic recovery

1.1 Introduction

There is now experimental evidence that the strength of relatively warm, polycrystalline ice, whether measured under tension, under compression or by flexing, increases upon cyclic loading. For instance, (Cole 1990) found through direct tension–compression tests at −10 °C (0.96 homologous temperature) on equiaxed and randomly oriented polycrystalline aggregates of freshwater ice of 2.5 mm grain size that the tensile strength more than doubled upon reverse cycling at 0.1 Hz. Similarly, (Picu and Gupta 1995) and (Iliescu and Schulson 2002) found through compression tests at the same temperature on columnar-grained freshwater ice of ~ 5 mm grain diameter (that possessed the S2 growth texture in which the crystallographic c-axes of the ice Ih grains were randomly oriented in a plane perpendicular to the long axis of the grains) that the across-column, brittle compressive strength of material moderately confined across the columns increased by a factor of ~ 1.5 upon cyclic loading. And more recently, (Iliescu et al. 2017) and (Murdza 2019) found, again from experiments on S2 columnar-grained ice loaded across the columns at −10 °C, that the flexural strength increased by a factor of two or more upon cyclic loading at 0.1 Hz.

In each of these instances, strengthening was attributed to the development of an internal back stress that opposes the initiation of micro-cracks, thereby making

E. M. Schulson (✉) · A. Murdza · C. E. Renshaw
Thayer School of Engineering, Dartmouth College, Hanover, NH 03755, USA
e-mail: erland.m.schulson@dartmouth.edu

J. Tuhkuri and A. Polojärvi (eds.), *IUTAM Symposium on Physics and Mechanics of Sea Ice*, IUTAM Bookseries 39, https://doi.org/10.1007/978-3-030-80439-8_1

necessary the application of a greater applied stress to trigger the failure process. Given that internal stress is thermodynamically unstable, the implication of this explanation is that cyclic-induced strength is recoverable, a hint of which is evident in the observations of (Iliescu et al. 2017) and now confirmed through systematic experiments by (Murdza 2019).

1.2 Principal Observations of Strengthening and Weakening

To set the scene, Fig. 1.1 shows the effect of stress amplitude on the flexural strength of S2 columnar-grained freshwater ice and saline ice that had been subjected to ~ 300 or more cycles of four-point reverse bending under load control at −10 °C at 0.1 Hz to; i.e., to enough cycles to impart a leveling off of strength (Murdza 2019). The flexural strength σ_{fc} increases linearly with stress amplitude σ_a and may be described by the relationship:

$$\sigma_{fc} = \sigma_{fo} + k\sigma_a \tag{1.1}$$

where, for the freshwater ice, the non-cycled strength $\sigma_{fo} = 1.67 \pm 0.22$ MPa, in agreement with (Timco and O'Brien 1994) average value of 1.73 ± 0.25 MPa, and where $k = 0.68$ under the conditions of the experiments. Saline ice (3.0 ± 0.9 ppt

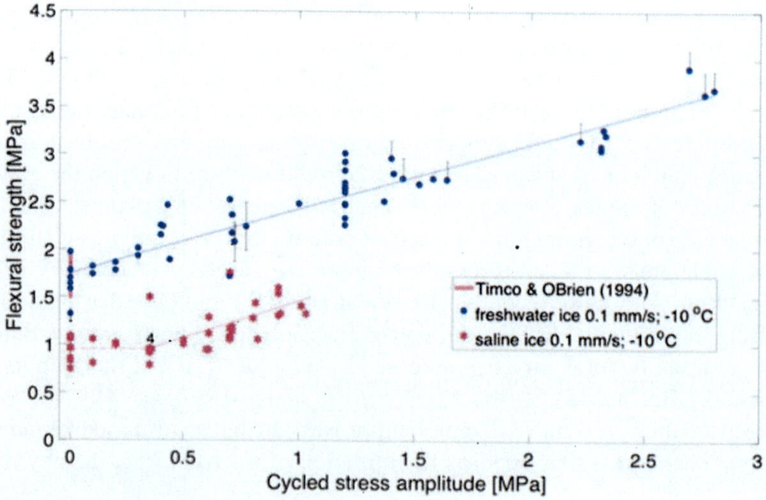

Fig. 1.1 Flexural strength of columnar-grained S2 ice loaded across the columns by cycling under load control at −10 °C at 0.1 Hz, at a rate of center-point displacement of 0.1 mm s^{-1}. (Murdza 2019)

salinity) of the same S2 texture shows similar behavior, although the non-cycled strength is lower by about a factor of two owing to the stress concentrating effect of brine pockets. To obtain these results, the amplitude of the cycled stress was gradually increased to the level noted, as detailed elsewhere (Murdza 2019), strengthening the ice in the process and thus accounting for amplitudes greater than the strength of non-cycled ice.

Figure 1.2 shows the recovery of the freshwater ice after having been cyclically strengthened under an amplitude of $\sigma_a = 2.5$ MPa and then unloaded and annealed for up to 88 h at either −10 or −25 °C before being bent to failure. The strength is expressed in terms of its relative retention $f(t)$, defined as:

$$f(t) = \frac{\sigma_{fa} - \sigma_{fo}}{\sigma_{fc} - \sigma_{fo}} \tag{1.2}$$

where σ_{fa} denotes the strength that is retained upon annealing. Recovery is essentially complete *(f(t) = 0)* after about two days. Similar behavior is found for ice that had been strengthened under lower stress amplitudes of 0.7, 2.3, and 2.5 MPa and under a higher amplitude of 2.7 MPa. Analysis by (Murdza 2019) of the recovery kinetics in terms of stress-relaxation theory that is based upon the rate of power-law or dislocation creep $\dot{\varepsilon} = B(T)\sigma_{fa}^n$, where B is a temperature dependent materials constant and n denotes the sensitivity of creep to applied stress, yielded the values B (−10 °C) = 4.0×10^{-25} and B(−25 °C) = 2.4×10^{-26} Pa^{-n} s^{-1} and n(−10 °C) = 2.7 and n(−25 °C) = 2.9. These value compare favorably with the values B(−10 °C) = 1.5×10^{-25} and B(−25 °C) = 1.7×10^{-26} Pa^{-n} s^{-1} and $n = 3.0$ at both temperatures

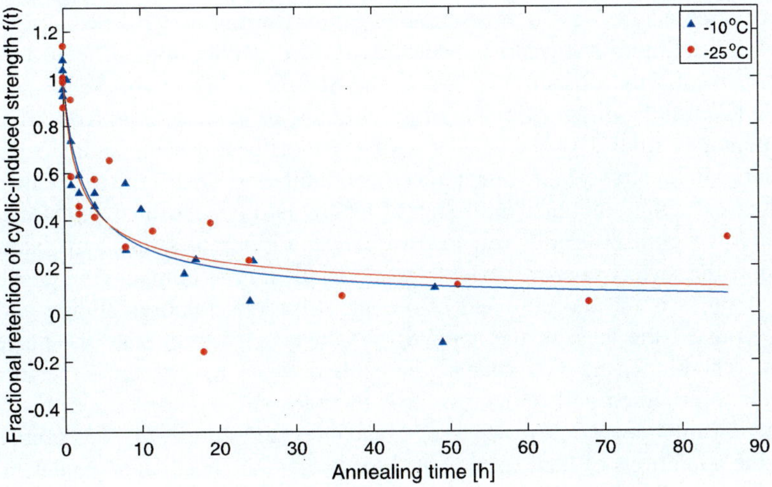

Fig. 1.2 Fraction of flexural strength retained upon annealing (Murdza 2019)

that were obtained by (Barnes et al. 1971) from creep experiments on freshwater ice. In other words, the kinetics underlying recovery appear to be similar to the kinetics of dislocation creep.

1.3 Physical Mechanisms

To understand this behavior, it is important to recognize two points: that flexural strength is governed by tensile strength, although greater by a factor of about 1.7 (Ashby and Jones 2013); and tensile strength is governed either by the stress to nucleate a crack or by the stress to propagate a crack (Schulson and Duval 2009) (and references therein). In the present instances, crack nucleation governs, at least in freshwater ice. (Saline ice too opaque to discern cracks.) This is evident from the fact that broken plates of freshwater ice contain no remnant cracks, just as cylindrical specimens of granular freshwater ice when broken under direct tension at −10 °C at a strain rate of 10^{-3} s^{-1} do not contain remnant cracks (Schulson and Duval 2009); see Fig. 1.4 of Schulson (1987). So, why and how do cracks nucleate in ice Ih and what accounts for the linear dependence of flexural strength on amplitude of the cyclic stress?

Cracks nucleate to relieve stress concentration that develops within the vicinity of grain boundaries (Gold 1972) following the imposition of a small amount of inelastic deformation. The process involves either grain boundary sliding or the piling up of dislocations against other boundaries following glide on the basal slip planes. Evidence exists for both mechanisms. From cyclic loading experiments on S2 freshwater ice, (Iliescu et al. 2017) and (Murdza 2019) observed grain boundary features oriented at ~ 45° to the direction of maximum principal stress—i.e., along planes of maximum shear stress—features termed decohesions. The features are reminiscent of those observed by Weiss and Schulson (Weiss and Schulson 2000) and are taken to be direct evidence of grain boundary sliding. And from in-situ X-ray topography studies, (Liu et al. 1995) observed dislocations emanating from grain boundaries in ice and forming pileups at other boundaries. Thus, both mechanisms are possible, and at this juncture it is difficult to distinguish between the two. The common factor is that both processes can lead to concentrations of stress high enough to overcome the surface energy barrier to crack nucleation. Accompanying the buildup of local stress is the development of a back stress that opposes sliding and slip, thereby raising the level of the applied stress necessary to operate the nucleation process. The higher the amplitude of the applied stress, the greater is the attendant inelastic deformation and so the greater is the back stress. Based upon this model, the relationship between flexural strength and stress amplitude, Eq. 1.1, implies that under the conditions of the experiments noted above the back stress scales linearly with stress amplitude.

Recovery opposes strengthening. At the same time that cycling is strengthening the ice through the development of an internal back stress, the internal stress tends to relax. A dynamic competition thus exists between strengthening and recovery. The fact that the parameters B and n that describe the kinetics of recovery are close in magnitude to values derived from creep experiments on polycrystalline ice at the same temperatures might seem to argue in favor of the pileup mechanism being the more important, for at temperatures as high as those under which the above experiments were performed (>0.9 homologous temperature) one can imagine dislocations climbing out of such arrangements and, in the process, reducing the magnitude of the back stress.

We recognize that recovery could originate not only from the rearrangement of a thermodynamically unstable dislocation array or from stress relaxation at a triple point, but also from a different restorative process. Dynamic recrystallization is one such process, but is one that we are inclined to rule out because examination of thin sections of cycled material did not reveal any evidence of new grains (Murdza 2019).

An implication of this interpretation is that ice poorly oriented for basal slip and/or free from grain boundaries may not harden as much as noted here. Another is that recovery is expected to occur more slowly in colder ice, although the scatter in the results and the relatively small difference in homologous temperature in the experiments mentioned above (Fig. 1.2) prevent comment on this point. The other implication is that cycling under conditions where recovery dominates hardening, either little or no cyclic strengthening is expected.

1.4 Conclusion

The strengthening and subsequent recovery of relatively warm polycrystalline ice favorably oriented for slip on basal planes and subjected to slow cyclic loading is attributed to a dynamic competition between the buildup of an internal back stress that originates from either dislocation pileups or grain boundary sliding and a process of creep-driven stress relaxation.

References

D.M. Cole, Reversed direct-stress testing of ice: initial experimental results and analysis. Cold Reg. Sci. Tech. **18**, 303–321 (1990)

R.C. Picu, V. Gupta, Crack nucleation in columnar ice due to elastic anisotropy and grain boundary sliding. Acta Metall. Mater. **43**, 3783–3789 (1995)

D. Iliescu, E.M. Schulson, Brittle compressive failure of ice: monotonic versus cyclic loading. Acta Mater. **50**, 2163–2172 (2002)

D. Iliescu, A. Murdza, E.M. Schulson, C.E. Renshaw, Strengthening ice through cyclic loading. J. Glac. **63**(240), 663–669 (2017)

A. Murdza PhD thesis, Dartmouth College, Hanover, NH, USA, (2019)

G.W. Timco, S. O'Brien, Flexural strength equation for sea ice. Cold Reg. Sci. Tech. **22**(3), 285–298 (1994)

P. Barnes, D. Tabor, J.C.F. Walker, The friction and creep of polycrystalline ice. Proc. Roy. Soc. Lond. A **324**, 127–155 (1971)

M.F. Ashby, D.R.H. Jones, *Engineering Materials 2: An Introduction to Microstructures and Processing* (Elsevier/ Butterworths-Heinemann, Oxford, 2013)
E.M. Schulson, P. Duval, *Creep and Fracture of Ice* (Cambridge Univ, Press, 2009)
E.M. Schulson (1987) The fracture of ice Ih. J. de Phys. Coll. C1 Suppl. **48**, C1–207-C1–220
L.W. Gold, The process of failure of columnar-grained ice. Phil. Mag. **26**, 311–328 (1972)
J. Weiss, E.M. Schulson, Grain-boundary sliding and crack nucleation in ice. Phil. Mag. **A80**(2), 279–300 (2000)
F. Liu, I. Baker, M. Dudley Dislocation-grain boundary interactions in ice crystals. Phil. Mag. A **71**, 15–42 (1995)

Chapter 1
Mechanisms of Cyclic Strengthening and Recovery of Polycrystalline Ice

Erland M. Schulson, Andrii Murdza, and Carl E. Renshaw

Abstract Strengthening and subsequent recovery of relatively warm polycrystalline ice cyclically loaded is attributed to a dynamic competition between the buildup of an internal back stress that originates either from dislocation pileups or grain boundary sliding and a process of creep-driven stress relaxation.

Keywords Flexural strength · Strengthening upon cyclic loading · Dynamic recovery

1.1 Introduction

There is now experimental evidence that the strength of relatively warm, polycrystalline ice, whether measured under tension, under compression or by flexing, increases upon cyclic loading. For instance, (Cole 1990) found through direct tension–compression tests at −10 °C (0.96 homologous temperature) on equiaxed and randomly oriented polycrystalline aggregates of freshwater ice of 2.5 mm grain size that the tensile strength more than doubled upon reverse cycling at 0.1 Hz. Similarly, (Picu and Gupta 1995) and (Iliescu and Schulson 2002) found through compression tests at the same temperature on columnar-grained freshwater ice of ~ 5 mm grain diameter (that possessed the S2 growth texture in which the crystallographic c-axes of the ice Ih grains were randomly oriented in a plane perpendicular to the long axis of the grains) that the across-column, brittle compressive strength of material moderately confined across the columns increased by a factor of ~ 1.5 upon cyclic loading. And more recently, (Iliescu et al. 2017) and (Murdza 2019) found, again from experiments on S2 columnar-grained ice loaded across the columns at −10 °C, that the flexural strength increased by a factor of two or more upon cyclic loading at 0.1 Hz.

In each of these instances, strengthening was attributed to the development of an internal back stress that opposes the initiation of micro-cracks, thereby making

E. M. Schulson (✉) · A. Murdza · C. E. Renshaw
Thayer School of Engineering, Dartmouth College, Hanover, NH 03755, USA
e-mail: erland.m.schulson@dartmouth.edu

J. Tuhkuri and A. Polojärvi (eds.), *IUTAM Symposium on Physics and Mechanics of Sea Ice*, IUTAM Bookseries 39, https://doi.org/10.1007/978-3-030-80439-8_1

necessary the application of a greater applied stress to trigger the failure process. Given that internal stress is thermodynamically unstable, the implication of this explanation is that cyclic-induced strength is recoverable, a hint of which is evident in the observations of (Iliescu et al. 2017) and now confirmed through systematic experiments by (Murdza 2019).

1.2 Principal Observations of Strengthening and Weakening

To set the scene, Fig. 1.1 shows the effect of stress amplitude on the flexural strength of S2 columnar-grained freshwater ice and saline ice that had been subjected to ~ 300 or more cycles of four-point reverse bending under load control at −10 °C at 0.1 Hz to; i.e., to enough cycles to impart a leveling off of strength (Murdza 2019). The flexural strength σ_{fc} increases linearly with stress amplitude σ_a and may be described by the relationship:

$$\sigma_{fc} = \sigma_{fo} + k\sigma_a \tag{1.1}$$

where, for the freshwater ice, the non-cycled strength $\sigma_{fo} = 1.67 \pm 0.22$ MPa, in agreement with (Timco and O'Brien 1994) average value of 1.73 ± 0.25 MPa, and where $k = 0.68$ under the conditions of the experiments. Saline ice (3.0 ± 0.9 ppt

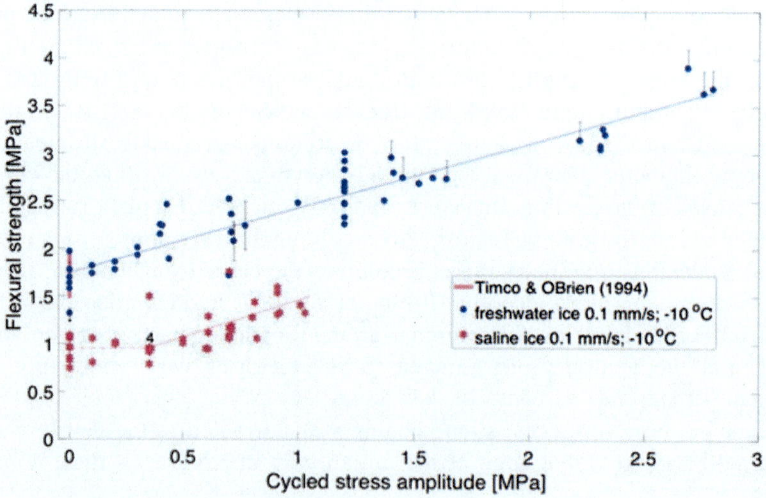

Fig. 1.1 Flexural strength of columnar-grained S2 ice loaded across the columns by cycling under load control at −10 °C at 0.1 Hz, at a rate of center-point displacement of 0.1 mm s^{-1}. (Murdza 2019)

salinity) of the same S2 texture shows similar behavior, although the non-cycled strength is lower by about a factor of two owing to the stress concentrating effect of brine pockets. To obtain these results, the amplitude of the cycled stress was gradually increased to the level noted, as detailed elsewhere (Murdza 2019), strengthening the ice in the process and thus accounting for amplitudes greater than the strength of non-cycled ice.

Figure 1.2 shows the recovery of the freshwater ice after having been cyclically strengthened under an amplitude of $\sigma_a = 2.5$ MPa and then unloaded and annealed for up to 88 h at either -10 or -25 °C before being bent to failure. The strength is expressed in terms of its relative retention $f(t)$, defined as:

$$f(t) = \frac{\sigma_{fa} - \sigma_{fo}}{\sigma_{fc} - \sigma_{fo}} \tag{1.2}$$

where σ_{fa} denotes the strength that is retained upon annealing. Recovery is essentially complete *(f(t)* = *0)* after about two days. Similar behavior is found for ice that had been strengthened under lower stress amplitudes of 0.7, 2.3, and 2.5 MPa and under a higher amplitude of 2.7 MPa. Analysis by (Murdza 2019) of the recovery kinetics in terms of stress-relaxation theory that is based upon the rate of power-law or dislocation creep $\dot{\varepsilon} = B(T)\sigma_{fa}^n$, where B is a temperature dependent materials constant and n denotes the sensitivity of creep to applied stress, yielded the values B (-10 °C) = 4.0×10^{-25} and $B(-25\ °\text{C}) = 2.4 \times 10^{-26}$ Pa^{-n} s^{-1} and $n(-10\ °\text{C}) = 2.7$ and $n(-25\ °\text{C}) = 2.9$. These value compare favorably with the values $B(-10\ °\text{C}) = 1.5 \times 10^{-25}$ and $B(-25\ °\text{C}) = 1.7 \times 10^{-26}$ Pa^{-n} s^{-1} and $n = 3.0$ at both temperatures

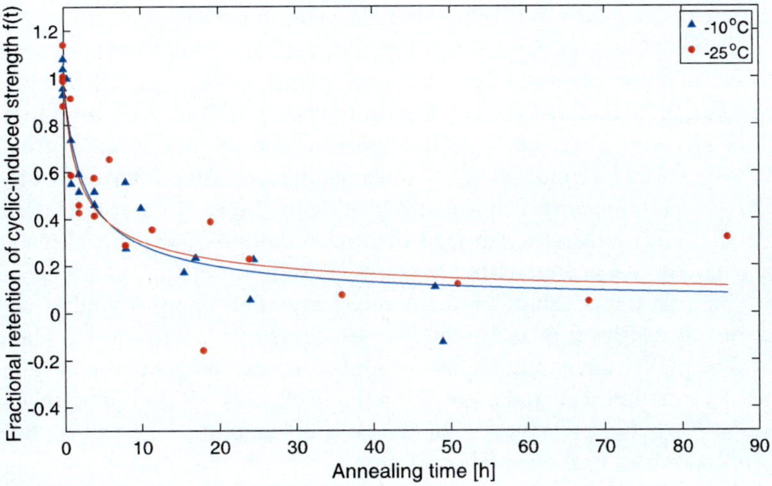

Fig. 1.2 Fraction of flexural strength retained upon annealing (Murdza 2019)

that were obtained by (Barnes et al. 1971) from creep experiments on freshwater ice. In other words, the kinetics underlying recovery appear to be similar to the kinetics of dislocation creep.

1.3 Physical Mechanisms

To understand this behavior, it is important to recognize two points: that flexural strength is governed by tensile strength, although greater by a factor of about 1.7 (Ashby and Jones 2013); and tensile strength is governed either by the stress to nucleate a crack or by the stress to propagate a crack (Schulson and Duval 2009) (and references therein). In the present instances, crack nucleation governs, at least in freshwater ice. (Saline ice too opaque to discern cracks.) This is evident from the fact that broken plates of freshwater ice contain no remnant cracks, just as cylindrical specimens of granular freshwater ice when broken under direct tension at -10 °C at a strain rate of 10^{-3} s^{-1} do not contain remnant cracks (Schulson and Duval 2009); see Fig. 1.4 of Schulson (1987). So, why and how do cracks nucleate in ice Ih and what accounts for the linear dependence of flexural strength on amplitude of the cyclic stress?

Cracks nucleate to relieve stress concentration that develops within the vicinity of grain boundaries (Gold 1972) following the imposition of a small amount of inelastic deformation. The process involves either grain boundary sliding or the piling up of dislocations against other boundaries following glide on the basal slip planes. Evidence exists for both mechanisms. From cyclic loading experiments on S2 freshwater ice, (Iliescu et al. 2017) and (Murdza 2019) observed grain boundary features oriented at ~ 45° to the direction of maximum principal stress—i.e., along planes of maximum shear stress—features termed decohesions. The features are reminiscent of those observed by Weiss and Schulson (Weiss and Schulson 2000) and are taken to be direct evidence of grain boundary sliding. And from in-situ X-ray topography studies, (Liu et al. 1995) observed dislocations emanating from grain boundaries in ice and forming pileups at other boundaries. Thus, both mechanisms are possible, and at this juncture it is difficult to distinguish between the two. The common factor is that both processes can lead to concentrations of stress high enough to overcome the surface energy barrier to crack nucleation. Accompanying the buildup of local stress is the development of a back stress that opposes sliding and slip, thereby raising the level of the applied stress necessary to operate the nucleation process. The higher the amplitude of the applied stress, the greater is the attendant inelastic deformation and so the greater is the back stress. Based upon this model, the relationship between flexural strength and stress amplitude, Eq. 1.1, implies that under the conditions of the experiments noted above the back stress scales linearly with stress amplitude.

Recovery opposes strengthening. At the same time that cycling is strengthening the ice through the development of an internal back stress, the internal stress tends to relax. A dynamic competition thus exists between strengthening and recovery. The fact that the parameters *B* and *n* that describe the kinetics of recovery are close in magnitude to values derived from creep experiments on polycrystalline ice at the same temperatures might seem to argue in favor of the pileup mechanism being the more important, for at temperatures as high as those under which the above experiments were performed (>0.9 homologous temperature) one can imagine dislocations climbing out of such arrangements and, in the process, reducing the magnitude of the back stress.

We recognize that recovery could originate not only from the rearrangement of a thermodynamically unstable dislocation array or from stress relaxation at a triple point, but also from a different restorative process. Dynamic recrystallization is one such process, but is one that we are inclined to rule out because examination of thin sections of cycled material did not reveal any evidence of new grains (Murdza 2019).

An implication of this interpretation is that ice poorly oriented for basal slip and/or free from grain boundaries may not harden as much as noted here. Another is that recovery is expected to occur more slowly in colder ice, although the scatter in the results and the relatively small difference in homologous temperature in the experiments mentioned above (Fig. 1.2) prevent comment on this point. The other implication is that cycling under conditions where recovery dominates hardening, either little or no cyclic strengthening is expected.

1.4 Conclusion

The strengthening and subsequent recovery of relatively warm polycrystalline ice favorably oriented for slip on basal planes and subjected to slow cyclic loading is attributed to a dynamic competition between the buildup of an internal back stress that originates from either dislocation pileups or grain boundary sliding and a process of creep-driven stress relaxation.

References

D.M. Cole, Reversed direct-stress testing of ice: initial experimental results and analysis. Cold Reg. Sci. Tech. **18**, 303–321 (1990)

R.C. Picu, V. Gupta, Crack nucleation in columnar ice due to elastic anisotropy and grain boundary sliding. Acta Metall. Mater. **43**, 3783–3789 (1995)

D. Iliescu, E.M. Schulson, Brittle compressive failure of ice: monotonic versus cyclic loading. Acta Mater. **50**, 2163–2172 (2002)

D. Iliescu, A. Murdza, E.M. Schulson, C.E. Renshaw, Strengthening ice through cyclic loading. J. Glac. **63**(240), 663–669 (2017)

A. Murdza PhD thesis, Dartmouth College, Hanover, NH, USA, (2019)

G.W. Timco, S. O'Brien, Flexural strength equation for sea ice. Cold Reg. Sci. Tech. **22**(3), 285–298 (1994)

P. Barnes, D. Tabor, J.C.F. Walker, The friction and creep of polycrystalline ice. Proc. Roy. Soc. Lond. A **324**, 127–155 (1971)

M.F. Ashby, D.R.H. Jones, *Engineering Materials 2: An Introduction to Microstructures and Processing* (Elsevier/ Butterworths-Heinemann, Oxford, 2013)

E.M. Schulson, P. Duval, *Creep and Fracture of Ice* (Cambridge Univ, Press, 2009)

E.M. Schulson (1987) The fracture of ice Ih. J. de Phys. Coll. C1 Suppl. **48**, C1–207-C1–220

L.W. Gold, The process of failure of columnar-grained ice. Phil. Mag. **26**, 311–328 (1972)

J. Weiss, E.M. Schulson, Grain-boundary sliding and crack nucleation in ice. Phil. Mag. **A80**(2), 279–300 (2000)

F. Liu, I. Baker, M. Dudley Dislocation-grain boundary interactions in ice crystals. Phil. Mag. A **71**, 15–42 (1995)

Chapter 2
Strength of Ice in Brittle Regime—Multiscale Modelling Approach

Kari Kolari and Reijo Kouhia

Abstract A three dimensional constitutive model based on micromechanical behaviour of granular and columnar ice in the brittle regime is proposed. The model exploits the experimentally observed crack initiation mechanism; grain boundary sliding. The growth of the initiated cracks is assumed to follow the sliding wing crack approach. The model predicts qualitatively and quantitatively the major failure modes and strength of columnar and granular ice under biaxial compression. As shown in the numerical examples, the model captures for example the anisotropic strength of columnar ice under biaxial loading both across and along column loading. Although the model is able to predict splitting and spalling of columnar ice, the shear-like fault is not captured by the current model. The model is implemented into the Abaqus/Explicit FE-software.

2.1 Introduction

Crushing of brittle ice is an important failure mode during impact and interaction with vertical offshore structures. Ice loads exerted on structures are limited by the failure process of the ice during the interaction. Simulation of the failure process based on observed physical mechanisms is of importance in the safe design of offshore structures. The transition of failure modes and their dependence on the micro-structure and stress state play a significant role in the process. The development of a physically based three-dimensional continuum model for both granular and columnar fresh water ice is considered here. Beside the correct failure stress, it is highly important that material models predict correct failure modes. This is especially important in ice-structure interaction simulation which consists of sequential fragmentation of intact

K. Kolari (✉)
VTT Technical Research Centre of Finland, Kemistintie 3, 02150 Espoo, Finland
e-mail: Kari.Kolari@vtt.fi

R. Kouhia
Tampere University, P.O.Box 600, 33101 Tampere, Finland
e-mail: Reijo.Kouhia@tuni.fi

J. Tuhkuri and A. Polojärvi (eds.), *IUTAM Symposium on Physics and Mechanics of Sea Ice*, IUTAM Bookseries 39, https://doi.org/10.1007/978-3-030-80439-8_2

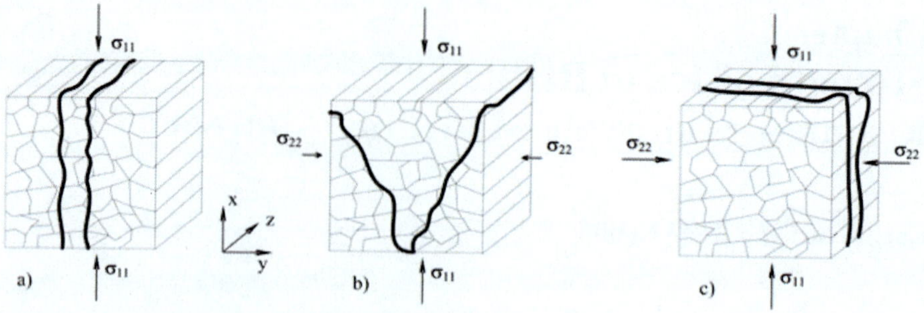

Fig. 2.1 Failure modes of uniaxially and biaxially compressed columnar ice. **a** Splitting; **b** Shear-kind fault; **c** Spalling (splitting across the columns) (From Kolari 2019)

ice. If failure modes were faulty the simulation of the interaction process would be incorrect.

A number of measurements of uniaxial compressive strength of sea ice and fresh water ice have shown that the strength increases with decreasing temperature (Arakawa and Maeno 1997; Schulson 1990) and decreasing grain diameter (Cole 1986; Schulson 1990). Naturally grown ice is either granular or columnar. The columnar ice exhibits anisotropic behavior under applied stress. Under confined compression, the failure modes of the columnar ice change with the confinement ratio (Iliescu and Schulson 2004; Renshaw and Schulson 2001; Smith and Schulson 1993) as discussed later.

Objectives. The objective is to introduce a new three-dimensional model for granular and columnar ice to capture the major failure modes in the brittle regime based on observed mechanisms in micro- and macroscale. The failure modes of columnar ice are illustrated in Fig. 2.1. Modelling concentrates on the uniaxial tension and biaxial compression based on the sliding wing crack model as described by Kolari (2019, 2017).

Method. The approach is based on the hypothesis that the initiation, growth and coalescence of microcracks is the mechanisms behind the macroscopic failure. The micromechanical model is based on the sliding wing crack approach proposed by Ashby and Sammis (1990). Their model has been extended, modified and further developed to formulate a 3D anisotropic continuum damage model suitable for FE-implementation (Kolari 2017, 2019). The use of wing crack approach enables to link crack initiation in grain scale with macroscopic failure modes.

2.2 Continuum Model—Crack Weakened Solids

Crack initiation and growth plays significant role in the brittle compressive fracture of ice. The diameter of the initiated cracks are the order of grain diameter (Cole 1986; Gold 1966; Schulson 1991). The first cracks initiate under compressive stress of about one-third to one-half of the fracture stress (Schulson 1990). The density of cracks increases with loading.

Micromechanical continuum damage models are often formulated using kinematic algorithm to describe inelastic strains induced by cracks (Basista and Gross 1998). In the Algorithm, the total strain tensor ($\varepsilon = \varepsilon^e + \varepsilon^i$) is decomposed to elastic (ε^e) and inelastic (ε^i) parts. Consider continuum with open cracks shown in Fig. 2.2a. The kinematic formulation for inelastic strains reads:

$$\varepsilon^i = \frac{1}{V_0} \sum_{j=1}^{M} \int_{A_j} (\mathbf{g} \otimes \mathbf{b} + \mathbf{b} \otimes \mathbf{g})_j dA \tag{2.1}$$

where V_0 is the volume of the representative volume element, **g** in the unit normal vector of the crack,**b** is the displacement jump vector,A_j is the area of the jth crack, M is the number of cracks in the representative volume element (RVE) and $\otimes$ represents dyadic product. Stress–strain relation can be expressed as

$$\sigma = \mathbf{C} : \varepsilon^e = \mathbf{C} : (\varepsilon - \varepsilon^i) \tag{2.2}$$

where σ is the Cauchy stress tensor and **C** is the elastic constitutive tensor. Thus, the macroscopic stress state σ of the representative volume element can be defined when the configuration and the opening of all cracks are known.

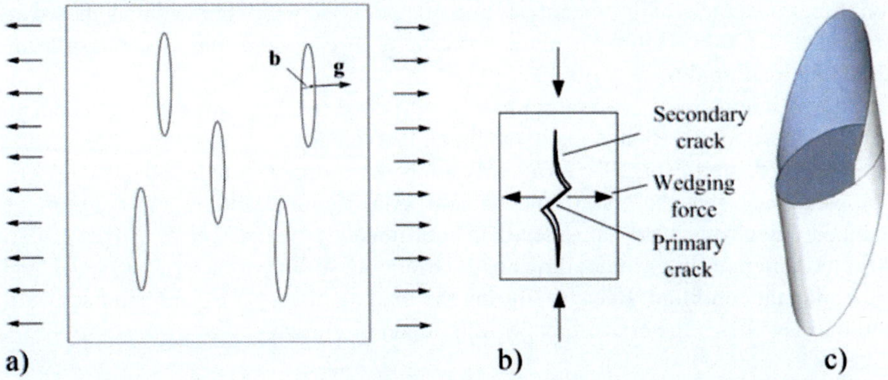

Fig. 2.2 Cracks in a continuum. **a** Opening of cracks in a continuum. **b** Is the displacement jump vector and **g** is normal vector of the crack surface. **b** Formation of a 2D-wing crack; **c** Illustration of 3D-crack (From Kolari 2017)

2.3 Sliding Wing-Crack Model

The axial splitting under uniaxial compression characterises the macroscopic compressive failure mode of many brittle materials like concrete, rock, ceramics and natural ice. A mechanism proposed for the modelling of the splitting is the wing-crack approach introduced by Brace and Bombolakis (1963) and extensively studied by Horii and Nemat-Nasser (Horii and Nemat-Nasser 1986, 1985, 1982; Nemat-Nasser and Obata 1988). Wing cracks has been observed in ice, both in small and large scale (Cannon et al. 1990; Hawkes and Mellor 1970; Iliescu and Schulson 2004; Schulson 2004; Schulson et al. 2006; Smith and Schulson 1994).

As illustrated in Fig. 2.2b the failure begins when a primary crack undergoes frictional sliding, creating secondary cracks at the tips of the primary crack. The macroscopic failure occurs when series of cracks extend and finally link together and split the material. It has been shown, that the size of the primary crack is directly related to the grain size (Cole 1986; Picu and Gupta 1995a) therefore the wing crack model can be linked both to the grain size and to the temperature dependent friction of ice as proposed by Schulson (1990).

The wing-crack mechanism in 3D is different to that of 2D. Under uniaxial loading of single, inclined primary crack the secondary crack wraps (curves) as illustrated in Fig. 2.2c (Adams and Sines 1978; Cannon et al. 1990; Dyskin et al. 1994). When the length of the wrapped wings is 1.0–1.5 times the diameter of the primary crack, the growth stops. As shown by Dyskin et al. (1994) the wings do not grow under further loading. When the number of cracks increases, the interaction of cracks has to be considered. When the centre-to-centre distance between primary cracks is less than four times the crack radius, the interaction of cracks becomes significant as shown by Dyskin et al. (2003, 1999). Then, the resulting failure mode is macroscopic splitting along the loading direction (ibid.).

Three-dimensional continuum models motivated by the wing-crack mechanism have been proposed (Bhat et al. 2011; Jin and Arson 2017; Pensée et al. 2002; Yang et al. 1999; Yu and Feng 1995). But only a few three-dimensional models have been proposed, such that the growth mechanism with frictional sliding and the damage induced anisotropy were considered in the models (Ayyagari et al. 2018; Kolari 2017). Challenges in the modelling are the frictional sliding of the primary crack and the unilateral condition. Because, during the loading history, forward and backward sliding may occur; the crack may be either open or closed.

2.4 Model Formulation

Formulation of the model described in Chap. 3 and illustrated in Fig. 2.3 consists the following steps:

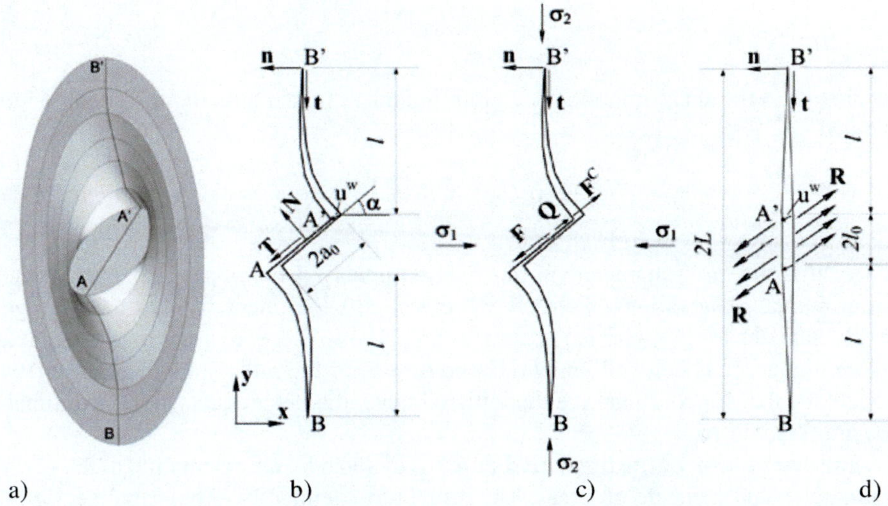

Fig. 2.3 Sketch of 3D crack and its section views. (a) 3D view; (b) Section view and notations; (c) Forces acting under sliding of the primary crack; (d) the representative crack model for crack growth calculations (From Kolari 2017)

1. Nucleation of the primary cracks.
   - Solve the orientation **N** and diameter $2a_0$ of the primary crack.
2. Sliding of the primary crack.
   - Formulate a sliding condition for the primary crack.
3. Growth and interaction of cracks.
   - Solve the secondary crack orientation **n** and the length LL.
     - Based on the linear elastic fracture mechanics.
   - Introduce crack interaction formulation.
4. Opening of the cracks.
   - Solve average crack opening displacements $\mathbf{u}^w$ and $\mathbf{u}^F$.
     - Based on the energy release rate and Castigliano's second theorem.

2.4.1 Nucleation of the Primary Crack

As observed by Picu and Gupta (1995a, b) in the brittle regime, cracks nucleate by grain boundary sliding mechanism. The nucleation is driven by the shear stress on the grain boundary. The crack is nucleated when the nucleation threshold value is exceeded:

$$\left\|\tau^{nz}\right\| > \tau_{ini}[\text{MPa}] \tag{2.3}$$

where τ^{nz} is the shear traction on a grain boundary, and τ_{ini} is the threshold value defined as:

$$\tau_{ini} = 0.18 + 0.76\,d^{-1/2}[\text{MPa}] \tag{2.4}$$

where d is the grain diameter (in mm). **The nucleated crack size** is based on the experimental observations. Cole (1986) observed experimentally that the average crack diameter for granular ice is about $0.65 \cdot d$. But, the size of the initiated crack for columnar ice is not well known. Based on a brief literature study, Kolari (2019) concluded that for columnar ice the initiated crack diameter equals about two times the grain diameter.

The orientation of the nucleated crack is dictated by the orientation of the grain boundaries and the state of stress. The orientation defined by the normal vector of the boundary $\mathbf{n}^z$ shown in Fig. 2.4. The sliding plane is considered as a mode II crack from which the primary crack nucleates as described for example by Cole (1986) and Picu and Gupta (1995b). The angle Ψ between the nucleated crack and sliding plane is 70.5° as proposed by Stroh (1957). The angle is obtained by maximising the energy release. The normal of the primary crack is formulated as follows:

$$\mathbf{N} = \frac{\tau^{nz}}{\|\tau^{nz}\|}\sin\Psi + \mathbf{n}^z\cos\Psi \tag{2.5}$$

where τ^{nz} is the shear traction on the boundary is defined as:

$$\tau^{nz} = \mathbf{n}^z \cdot \sigma - \sigma^{nz} \tag{2.6}$$

and

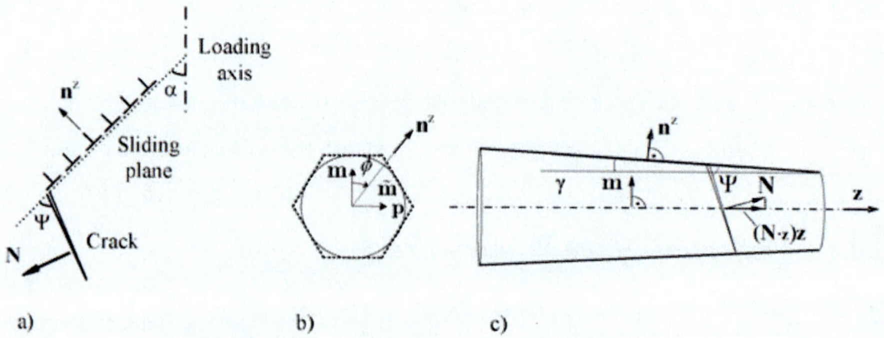

Fig. 2.4 The initiation of primary crack in granular ice and columnar ice. **a** Sliding plane; **b** Cross-section of a grain; **c** columnar grain with the taper angle of γ (From Kolari 2019)

$$\begin{aligned}\sigma^{nz} &= \sigma^{nz}\mathbf{n}^z \\ \sigma^{nz} &= \mathbf{n}^z \cdot \sigma \cdot \mathbf{n}^z\end{aligned} \tag{2.7}$$

The actual orientation of grain boundaries is not known. Thus, for **granular ice** the orientation of $\mathbf{n}^z$ is defined such that the shear traction τ^{nz} is maximized.

The behavior is **columnar ice** is different. As suggested by Iliescu and Schulson (2004) thc taper angle γ is of importance especially in the initiation of across column cracks under biaxial compression across the columns. To define the maximum shear traction τ^{nz} on a columnar grain boundary, the cross section is idealized to be circular and tapered with the taper angle of γ as shown in Fig. 2.4a and 4b. The normal vector of the grain boundary $\mathbf{n}^z$ is given as follows

$$\mathbf{n}^z = \tilde{\mathbf{m}} \cos\gamma + \mathbf{z} \sin\gamma \tag{2.8}$$

where

$$\begin{aligned}\tilde{\mathbf{m}} &= \mathbf{m}\cos\phi + \mathbf{p}\sin\phi \\ \mathbf{p} &= \mathbf{m} \times \mathbf{z}\end{aligned} \tag{2.9}$$

The arbitrary unit vector $\mathbf{m}$ in Eq. (2.9) is chosen such that it is normal to the symmetry axis $\mathbf{z}$ as shown in Fig. 2.4c; thus, $\mathbf{m} \cdot \mathbf{z} = 0$. The angle ϕ of Eq. $(2.9)_1$ is defined such that the shear traction is maximized. Then the primary crack orientation vector $\mathbf{N}$ is obtained from Eq. (2.5).

2.4.2 Sliding Condition

One of the important features of the wing-crack model is the cohesion and friction controlled sliding of the primary crack. The ice-ice friction is known to be a function of the temperature and the velocity (Kennedy et al. 2000; Maeno and Arakawa 2004; Schulson 1990). As shown later the impact of the ice-ice friction to the compressive strength is significant.

As described by Nemat-Nasser and Obata (1988) and Basista and Gross (1998) forward and backward sliding with or without crack growth may occur during loading–unloading process. The proposed model is intended for the simulation of tension also, thus the opening of the primary crack has to be taken into account.

Consider the forces described in Fig. 2.3c. Driving traction $\mathbf{F}$ is either opening or closing the wing crack. The traction is formulated as follows

$$\mathbf{F} = \pi a_0^2 \left[\mathbf{N} \cdot \sigma - \sigma^N \mathbf{N}\left(1 - H(\sigma^N)\right)\right] \tag{2.10}$$

where a_0 is the radius of the primary crack, $\mathbf{N}$ is the unit normal vector of the crack and $\sigma^N = \mathbf{N} \cdot \sigma \cdot \mathbf{N}$. The Heaviside function $H(\cdot)$ is applied to cover both the compression and the tension of the primary crack. The other forces to be considered are the dissipative force $\mathbf{Q}$ induced by friction and cohesion and the elastic crack closing force $\mathbf{F}^C(\mathbf{u}^w, l)$. The crack closing force is a function of the opening and the length of the crack as illustrated in Fig. 2.3d. The closing force $\mathbf{F}^C$ resists the opening of the wing crack. The "direction" of the other forces may vary: the direction of the driving traction $\mathbf{F}$ depends on the state of stress; the dissipative force $\mathbf{Q}$ is opposite to the sliding direction. The potential sliding direction $\tilde{\mathbf{n}}$ is defined as a function of the actual forces $\mathbf{F}$ and $\mathbf{F}^C$:

$$\tilde{\mathbf{n}} = \frac{\mathbf{F} + \mathbf{F}^C}{\left\|\mathbf{F} + \mathbf{F}^C\right\|} \tag{2.11}$$

The dissipative, resisting force is:

$$\mathbf{Q} = (Q^F + Q^C)\,\tilde{\mathbf{n}} \tag{2.12}$$

where Q^F is the Coulomb friction and Q^C is the cohesion described as follows:

$$\begin{aligned} Q^F &= \pi\, a_0^2 \mu\, \sigma^N (1 - H(\sigma^N)) \\ Q^C &= -\pi\, a_0^2 \tau^c \end{aligned} \tag{2.13}$$

where μ is the friction coefficient and τ^c is the cohesion. Note that under tension ($\sigma^N > 0$) the frictional force Q^F equals zero. Then the sliding condition f can be written as follows:

$$f = \left\|\mathbf{F} + \mathbf{F}^C\right\| + Q^F + Q^C \tag{2.14}$$

If $f > 0$, sliding to the direction defined by $\tilde{\mathbf{n}}$ occurs until $f = 0$. For $f \leq 0$ there is no sliding.

2.4.3 *Growth and Interaction of Cracks*

The **growth of cracks** is based on the model described by Horii and Nemat-Nasser (1986) and Ashby and Sammis (1990). They proposed that the wing crack can be approximated by a straight representative crack shown in Fig. 2.3d. Horii and Nemat-Nasser (ibid.) showed that in 2D a straight representative crack estimates the stress intensity factor K_I^0 relatively well. The effective length L of the representative crack is defined as

$$L = l + l_0 \tag{2.15}$$

where $l_0 = \beta a_0$ and β is constant (Ashby and Sammis 1990; Horii and Nemat-Nasser 1986). The stress intensity factor for non-interacting 3D representative crack is given as

$$K_I^0 = \frac{2\mathbf{R} \cdot \mathbf{n}\left(L - \sqrt{L^2 - l_0^2}\right)}{\sqrt{\pi L}} + 2\sigma^n \sqrt{\frac{L}{\pi}} \tag{2.16}$$

where $\mathbf{R} = \mathbf{F}^C(\mathbf{u}^w, l)/(\pi l_0^2)$ is the surface traction, $\mathbf{n}$ is the normal of the secondary crack and $\sigma^n = \mathbf{n} \cdot 1\sigma \cdot \mathbf{n}$.

As stated before the **interaction and coalescence** of cracks are considered to be the mechanisms behind the macroscopic brittle failure. Under uniaxial compression the failure mode is axial splitting, while under confined compression crack arrays may become tilted and the failure mode is shear kind as described for example by Horii and Nemat-Nasser (1986) and Smith and Schulson (1993). Detailed analysis of crack interaction is out of the scope of this paper. Readers interested in detailed analysis of crack interaction are advised to consult the papers of e.g. Kachanov (1987), Horii and Nemat-Nasser (1986, 1985), Dyskin et al. (1999) and Kuutti and Kolari (2018). The ad-hoc interaction function is applied in this approach. It is assumed that the interaction is only a function of crack spacing. Thus, shear kind failure modes cannot be predicted with the model. The interaction is described with the function

$$g(L/b) = \left(1 - \left(\frac{L}{b}\right)^2\right)^{-1/2} \tag{2.17}$$

where b is the average centre-to-centre distance of cracks shown in Fig. 2.5. The stress intensity factor for interacting crack is then:

$$K_I = g(L/b)K_I^0 \tag{2.18}$$

As shown in Fig. 2.5 the proposed interaction function predicts weaker interaction than the function proposed by Ashby and Sammis (1990). The growth criterion is based on the linear elastic fracture mechanics and defined to be the difference of the stress intensity factor for mode I and the fracture toughness[1] K_{IC}:

$$Z = K_I - K_{IC} \tag{2.19}$$

[1] Note: Fracture toughness testing standard does not exist for ice. Thus, Dempsey (1991) recommends applying notation K_Q for the "apparent fracture toughness" of ice.

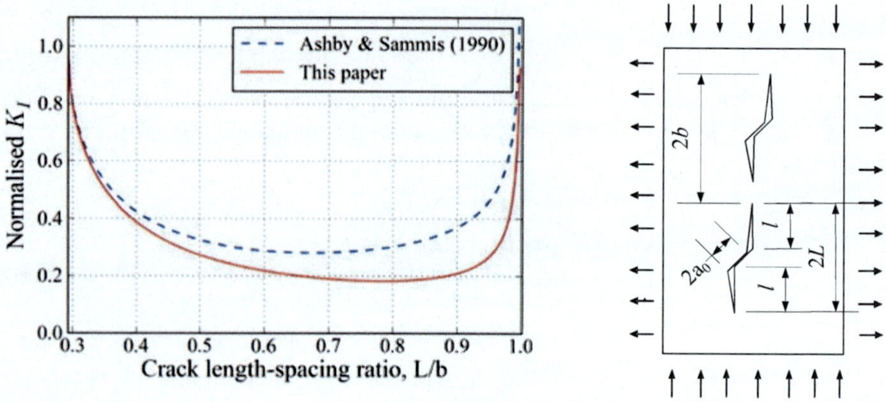

a) b)

Fig. 2.5 Interaction of cracks. **a** The proposed interaction function compared to the interaction derived by Ashby & Sammis (1990) under uniaxial compression. **b** Notations applied for cracks

when $Z > 0$, crack grows until $Z = 0$. If $Z \leq 0$ crack does not grow. The growth of isolated wing crack under uniaxial compression is stable: to propagate a crack, load must be increased. Due to interaction of wing cracks under uniaxial compression, crack propagation can become unstable. As shown in the Fig. 2.5 (LHS) the growth under uniaxial compression is stable up to the *L/b*-ratio of about 0.8. In the unstable regime the growth is limited by the maximum growth rate $\dot{l}_{\text{max}}$ as follows:

$$\Delta l = \Delta t \, \dot{l}_{\text{max}} \tag{2.20}$$

where Δt is time increment.

As proposed by Horii and Nemat-Nasser (1986) the orientation **n** of the secondary crack is defined by maximising K_I. The orientation of the secondary crack is known to change during the growth process. But, the rotation of the crack would cause directional stiffness recovery, which is not acceptable. Therefore "a long wing approach" is applied. When secondary crack grows for the first time, the orientation is defined as follows; K_I is maximised by setting $L = 10\,a_0$ and $b = \infty$ in Eq. (2.18), then the defined orientation **n** remains fixed during the analysis.The opening of cracks.

2.5 The Opening of Cracks

The opening of cracks is needed both in the sliding condition (Eq. 2.14) and in the calculation of inelastic strains induced by the crack openings as described in Eq. (2.1). As illustrated in Fig. 2.6 the opening displacements is the sum of the sliding displacement $\mathbf{u}^{\mathrm{w}}$ and the displacement $\mathbf{u}^F$ induced by the far field stresses.

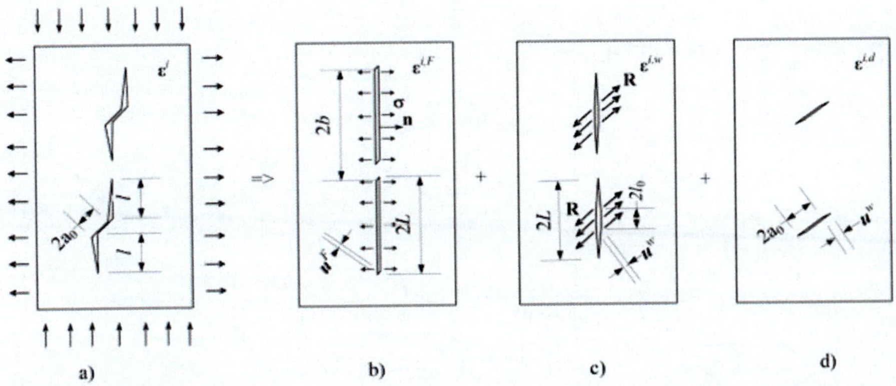

Fig. 2.6 Superposition of crack opening displacements; **a** Open wing crack; **b** Opening of the representative crack induced by the far-field stresses; **c** Opening of the representative crack induced by the wedging forces; **d** opening of the primary crack induced by the wedging (From Kolari 2017)

The displacements are derived from the Castigliano's theorem and energy release by the crack formation as described in Kolari (2017). The energy release $\Delta\Pi$ is

$$\Delta\Pi = \int_0^L \int_0^{2\pi} G(\theta)\, L d\theta\, dL \tag{2.21}$$

where $G(\theta)$ is the release rate along the crack edge as a function of the polar coordinate θ:

$$G(\theta) = \frac{(1-\nu^2)}{E}\left(K_I^2 + K_{II}(\theta)^2 + \frac{1}{1-\nu}K_{III}(\theta)^2\right) \tag{2.22}$$

where is ν the Poisson's ratio. The average crack opening displacement u_x can be derived using the Castigliano's second theorem (Kemeny and Cook 1991):

$$u_x = \frac{\partial \Delta\Pi}{\partial F_x} \tag{2.23}$$

where F_x is the force conjugate to the average opening u_x. For example, in the loading case illustrated in Fig. 2.6b the normal and shear forces are $\pi L^2 \sigma^n$ and $\pi L^2 \tau^n$ respectively. The corresponding opening displacement obtained from Eqs. (2.18) and (2.23) is as follows:

$$\begin{aligned} \mathbf{u}^F &= \frac{\partial \Delta\Pi}{\pi L^2 \partial \boldsymbol{\sigma}^n} + \frac{\partial \Delta\Pi}{\pi L^2 \partial \boldsymbol{\tau}^n} \\ &= \frac{16 b^2}{\mathrm{E} L^2 \pi}(1-\nu^2)\left(b \tan h^{-1}\left(\frac{L}{b}\right) - L\right)\left[\boldsymbol{\sigma}^n + \frac{\boldsymbol{\tau}^n}{(1-\nu/2)}\right] \end{aligned} \tag{2.24}$$

Similarly we can derive the function for the crack closing force $\mathbf{F}^C$ described in Eq. (2.14) and shown in Fig. 2.3:

$$\mathbf{F}^C = -\frac{\pi l_0^2}{C_P} 1 u^w = -\pi l_0^2 \mathbf{R} \tag{2.25}$$

where

$$\begin{aligned} C_P = \frac{16(1-\nu^2)}{\pi l_0^2 E} \frac{(1-\nu/4)}{(1-\nu/2)} \frac{b}{2} \Bigg[& \log\left(\frac{(b-L)(b+l_0)}{(b+L)(b-l_0)}\right) l_0^2 \\ & 4b\sqrt{L^2 - l_0^2} + 2b\left(2b \tanh^{-1}\left(\frac{L}{b}\right) - b \tanh^{-1}\left(\frac{l_0}{b}\right) + l_0 - 2L\right) \\ & 2b\sqrt{b^2 - l_0^2} \log\left(\frac{b^2 - L^2}{b^2 + L^2 - 2l_0^2 + 2\sqrt{(b^2 - l_0^2)(L^2 - l_0^2)}}\right)\Bigg] \end{aligned} \tag{2.26}$$

When the displacements $\mathbf{u}^F$ and $\mathbf{u}^w$ are known, the inelastic strains can be calculated from Eq. (2.1). The number of cracks (M) in the Equation is a function of the crack spacing and the size of the element: $M = V_0/(2b)^3$. The derivation the opening displacements and corresponding strains is described more in detail by Kolari (2017).

2.6 Numerical Verification

2.6.1 *Implementation*

The model introduced above was implemented into Abaqus/explicit finite element software as a user material subroutine VUMAT. The following assumptions were applied in the implementation:

- Both the primary and secondary crack size and orientation are assumed to be the same in the volume related to a single integration point.
- To introduce material inhomogeneity in the model, the size and orientation of the grains are assumed to be lognormally distributed between the integration points. Therefore, also the size (a_0, L), taper angle (γ) and the orientation of cracks ($\mathbf{N}$, $\mathbf{n}$) vary between the integration points.

Widely used one-step forward Euler approach is applied to compute the inelastic strains.

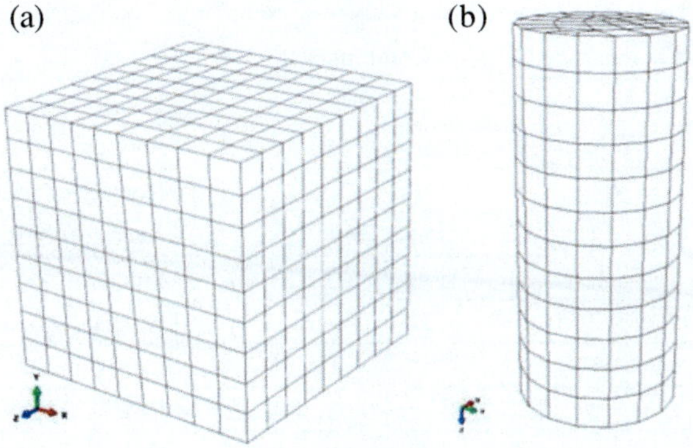

Fig. 2.7 FE mesh; **a** Cube (LHS) and **b** Cylinder

2.6.2 FE-Model

Cubical specimens of side length of 125 mm were used in the simulation of biaxial compression tests, while cylindrical specimen (96 mm in diameter and 230 mm in length) were used in pure uniaxial compression and uniaxial tension tests of granular ice. The size of the cylindrical specimens were the same as in the experiments of Schulson (1990). The ends of all the specimens were assumed to be frictionless. Loading rate was set to coincide with the average strain rate of 10^{-3} s^{-1}. To simulate natural variation of ice, simulations were repeated five to six times with random crack size and random taper angle.

The specimens were modelled using eight-node linear brick elements with reduced integration (C3D8R). The cube was modelled with 729 elements while 384 elements were used in the modelling of the cylinder as shown in Fig. 2.7. The material parameters used in the simulations are given in Table 2.1.

2.6.3 Granular Ice—Compression and Tension

The model was tested against the experimental results to verify whether the model is capable to predict grain size and temperature dependent uniaxial compressive and uniaxial tensile strength. In addition, biaxial compression test was also simulated.

In the uniaxial compression and tension test simulations the grain size were varied in the range of about 2–10 mm at −10 °C and −50 °C. The test results and simulated results shown in Figs. 2.9 and 2.10 coincide relatively well both under compression and tension; the strength increases with decreasing grain diameter (Fig. 2.11). The simulated failure modes illustrated in Fig. 2.12 correspond the observations of Schulson (1990) Lee and Schulson (1988) and Weiss and Schulson (1995).

The failure mode observed in the biaxial compression experiments of Weiss and Schulson (1995) changed with increasing stress ratio $R(R = \sigma_{11}/\sigma_{22})$ from multiple longitudinal splitting to formation of slabs under biaxial loading. The normal of the

Table 2.1 Material parameters for granular and columnar ice

Granular ice	Columnar ice	Description
$E = 9.0$	$E = 9.0$	Young's modulus, GPa
$\nu = 0.3$	$\nu = 0.3$	Poisson's ratio
$\rho = 916$	$\rho = 916$	Density of ice,kg/m^3
$a_{0,m} = 0.65\,d/2$	$a_{0,m} = d$	Average crack radius versus grain diameter, mm
$a_{0,s} = 0.65 \cdot a_{0,m}$	$a_{0,s} = 0.667 \cdot a_{0,m}$	Standard deviation of crack radius, mm
$b = d/2$	$2b = 3\,a_{0,m}$	Average spacing of cracks, mm
$\tau^c = 0.01$	$\tau^c = 0.01$	Cohesion, set to negligible, kPa
$\mu = 0.06...0.8$	$\mu = 0.06...0.8$	Kinetic friction (see Fig. 2.8)
$K_{IC} = 100$	$K_{IC} = 100$	Fracture toughness (kPa$\sqrt{\text{m}}$)
$\dot{L}_{\max} = 400$	$\dot{L}_{\max} = 400$	Maximum crack velocity, m/s
$\beta = 0.45$	$\beta = 0.45$	Parameter,$l_0 = \beta\, a_0$
$\Psi = 70.5°$	$\Psi = 70.5°$	Angle between parent crack and sliding plane
	$\gamma_m = 0°$	Mean taper angle
	$\gamma_m = 9°$	Standard deviation, taper angle

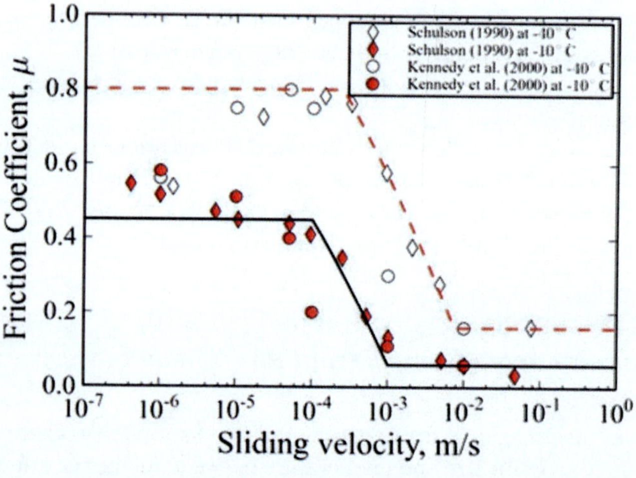

Fig. 2.8 Ice-ice kinetic friction. The solid and dashed lines represent values applied in the simulations. Experimental values are from Kennedy et al. (2000) and Schulson (1990) (From Kolari 2017)

slabs pointed to the no-load direction (Schulson and Duval 2009, p 270) as illustrated in Fig. 12c. Simulated failure modes correspond well the experimentally observed modes; the normal vectors of the wings rotate with increasing confinement. Under uniaxial compression the wings were randomly oriented to the no-load direction. At

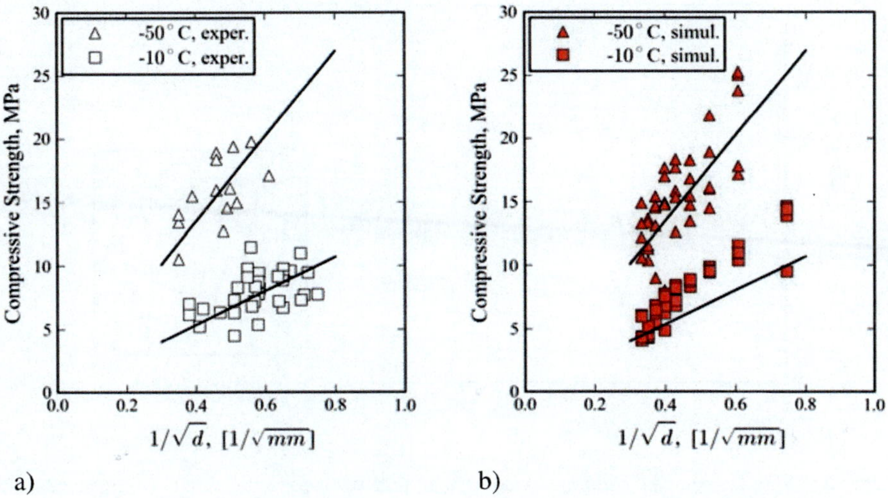

Fig. 2.9 Uniaxial compression of granular ice. Test results (**a**) and Simulation (**b**). Test results have been redrawn from Schulson (1990). Solid lines represent fit to the experimentally observed values

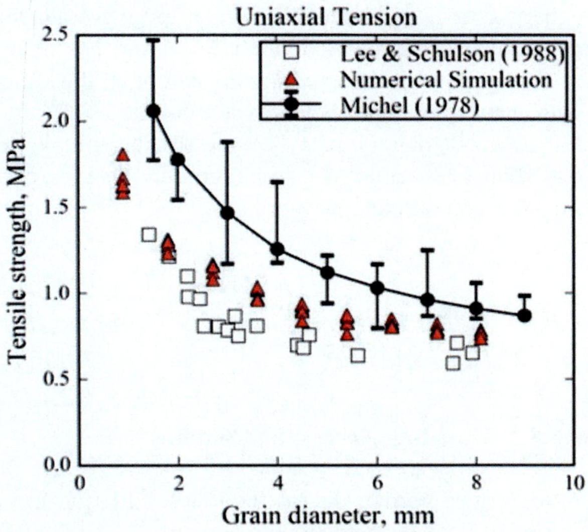

Fig. 2.10 Granular ice under uniaxial tension. Test results are from Lee and Schulson (Lee and Schulson 1988) and Michel (1978) (From Kolari 2017)

small stress ratio of $\sigma_{11}/\sigma_{22} = 0.01$ normals become parallel and point to the no-load direction illustrating formation of slabs.

Weiss and Schulson (1995) observed that *the biaxial compressive strength* is essentially independent of the confinement ratio and "*is about equal to the uniaxial failure stress*". The observed uniaxial compressive strength was 7 ± 1 MPa while the biaxial compressive strength was 8 ± 1 MPa at $\sigma_{11}/\sigma_{22} = 1$.

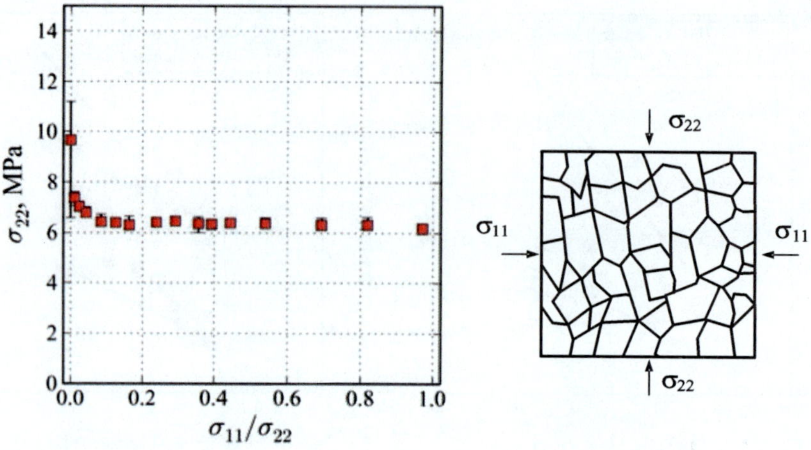

Fig. 2.11 Simulated biaxial compressive strength of granular ice as a function of confinement ratio σ_{11}/σ_{22}. The average grain diameter of granular ice was 7 mm. Error bars represent maximum and minimum values observed in the simulations

In the simulations the average grain diameter was set to be the same as in the experiments of Weiss and Schulson (1995): average diameter of 7 mm with standard deviation of 2.5 mm was applied. The simulated biaxial compressive strength corresponds quite well the experiments, although the simulations shows small decrease with increasing confinement when confinement ratio $\sigma_{11}/\sigma_{22} < 0.1$ as shown in Fig. 2.11. At the ratio of $\sigma_{11}/\sigma_{22} = 0.1$ the strength levels off to the value of about 6.4 MPa. The scatter at uniaxial compressive strength is notable but negligible at low-to-high confinement ratio.

2.6.4 Columnar Ice—Biaxial Compression Across and Along the Columns

2.6.4.1 Loading Across the Columns

As described earlier, due to the crack initiation mechanism the behaviour of the columnar ice in the along column direction is totally different from the across column direction. Although the biaxial compressive strength of granular ice is almost insensitive to the confined compression, the columnar ice is very sensitive to the confinement ratio as is be shown below.

The simulations were verified against the test results of Iliescu and Schulson (2004), Schulson and Nickolayev (1995), Smith and Schulson (1993) and Gratz and Schulson (1997). In the tests, the average grain diameter varied in the range of 4–8.5 mm. There is not much information available about the taper angle γ; Iliescu and

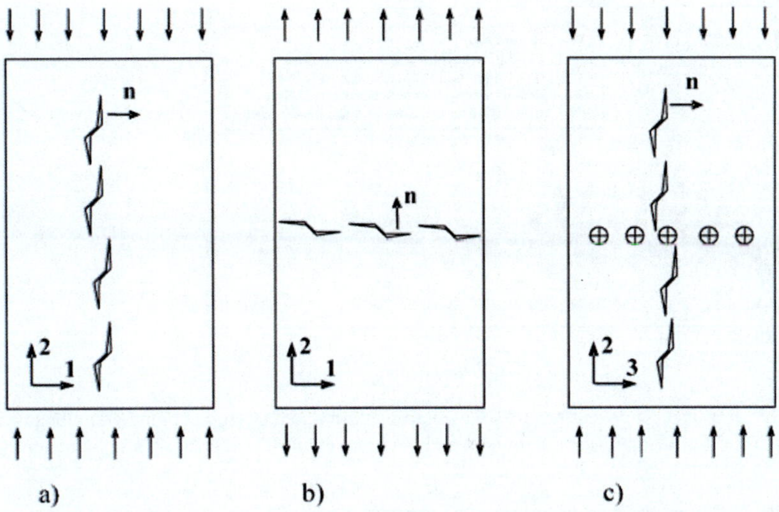

Fig. 2.12 Illustration of the primary failure modes of granular ice. **a** Splitting along uniaxial compression direction; **b** Tensile failure; **c** Spalling to the stress free direction under biaxial compression ($\sigma_{11} \neq 0,\ \sigma_{22} \neq 0,\ \sigma_{33} = 0$)

Schulson (2004) concluded that the angle is about 5° while and Gratz and Schulson (1997) observed the angle deviation to be ± 9°.

In the simulations the taper angle was assumed to be normally distributer with the mean of 0° and standard deviation of 9°. The average grain diameter was set to 6 mm with the standard deviation of 4 mm. The diameter was assumed to be lognormally distributed.

The failure modes observed in the biaxial compression tests across the columns (Iliescu and Schulson 2004; Schulson and Nickolayev 1995; Smith and Schulson 1993) were changing with increasing stress ratio R as follows: splitting ($R < 0.1\ldots0.2$), shear faulting ($0.1\ldots0.2 < R < 0.3\ldots0.5$) and mixed shear faulting and spalling at higher confinement ratio ($0.3\ldots0.5 < R < 1.0$).

As shown in the Fig. 2.13 simulated strength corresponds reasonable well the experimental results, although the model overestimates the strength in the confinement range of $0.05 < R < 0.4$. In that range the primary failure mode observed in the tests was shear faulting, while the failure mode in the simulations was mixed spalling and splitting. Both the simulated strength and failure modes correspond experiments outside of the confinement range of $0.05 < R < 0.4$. It is worth to note the effect of end conditions in the experiments; the strength is clearly higher when solid platens instead of brush platens were used (Fig. 2.13). The ends in the simulations were frictionless, thus the results should be closer to the experiments done with brush platens. As illustrated in Fig. 2.14 the orientations of primary cracks (**N**) and secondary cracks (**N**) give information about the orientation of the macroscopic failure plane. When both **N** and **n** point to the normal of the column axis the mode

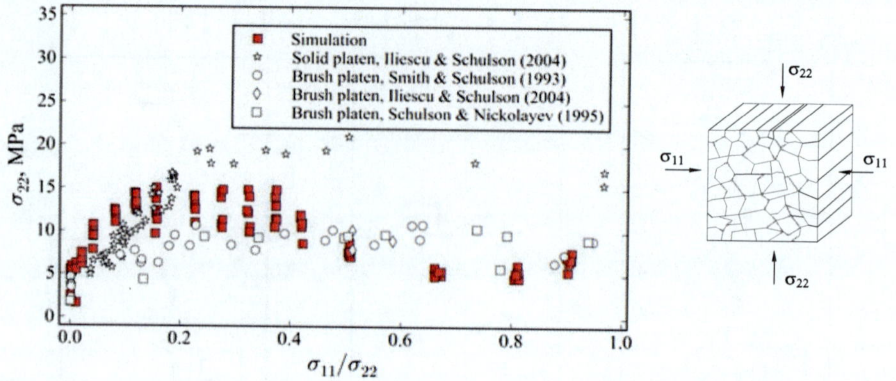

Fig. 2.13 Biaxial compression across the columns. Simulated strength compared to the test results of Iliescu and Schulson (2004), Schulson and Nickolayev (1995) and Smith and Schulson (1993)

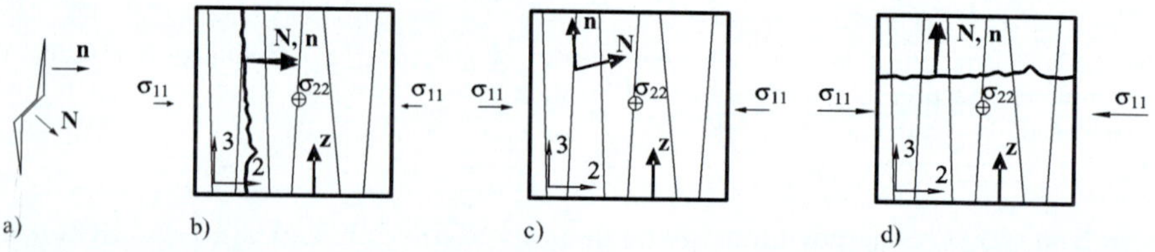

Fig. 2.14 Failure modes observed in the simulations of biaxial compression across the columns. The modes are based on the orientation of the primary crack (**N**) and secondary crack (**n**) normals: **a** Notations; **b** Splitting at low confinement; both vectors are normal to the columns; **c** Mixed mode, secondary crack normal (**n**) points to the unstressed direction but the parent crack normal is pointing normal to the columns; **d** Spalling, both vectors are pointing to the unstressed direction

is splitting. When both **N** and **n** point to the direction of the columns (**z**) the mode is spalling. Otherwise the mode was "mixed", between splitting and spalling.

Fig. 2.15 Spalling under confinement ratio of 0.8. Red colour represents fully damaged elements

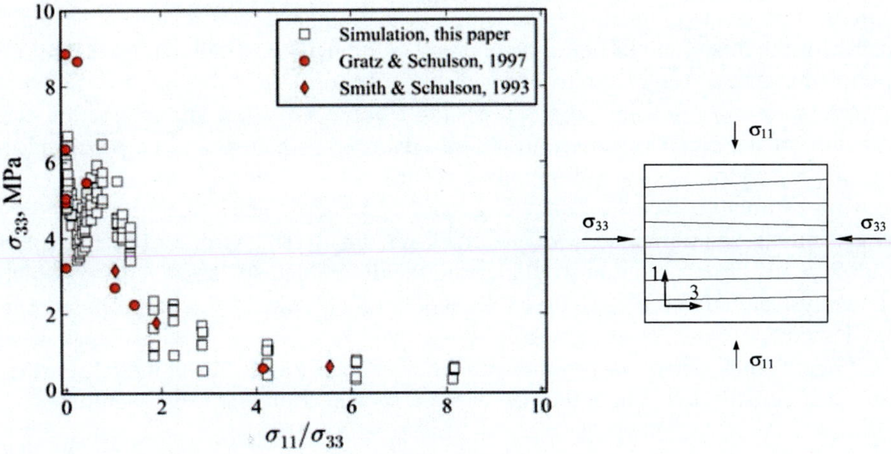

Fig. 2.16 Biaxial compression along the columns. Simulated strength compared to the test results of Smith and Schulson (1993) and Gratz and Schulson (1997)

2.6.4.2 Loading Along the Columns

The material parameters used were the same as in the across column compression simulations. The simulations were compared to the test results of Smith and Schulson (1993) and Gratz and Schulson (1997). Although Smith and Schulson used both steel platen and brush platen, only those where the brush platen were used were considered in the verification. As shown in Fig. 2.16 the simulated strengths correspond reasonable well the experiments, although the scatter at low or zero confinement was clearly higher in the experiments.

In the experiments the observed failure mode was splitting along the colums. Under zero or low confinement specimens failed to column-like shards that were paralled to the columns. Biaxial loading led to formation of planar cracks oriented perpendicular to the no-load direction. The simulated failure modes correspond the experimental observations. At zero or low confinement the secondary crack normals (**n**) were randomly oriented, but were perpendicular to the loading direction. At higher confinement the normals were parallel to the no-load direction, simulating planar-like failure mode.

2.7 Discussion

Compared to the experiments, the proposed method is able to predict qualitatively and quantitatively both the major failure modes and failure stresses. The grain boundary sliding approach combined with the wing-crack sliding model was show to be applicable for the modelling of both granular and columnar ice. Although the results are promising, further research required is discussed here.

Interaction of cracks is based on a simple approach where the interaction is a function of the crack spacing only. More advanced model should be proposed to tackle for example shear-faulting failure mode.

Crack growth is based on the "long crack approach". Thus, once the secondary crack initiates, the orientation of the crack remains fixed. As described by Kolari (2007, p. 132) crack rotation would lead to the directional stiffness recovery, which is unphysical. Therefore, a "curved secondary crack" model should proposed and implemented.

The size and density of primary cracks is of importance in the model. Further information especially about the size of the cracks in columnar ice is required.

References

M. Adams, G. Sines, Crack extension from flaws in a brittle material subjected to compression. Tectonophysics **49**, 97–118 (1978). https://doi.org/10.1016/0040-1951(78)90099-9

M. Arakawa, N. Maeno, Mechanical strength of polycrystalline ice under uniaxial compression. Cold Reg. Sci. Technol. **26**, 215–229 (1997). https://doi.org/10.1016/S0165-232X(97)00018-9

Ashby, M.F., Sammis, C.G., 1990. The Damage Mechanics of Brittle Solids in Compression. Pure Appl. Geophys. 133, 489–521. https://doi.org/10.1007/BF00878002

R.S. Ayyagari, N.P. Daphalapurkar, K.T. Ramesh, The effective compliance of spatially evolving planar wing-cracks. J. Mech. Phys. Solids **111**, 503–529 (2018)

M. Basista, D. Gross, The sliding crack model of brittle deformation: An internal variable approach. Int. J. Solids Struct. **35**, 487–509 (1998). https://doi.org/10.1016/S0020-7683(97)00031-0

H. Bhat, C. Sammis, A. Rosakis, The Micromechanics of Westerley Granite at Large Compressive Loads. Pure Appl. Geophys. **168**, 2181–2198 (2011)

W.F. Brace, E.G. Bombolakis, A Note on Brittle Crack Growth in Compression. J. Geophys. Res. **68**, 3709–3713 (1963)

N.P. Cannon, E.M. Schulson, T.R. Smith, H.J. Frost, Wing cracks and brittle compressive fracture. Acta Metall. Mater. **38**, 1955–1962 (1990)

D.M. Cole, Effect of grain size on the internal fracturing of polycrystalline ice, CRREL report (1986)

J.P. Dempsey, The fracture toughness of ice, in: Ice-Structure Interaction. Springer, Berlin Heidelberg, pp. 109–145 (1991). https://doi.org/10.1007/978-3-642-84100-2_8

A.V. Dyskin, R.J. Jewell, H. Joer, E. Sahouryeh, K.B. Ustinov, Experiments on 3-D crack growth in uniaxial compression. Int. J. Fract. **65**, R77–R83 (1994). https://doi.org/10.1007/BF00012382

A.V. Dyskin, L.N. Germanovich, K.B. Ustinov, A 3-D model of wing crack growth and interaction. Eng. Fract. Mech. **63**, 81–110 (1999)

A.V. Dyskin, E. Sahouryeh, R.J. Jewell, H. Joer, K.B. Ustinov, Influence of shape and locations of initial 3-D cracks on their growth in uniaxial compression. Eng. Fract. Mech. **70**, 2115–2136 (2003). https://doi.org/10.1016/S0013-7944(02)00240-0

L.W. Gold, Dependence of crack formation on crystallographic orientation for ice. Can. J. Phys. **44**, 2757–2764 (1966)

E.T. Gratz, E.M. Schulson, Brittle failure of columnar saline ice under triaxial compression. J. Geophys. Res. **102**, 5091–5107 (1997). https://doi.org/10.1029/96JB03738

V. Gupta, R.C. Picu, A model for the indentation-induced splitting ice floe experiments. Acta Metall. Mater. **43**, 1355–1362 (1995). https://doi.org/10.1016/0956-7151(94)00373-P
I. Hawkes, M. Mellor, Uniaxial testing in rock mechanics. Eng. Geol. **4**, 177–285 (1970)
H. Horii, S. Nemat-Nasser, Brittle failure in compression: splitting, faulting and brittle-ductile transition. Philos. Trans. R. Soc. London A Math. Phys. Sci. **319**, 337–374 (1986). https://doi.org/10.1098/rsta.1986.0101
H. Horii, S. Nemat-Nasser, Compression-induced microcrack growth in brittle solids: Axial splitting and shear failure. J. Geophys. Res., J. Geophys. Res. (USA) 90, 3105 (1985). https://doi.org/10.1029/JB090iB04p03105
H. Horii, S. Nemat-Nasser, Compression induced nonplanar crack extension with application to splitting, exfoliation, and rockburst. J. Geophys. Res. **87**, 6805–6821 (1982)
D. Iliescu, E.M. Schulson, The brittle compressive failure of fresh-water columnar ice loaded biaxially. Acta Mater. 52, 5723–5735 (2004). https://doi.org/10.1016/j.actamat.2004.07.027
W. Jin, C. Arson, Discrete equivalent wing crack based damage model for brittle solids. Int. J. Solids Struct. **110**, 279–293 (2017)
M. Kachanov, Elastic solids with many cracks: a simple method of analysis. Int. J. Solids Struct. Int. J. Solids Struct. (UK) **23**, 23–43 (1987)
J.M. Kemeny, N.G.W. Cook, Micromechanics of deformation in rocks, in *Toughening Mechanisms in Quasi-Brittle Materials*. ed. by S.P. Shah (Kluwer Academic Publishers, 1991), pp. 155–188
F.E. Kennedy, E.M. Schulson, D.E. Jones, The friction of ice on ice at low sliding velocities. Philos. Mag. A Phys. Condens. Matter. Struct. Defects Mech. Prop. **80**, 1093–1110 (2000). https://doi.org/10.1080/01418610008212103
K. Kolari, Modeling Splitting and Spalling of Columnar Ice Compressed Biaxially: The Role of Crack Nucleation. J. Geophys. Res. Solid Earth **124**, 3271–3287 (2019). https://doi.org/10.1029/2018JB017032
K. Kolari, A complete three-dimensional continuum model of wing-crack growth in granular brittle solids. Int. J. Solids Struct. **115–116**, 27–42 (2017). https://doi.org/10.1016/j.ijsolstr.2017.02.012
K. Kolari, Damage mechanics model for brittle failure of transversely isotropic solids - Finite element implementation, VTT Publications. Technical Research Center of Finland, Espoo 15, Finland (2007)
J. Kuutti, K. Kolari, Interaction of periodic arrays of wing cracks. Eng. Fract. Mech. **200**, 17–30 (2018). https://doi.org/10.1016/j.engfracmech.2018.03.030
R.W. Lee, E.M. Schulson, Strength and Ductility of ice under tension. J. Offshore Mech. Arct. Eng. **110**, 187–191 (1988)
N. Maeno, M. Arakawa, Adhesion shear theory of ice friction at low sliding velocities, combined with ice sintering. J. Appl. Phys. J. Appl. Phys. (USA) **95**, 134–139 (2004). https://doi.org/10.1063/1.1633654
B. Michel, Ice mechanics. University of Laval, Quebec. (1978). https://doi.org/10.1139/t71-004
S. Nemat-Nasser, M. Obata, A Microcrack Model of Dilatancy in Brittle Materials. J. Appl. Mech. Trans. ASME **55**, 24–35 (1988)
V. Pensée, D. Kondo, L. Dormieux, Micromechanical Analysis of Anisotropic Damage in Brittle Materials. J. Eng. Mech. **128**, 889–897 (2002). https://doi.org/10.1061/(ASCE)0733-9399(2002)128:8(889)
R.C. Picu, V. Gupta, Observations of crack nucleation in columnar ice due to grain boundary sliding. Acta Metall. Mater. **43**, 3791–3797 (1995a)
R.C. Picu, V. Gupta, Crack nucleation in columnar ice due to elastic anisotropu and grain boundary sliding. Acta Metall. Mater. **43**, 3783–3789 (1995b)
C.E. Renshaw, E.M. Schulson, Universal behaviour in compressive failure of brittle materials. Nature 412, 897–900 (2001). https://doi.org/10.1038/35091045
E.M. Schulson, Compressive shear faults within arctic sea ice: Fracture on scales large and small. J. Geophys. Res. **109**, C07016 (2004). https://doi.org/10.1029/2003JC002108

E.M. Schulson, The Tensile and Compresive Fracture of Ice, in *IUTAM / IAHR Symposium on Ice-Structure Interaction*. ed. by S.J. Jones, R.S. McKenna, J. Tillotson, I.J. Jordaan (Springer-Verlag, Berlin Heidelberg, 1991), pp. 165–183

E.M. Schulson, The brittle compressive fracture of ice. Acta Metall. Mater. **38**, 1963–1976 (1990). https://doi.org/10.1016/0956-7151(90)90308-4

E.M. Schulson, P. Duval, *Creep and Fracture of Ice* (Cambridge University Press, 2009)

E.M. Schulson, A.L. Fortt, D. Iliescu, C.E. Renshaw, Failure envelope of first-year Arctic sea ice: The role of friction in compressive fracture. J. Geophys. Res. Ocean., J. Geophys. Res., C. Oceans (USA) **111**, 11–25 (2006). https://doi.org/10.1029/2005JC003235

E.M. Schulson, O.Y. Nickolayev, Failure of columnar saline ice under biaxial compression: Failure envelopes and the brittle-to-ductile transition. J. Geophys. Res. Solid Earth **100**, 22383–22400 (1995)

T.R. Smith, E.M. Schulson, Brittle compressive failure of salt-water columnar ice under biaxial loading. J. Glaciol. **40**, 265–276 (1994). https://doi.org/10.1017/S0022143000007358

T.R. Smith, E.M. Schulson, The brittle compressive failure of fresh-water columnar ice under biaxial loading. Acta Metall. Mater. **41**, 153–163 (1993). https://doi.org/10.1016/0956-7151(93)90347-U

A.N. Stroh, A theory of the fracture of metals. Adv. Phys. **6**, 418–465 (1957)

J. Weiss, E.M. Schulson, The failure of fresh-water granular ice under multiaxial compressive loading. Acta Metall. Mater. **43**, 2303–2315 (1995)

Q. Yang, W.Y. Zhou, G. Swoboda, Micromechanical identification of anisotropic damage evolution laws. Int. J. Fract. **98**, 55–76 (1999)

S.W. Yu, X.Q. Feng, A micromechanics-based damage model for microcrack-weakened brittle solids. Mech. Mater. **20**, 59–76 (1995). https://doi.org/10.1016/0167-6636(94)00046-J

Chapter 3
Source Location and Dataset Incompleteness in Acoustic Emissions from Ice Tank Tests on Ice-Rubble-Ice Friction

Check for updates

Katerina Stavrianaki, Mark Shortt, and Peter Sammonds

Abstract Experiments in rock mechanics conducted in the laboratory have revealed that the generation of elastic waves during micro-fracturing provide a small-scale analogue to seismogenic processes. These elastic waves are called acoustic emissions (AE). In contrast to rock, the seismic behaviour of ice under applied stresses is relatively unstudied and a robust statistical categorisation of acoustic events has not yet been performed. In analogy with experiments from rock mechanics, where it has been proven that statistical laws of seismicity are obeyed in AE events, we aim to characterise seismic activity in ice. This was done by measuring acoustic emissions during ice-rubble-ice friction tests conducted at the HSVA ice tank. Specifically, we studied AE data from two tests which used different rubble geometries: large round and small angular. Using these datasets from we first conduct source location of the AE activity. Secondly, we investigate the possibility of incompleteness in the AE datasets during periods of increased activity. Our results from source location show that the round rubble geometry gave higher acoustic activity at the sliding interfaces. We observe potential incompleteness in both datasets. This analysis has applications in field of seismology as well as in ice mechanics.

K. Stavrianaki (✉) · M. Shortt · P. Sammonds
Institute for Risk and Disaster Reduction, University College London, London, UK
e-mail: k.stavrianaki@ucl.ac.uk

M. Shortt
e-mail: mark.shortt.12@ucl.ac.uk

P. Sammonds
e-mail: p.sammonds@ucl.ac.uk

K. Stavrianaki
Department of Statistical Science, University College London, London, UK

J. Tuhkuri and A. Polojärvi (eds.), *IUTAM Symposium on Physics and Mechanics of Sea Ice*, IUTAM Bookseries 39, https://doi.org/10.1007/978-3-030-80439-8_3

3.1 Introduction

Seismology is the study of the generation, propagation, and recording of elastic waves in the Earth (and other celestial bodies) and of the sources that produce them (Lay and Wallace 1995).

On Earth's scale, a portion of the energy that is released during frictional sliding along faults is converted to seismic waves that propagate outwards. On a laboratory scale, when a material is being deformed, elastic waves are emitted of similar nature to those generated by earthquakes (Scholz 1968). These waves are called acoustic emissions (AE). There are many similarities between earthquakes and AE events. Statistical laws that describe seismicity such as the Gutenberg-Richter and the Omori law are obeyed in AE events measured in laboratory samples of rock (Mogi 1963; Scholz 1968; Sammonds et al. 1992; Lockner 1993) and therefore AE provides a great tool to study the seismogenic process.

AE originates from strain localisation and deformation (e.g. crack nucleation/dislocation) in a material, where a portion of the locally accumulated energy is released in the form of the transient acoustic waves. Measurement of AE is common in experiments on rock mechanics, and provides an in-situ insight into the state of damage of a sample subjected to a given strain. Whilst AE measurements are less common in ice mechanics experiments, there are nevertheless a number of studies that have measured AE from various types of tests. The majority of these studies have taken place in the laboratory (e.g. Gold 1960; St Lawrence and Cole 1982; Sinha 1982, 1985; Rist and Murrell 1994; Weiss and Grasso 1997; Cole and Dempsey 2006; Li and Du 2016) where conditions are controlled and ambient noise is typically low. However, there have also been several (but fewer) studies documenting AE recorded from experiments on sea ice performed in-situ in the field (e.g. Langley 1989; Xie and Farmer 1994; Langhorne and Haskell 1996; Cole and Dempsey 2004; Lishman et al. 2019, 2020). Recently, Marchenko et al. (2019) and Lishman et al. (2019) have also documented results from AE in ice tank experiments.

A common trend in AE research on ice, particularly in earlier studies, is the focus on obtaining the temporal variation in AE activity, and its correlation with applied load. From this, potential source mechanisms can be inferred. In general, two primary source mechanisms have been identified for ice: dislocation breakaway and microcracking. Dislocation breakaway has been documented as a potential source mechanism in creep tests on ice under compressive loading (St Lawrence and Cole 1982; Weiss and Grasso 1997) and cyclic loading of cantilever beams (Langhorne and Haskell 1996). Furthermore, Weiss and Grasso (1997) recorded AE in the laboratory during creep deformation of single crystals of ice under uniaxial compression and torsion creep, and concluded that creep is a marginally stable state rather than a steady-stable state.

Microcracking is commonly identified in AE from tests involving ice fracture (Sinha 1982; Rist and Murrell 1994; Cole and Dempsey 2004, 2006; Li and Du 2016; Lishman et al. 2020) but has also been suggested to occur in creep tests when the stress level is high (St Lawrence and Cole 1982). Recently, Li and Du

(2016) measured AE in laboratory tests on fresh ice samples subjected to uniaxial compression and three-point bending. Analysing the waveforms of the individual AE signals, and plotting the average frequency against the rise amplitude, they were able to differentiate between tensile and shear cracking events. They found that shear cracking was the primary failure mode for ice under compression, whereas bending failure was controlled by tensile movement cracks. Lishman et al. (2019) used the same analysis on AE data obtained during field tests on sea ice under compression, tension, shear and indentation, but were unable to verify the results of Li and Du (2016), probably due to the 100 kHz low frequency cut-off of the instrumentation, which meant lower frequency shear cracks could not be recorded.

There exists a number of AE analyses used in rock mechanics and seismology that have yet to be conducted on ice. In this paper, we aim to implement two such analyses: (1) source location of AE events, and (2) short term aftershock incompleteness (STAI) of the AE datasets. We use AE data obtained from friction tests on saline ice conducted at Hamburgische Schiffbau-Versuchsanstalt (HSVA) in 2017. In these tests, numerous types of ice rubble were introduced between sliding interfaces with the aim of investigating the influence of fault gouge on the frictional behaviour.

To our knowledge, no previous studies on AE generated during ice-ice friction tests have been conducted, but we may postulate that in the static regime, the shear fracture of freeze-bonds is a probable source mechanism. This is particularly true in the case of a floating ice sheet, where surrounding water will act to amplify the bonding process. We can also make analogies with results from rock mechanics to gain an insight into other potential source mechanisms for ice friction. From data obtained from double-shear experiments on rock, Sammonds and Ohnaka (1998) suggest that the maximum AE rate coinciding with the maximum shear stress is a result of fracturing asperities at the fault surfaces. As the fault slides, a fractal length distribution of cracks grows as contacting asperities interact. Other studies (Michlmayr et al. 2012; Jiang et al. 2017) also suggest the breaking of asperities between solid surfaces as a potential source mechanism of AE generated during shearing of geological materials.

In the presence of a fault gouge within the sliding interface (analogous to the rubble region in our experiment), it is suggested by Michlmayr et al. (2012) that acoustic signals may be generated by the failure of buckling force chains and frictional slip between grains. Furthermore, it is noted by Mair et al. (2007) that the contact area between grains within the gouge is an important factor in determining the average AE occurrence rate.

If we wish to assess the evolution of damage in ice subjected to mechanical loading, it is desirable to map these deformation events so we may predict where or how the ice will fail. In rock mechanics this is done via source location (e.g. Brantut 2018). Although there are several methods of source location, the basic concept is the same—arrange an array of sensors across the surface area of the material to 'triangulate' the events given measurements of wave speed and arrival times.

For investigation of STAI, we introduce the Gutenberg-Richter (G-R) law (Ishimoto and Iida 1939; Gutenberg and Richter 1944) which gives the exponential relation between the frequency of earthquakes (N) and their magnitude (M) in a given region and period of time. This law is described mathematically as follows:

$$\log_{10}[N(m)] = \alpha - b(m - m_t) \tag{3.1}$$

In the equation above N(m) is the number of earthquakes with magnitude greater than or equal to m_t, α and b also known as the b-value are the ordinate intercept and slope, respectively, of the line that relates m_t and log10N(m).

The G-R law corresponds to a power law in terms of the energies which is a typical sign of scale-invariance (Main 1996; Turcotte, 1997) and it has been shown to hold for magnitudes as small as −4.4 (Kwiatek et al. 2010). The b-value is also linked with the magnitude of completeness of an earthquake catalogue. The magnitude of completeness (M_c) is the magnitude above which we are certain we have a complete dataset, and this is particularly important for any statistical analysis.

Here we use the equation from Sammonds and Ohnaka (1998) that relates the logarithm of AE hit frequency with the amplitudes. The amplitudes are measured in dB and divided by 20. This reproduces a G-R law for our experimental data and we can find the corresponding b-value. The b-value and its variations during a laboratory experiment is frequently used to monitor the fracturing process inside a specimen both for rock deformation (e.g. Mogi 1963; Scholz 1968; Lockner 1993; Sammonds and Ohnaka 1998) but also in ice.

Recent studies (Li and Du 2016; Lishman et al. 2019, 2020) have applied the concept of b-value to AE data from ice mechanics experiments. They calculated the temporal change in the b-value with the aim of inferring the evolution of damage within the material. Li and Du (2016) calculated this temporal change for laboratory tests on compression and three-point bending, and found that for both test types, the b-value gradually increased to above 1 as loading was applied, indicating that micro-cracks developed stability. During and after failure, the b-values dropped sharply, indicating a rapid expansion of cracks. From AE data obtained from compression and indentation tests on sea ice conducted in the field, Lishman et al. (2020) found respective b-values of 1.80 and 1.59. The temporal variation in b-value was also investigated for both types of tests. They found that as the ice evolved towards failure, the b-value decreased and tended towards 1. In the case of compression, the b-value increased again during unloading. Furthermore, Lishman et al. (2019) deduced the temporal change in b-value from AE data obtained in the HSVA Ice Tank from wave-induced deformation of a floating saline ice sheet, and speculated that cracks may have formed and healed repeatedly under cyclic motion. In each of the aforementioned studies, it is noted that analysis of b-values can provide a useful tool in predicting failure in ice under deformation.

STAI is a phenomenon observed in earthquake data. For a short period after large earthquakes the seismic activity is increased (we have aftershocks), and during this period smaller earthquakes can be lost in the coda of the larger events. The missingness of the small data has an effect on the b-value during this period. Since the b-value describes the ratio between large and small earthquakes, the lack of small events will cause an artificial decrease in the b-value.

Borrowing from seismology, we aim to look at these periods of increased activity and determine if we observe incompleteness. We use the method developed by Hainzl (2016), to find incompleteness. In his paper Hainzl demonstrates how the magnitude of completeness varies with seismicity rate. The b-value is estimated using a maximum likelihood method for different seismicity rates. Incomplete recordings will produce an apparent decrease in the b-value since we assume a constant frequency—magnitude distribution of the data.

3.2 Experimental Set-Up

AE data was gathered from a series of slide-hold-slide friction tests on saline ice conducted at the HSVA Large Ice Model Basin (LIMB) with the aim of investigating the influence of ice rubble gouge in between the contacting interfaces. A full description of the friction tests can be found in Scourfield (2019).

The HSVA LIMB measures 78 m x 12 m × 2.5 m and contains NaCl-doped water to a salinity of 6.8 ppt. The structure of S2-type columnar ice was replicated using 'artificial full-scale ice'. The air temperature was set to −8 °C for the duration of the tests and the average initial ice thickness (at the start of the experimental preparation) was 22 cm, and had reached 29 cm by the end of the test programme. The saline ice had bulk salinity of around 1.5 ppt and a bulk density of 919 kg/m^3.

The experimental set-up is shown in Fig. 3.1a. The experiment utilised a double-direct-shear configuration, comprising a central ice block measuring 3.5 m × 1.5 m and a side ice beam either side, both measuring 1.5 m × 0.8 m. The central block and the side beams were separated by a 0.5 m wide open channel of water containing rubble blocks, which were varied in size and shape over the experiment. A nominal normal load of 1.5kN was applied on each side beam by a set of side load frames. The central block was pushed under the normal load using a pusher plate attached to the moveable main carriage. The rate- and state-dependencies were investigated by varying the sliding velocity and hold time respectively.

Nine AE transducers were deployed on the central ice block by making 1 cm deep holes and then freezing using fresh water. The spatial arrangement of the AE transducers on the central block is shown in Fig. 3.1b. The transducers comprised PZT-5 h compressional piezoelectric crystals measuring 15 mm in diameter and 5 mm in thickness, with a 500 kHz central frequency. These crystals were soldered onto copper discs and potted in epoxy. The signals from each transducer were amplified locally by 40 dB via Vallen preamplifiers. Figure 3.2 shows an AE transducer connected to a Vallen preamplifier. The amplified signals were then transmitted to a Vallen AMSY5 system, which recorded a hit (maximum amplitude and time) if the signal exceeded a threshold of 40 dB. For each registered hit, the Vallen system also records a 400 μs transient waveform (voltage vs time) beginning 50 μs prior to the triggering of the threshold. Sampling frequency of the transient is 5 MHz. Observations of the shape of these transients are useful in noise filtering, which is discussed in the next section.

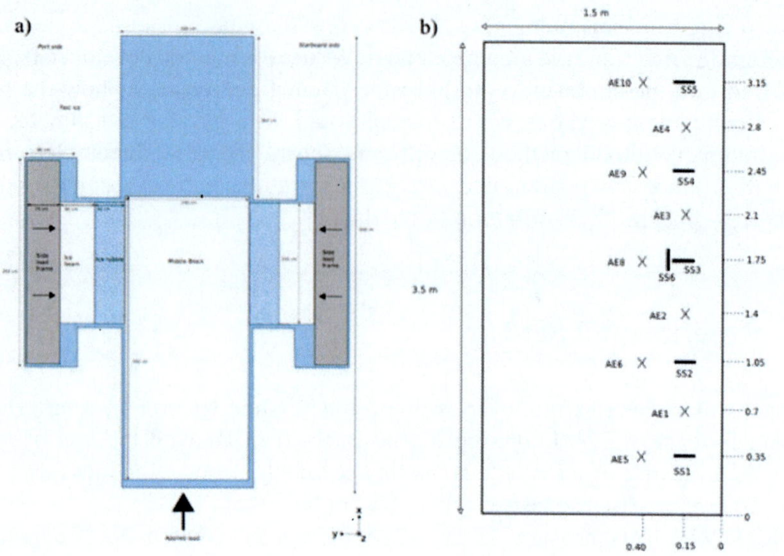

Fig. 3.1 **a** Overall experimental set-up for the friction tests, and **b** location of instruments on the central ice block – AE transducers are shown as crosses, mercury stress sensors (SS) are also shown (Scourfield 2019)

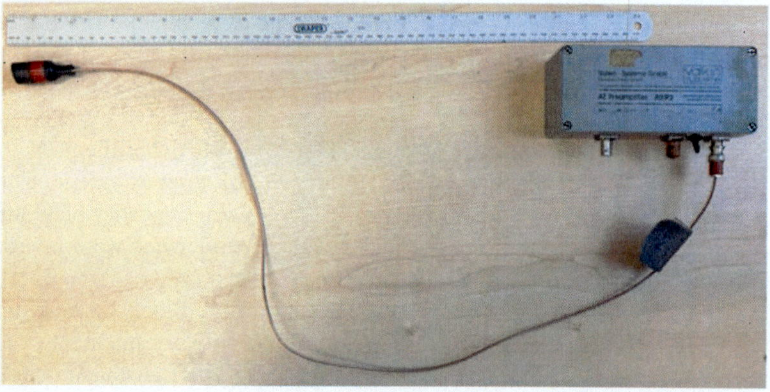

Fig. 3.2 Example of an AE transducer and Vallen pre-amplifier used in the friction tests

For the analysis in this paper, we use AE data from two selected tests, chosen on the basis of a clear spike in coefficient of friction under the initiation of motion of the central block. These tests were named 'Test 6b1' and 'Test 9b' respectively and conveniently shared the same hold time and sliding velocity, but had two different types of rubble—with varying size and shape. This enabled a direct investigation into the influence of rubble type on the acoustic activity. A summary of the properties of

Table 3.1 Summary of test properties for Test 6b1 and Test 9b

	Test 6b1	Test 9b
Rubble type	Large round	Small angular
Rubble dimensions	16.5 cm diameter, 9 cm depth	10 cm short axis, 17 cm long axis, 9 cm depth
Nominal normal load	1.5kN	1.5kN
Hold time	10,000 s	10,000 s
Nominal sliding velocity	0.3 cm/s	0.3 cm/s
Sliding distance	20 cm	20 cm

each test is given in Table 3.1. Plots of coefficient of friction against time for each test are given in Fig. 3.3.

In both tests there are clear spikes in the coefficient of friction at the initiation of sliding, corresponding to the static value and can be physically interpreted as the breaking of freeze-bonds developed between the ice blocks and the rubble during the hold period. Following this spike, the value of μ drops sharply to a relatively steady state corresponding to dynamic friction. The clear distinction between the static and dynamic friction regimes in these tests aids with the interpretation of the acoustic data. It should be noted that other tests conducted did not exhibit such a clear distinction.

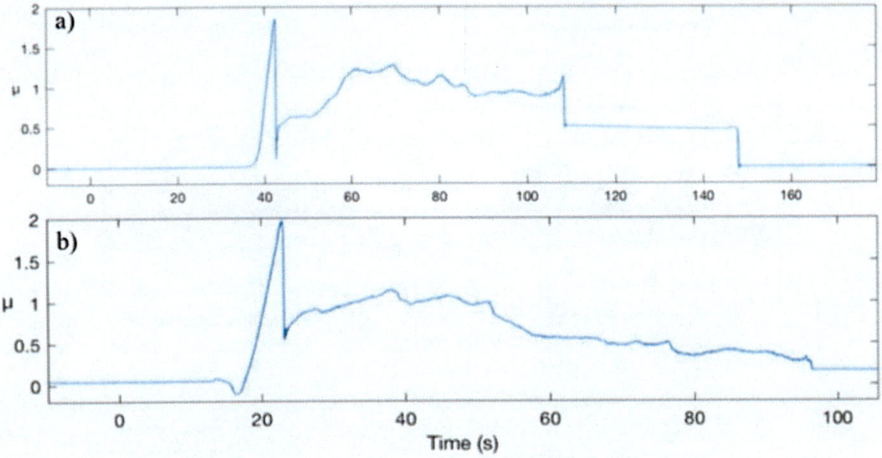

Fig. 3.3 Plots μ vs time for **a** Test 6b1, and **b** Test 9b. Clear spikes in coefficient of friction are visible at the initiation of sliding adapted from Scourfield (2019)

3.3 AE Datasets and Noise Filtering

The data acquisition period ran before and after the actual start and end time of the tests. Table 3.2 shows characteristics of the AE data from each test. In total there were 54,165 hits recorded in Test 6b1, and 70,471 hits recorded in Test 9b. The raw data from each test are given in the top two panels of Fig. 3.4. It can be seen that in both tests there is a large amount of noise, which needed to be filtered before any analysis could be performed. It is possible that this noise originated at the hardware level due to the inherent noise floor of the instrument chain and is thought to be of similar magnitude to the test source. Both the test signal and the noise floor were then subject to 40 dB gain provided locally by the Vallen preamplifiers.

Removal of noise was done in post-processing based on two criteria. The first criterion was time—the start and end times of the test, which were easily identified by viewing the overall AE activity. Any events occurring before or after these times were

Table 3.2 Characteristics of AE data from Test 6b1 and Test 9b

	Test 6b1	Test 9b
Total hits	54,165	70,471
Test start time t_{start}	43.5 s	76.9 s
Test end time t_{end}	139.5 s	159.0 s
Hits after filtering	5231	3325

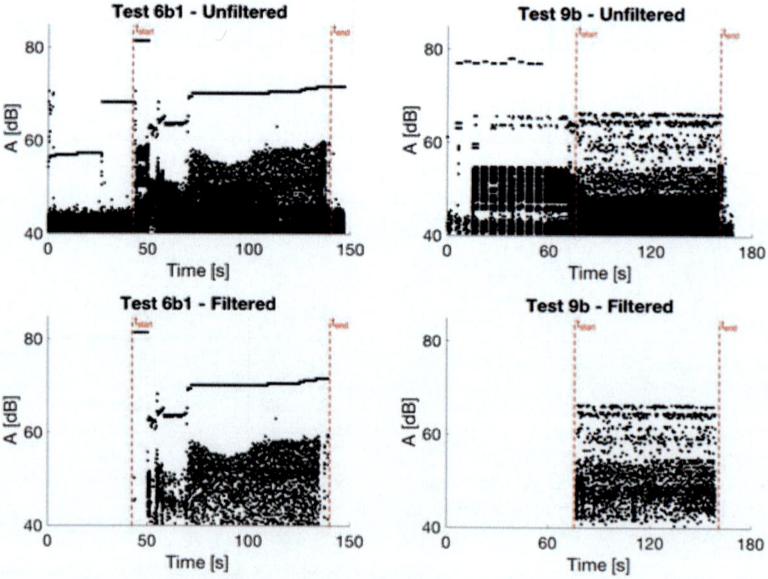

Fig. 3.4 Unfiltered and filtered datasets of maximum hit amplitude (A) with time for Test 6b1 and Test 9b. The start and end times of both tests are also identified

removed. The remaining events were then filtered based on a frequency criterion, similar to Lishman et al. (2019, 2020) and Marchenko et al. (2019). The frequency spectrum of a given hit is deduced by taking a fast Fourier transform of the transient waveform data. From this, the frequency at the maximum amplitude of the spectrum (FMXA) is calculated for each hit. By plotting FMXA against time, it is possible to deduce frequency bands of noise, which occur throughout the duration of the logging, irrespective of the test start and end times.

The FMXA against time data for both tests are plotted in the top panels of Fig. 3.5 for all hits. Focusing first on Test 6b1, it can be seen that a large number of hits occur across a narrow frequency band, between 420–450 kHz. Events within this frequency band occur across the duration of the logging period, and thus can be identified as noise. Now focusing on Test 9b, it can be seen that the noise is not confined to one frequency band and exists superimposed on real data. Therefore, to distinguish between the two we conducted further filtering based on analysis of hit duration (D) and counts (CNTS), which is the number of times a waveform crosses the threshold. Figure 3.6 gives a graphical description of these two variables. We conducted this filtering over the frequency band 100 < F < 200 kHz which appeared to show a high hit activity with noise superimposed. We found that the hits in this frequency region broadly fell into two regimes for D and CNTS. These corresponded to (1) D < 10 μs, CNTS = 1 and (2) D > 10 μs, CNTS ≠ 1.

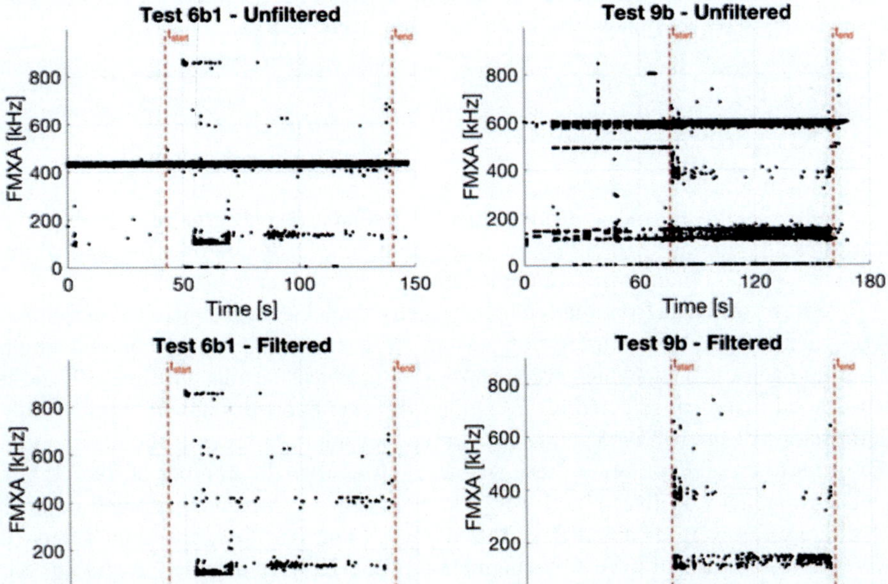

Fig. 3.5 Unfiltered and filtered datasets of frequency at maximum hit amplitude (FMXA) with time for Test 6b1 and Test 9b. The start and end times of both tests are also identified

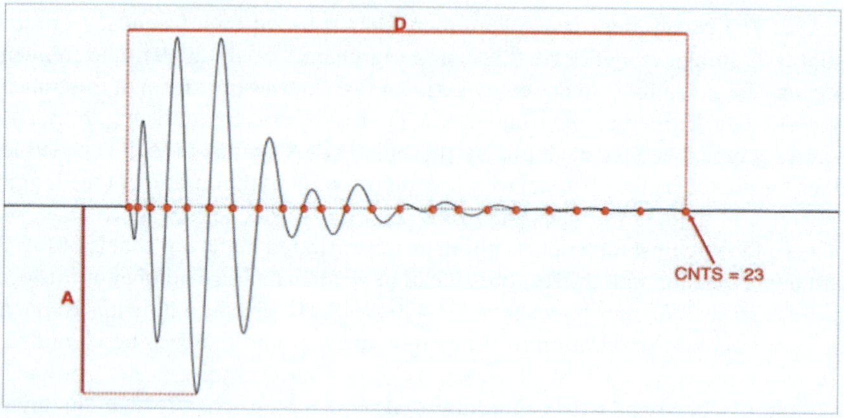

Fig. 3.6 Typical transient waveform recorded for an individual hit. The maximum amplitude (A), hit duration (D) and counts (CNTS) are also labelled

Figure 3.7 shows the hits filtered on these regimes. It can be seen in the latter criteria ($D > 10\,\mu s$, $CNTS \neq 1$), there exists concentrations of hits in two bands which are consistent with noise existing outside the test period. Therefore, we believe these criteria to be sufficient for filtering over this frequency region. The filtered data for amplitude and FMXA are given in the bottom panels of Figs. 3.4 and 3.5 respectively. The filtering gave less than 10% hits remaining from the total.

3.4 AE Source Location

The Vallen acquisition system measures the transient waveforms of each hit, and from this and due to the fact that we used multiple transducers over the area of the ice sheet we can obtain basic source location of microseismic events.

The waveform data from filtered hits on each channel were grouped into individual AE structures. Each structure (known as an AE event hereafter) contained a single hit from each channel, which were chosen on the basis similar hit times. Thus, it is assumed that each selected hit in a given AE event corresponds to an individual deformation event, received at different times for each transducer. It should be noted that prior to the generation of these AE event structures, the number of filtered hits on each channel was determined. Any channels with a comparatively small number of filtered hits were disabled from this analysis. For Test 6b1, channels 4 and 10 were disabled, whilst for Test 9b channels 4, 9 and 10 were disabled. It should also be noted that channel 7 was unused in all tests.

To conduct source location, it was necessary to obtain the p-wave velocity through the ice. This was done using the calibration function on the Vallen system, which emits four evenly spaced pulses (1 per second) on each channel consecutively. The frequency of the emitted pulses was approximately 150 kHz. These pulses are

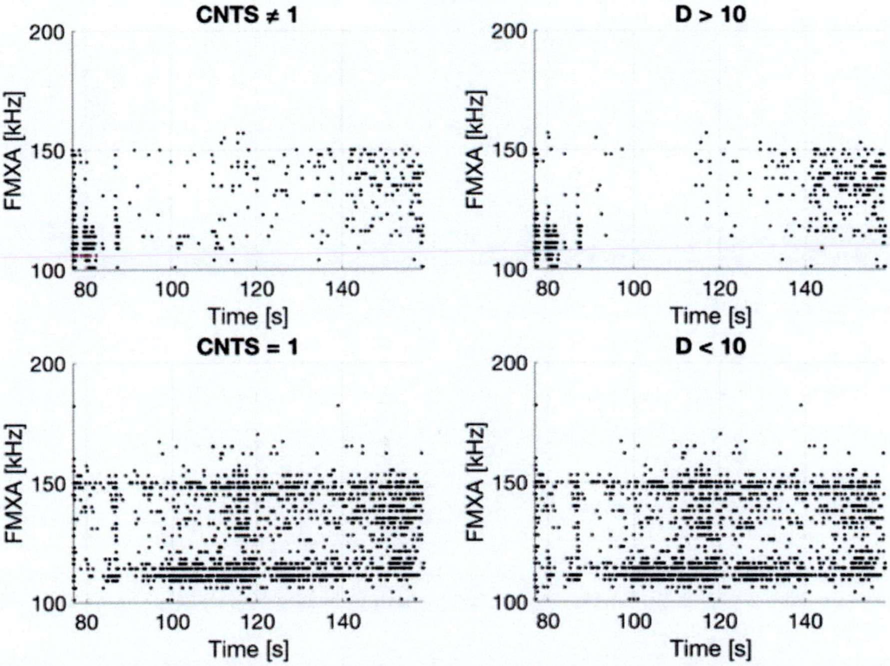

Fig. 3.7 Filtering of AE hits for Test 6b1 in the frequency range 100–200 kHz. We filter this data based on two regimes for D and CNTS: (1) D < 10 μs, CNTS = 1 and (2) D > 10 μs, CNTS ≠ 1. We believe that the latter conditions correspond to noise

received by surrounding channels at a lower amplitude a short time Δt after the emission pulse. A typical hits vs time graph for such a calibration test is shown in Fig. 3.8. Knowing the distance d between the emitting and receiving transducers it is possible to obtain an estimate for the p-wave velocity v_p via the basic distance-time relation:

$$v_p = \frac{d}{\Delta t}. \tag{3.2}$$

This calculation was conducted for each emitted hit with an obvious corresponding detected hit. The average velocity was then calculated over all emitted-received pairs and was found to be $v_p = 3213$ m/s. Our value of v_p is slightly higher than those measured in field sea ice (Xie and Famer 1994; Marchenko et al. 2020), but is slower than in freshwater bubble-free ice (Vogt et al. 2008). We believe that our intermediate value for v_p is a result of the low water salinity used in the tank. It can be possible for the underlying water to have an influence on the wave velocity, resulting in an artificial decrease in the measured value. Since our value for v_p lies within the range of previously measured values, and since the ice was relatively thick, we do not believe that water had a significant influence in our study.

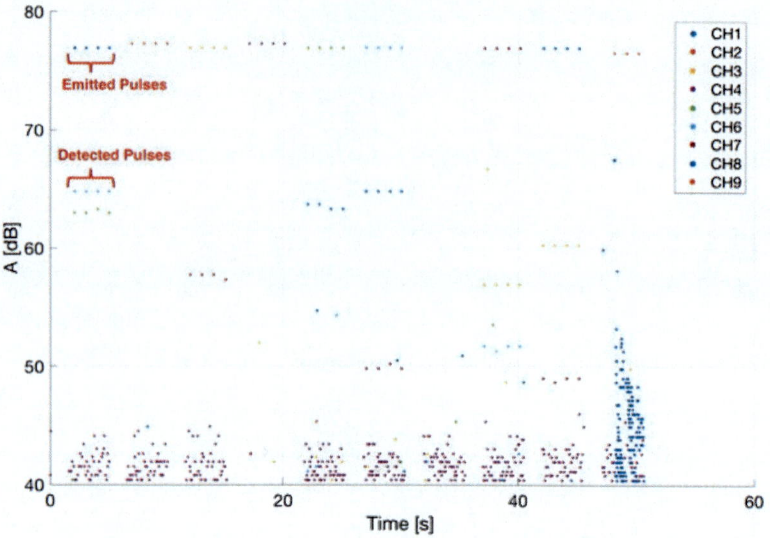

Fig. 3.8 Typical hits versus time graph for a calibration run, which we use to identify the p-wave velocity. Four evenly spaced pulses are emitted by each channel consecutively. These pulses are detected on surrounding channels

It should be noted that for a complete analysis of source location, the degree of anisotropy of p-wave velocity should be determined by deploying transducers in all directions across the area and depth of the ice block. However, in this analysis, we use an isotropic velocity model, since the positioning of the transducers meant we could only determine p-wave velocity in two directions.

The next step was to determine the p-wave arrival times at each transducer. This was done via the RMS amplitude method, which calculates an autopicking function via a moving window approach (ASC 2014). At each waveform datapoint *i*, two windows are generated—a front window of length (in datapoints) *NF* and a back window of length *NB*. The value of the autopicking function is given by Eq. (3.3), and represents the difference in energy contained in the front window compared with the back window:

$$F_i = \frac{\sum_{j=i+1}^{i+NF} A_j^2}{\sum_{j=i-1}^{i-NB} A_j^2} \tag{3.3}$$

where A_j is the amplitude. This gives an array of RMS values of the length of datapoints within the waveform. We use the following values: NF = 100, NB = 150, which were deduced by viewing transient waveforms. The p-wave arrival time was then then deduced by calculating the time in the waveform that coincides with the maximum value of RMS. Each event then has an arrival time for each channel and given these arrival times, in addition to the p-wave velocity and the relative

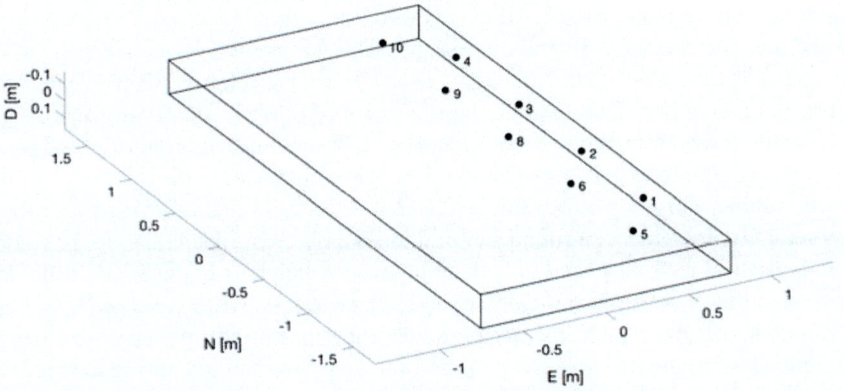

Fig. 3.9 Geometry for source location corresponding to the dimensions of the central ice block. The positions of the AE transducers are also shown

positions of the transducers, an estimate for the source location of a given event can be obtained. Events identified from Test 6b1 and Test 9b are plotted onto the central ice block geometry (see Fig. 3.9) in Fig. 3.10a, b respectively. It should be noted that, since the transducers were all positioned at the same depth within the central block, we can only estimate source location across a plane at this depth. Incorporation of depth effects requires positioning of transducers throughout the thickness of the ice block.

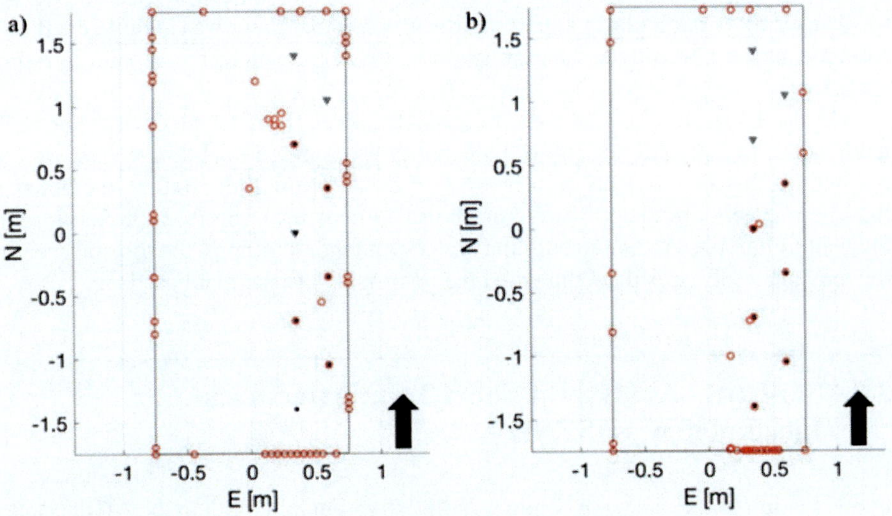

Fig. 3.10 Results from source location for **a** Test 6b1 and **b** Test 9b plotted on the geometry of the central ice block, The black arrows represent the direction in which the pushing load was applied. The triangles represent channels which were disabled from the analysis

For Test 6b1, a total of 4967 AE events were identified, which was reduced to 331 events after filtering. For Test 9b, a total of 6674 AE events were identified, which reduced to just 67 events after filtering. Thus there were relatively fewer AE events in Test 9b, which is reflective of the high amount of noise recorded during this test.

Overall, there were more events located at the sliding edges in Test 6b1, and thus we may conclude that large round rubble gave higher AE activity than the small angular rubble. This may be due to the different geometries in the two types of rubble. Theoretically, each piece of round rubble should only create a single point of contact with the central ice block (and other rubble pieces), whereas in the angular rubble it is possible for an entire face to contact and even tesselate with one another.

Acoustic activity at the interacting edges may be initially governed by freeze-bonding between the ice rubble and the central ice block forming during the previous 10,000 s hold period. We can thus assume that greater AE activity in the round rubble resulted (at least initially) from a greater number of individual freeze-bond failure events. At the initiation of sliding, it may have been easier for bonds in round rubble to break since in this geometry both rotational and linear forces will be acting in significant proportions. In the case of the angular rubble, rotational forces should be less apparent. Transmission of force chains (as noted by Michlmayr et al. 2012) is also easier with round rubble which may have further increased freeze-bond failures between individual blocks. Following initial freeze-bond failure, in the sliding regime, further AE may be generated from sliding between the central block and the (now freed) rubble pieces.

One interesting similarity between the two tests is the clustering of events at the near end of the ice block. We attribute this to indentation in the ice block caused by the pusher plate which was located at the same end. Indeed in Test 9b, where this clustering is particularly apparent, it was noted that there was audible creaking from the pusher plate due to the high load (7.7 kN) experienced at the point of static friction.

It is also interesting to note the events located away from the edges of the central ice block. Since the AE activity is expected to be confined to sliding edges, we do not believe that these events were generated by deforming ice. Rather, we believe that these events originated from other noise sources resulting from the motion of the central block, such as surrounding water that tended to flood the top surface, or the movement of wires that connected the instruments to the main carriage.

3.5 Statistical Analysis of Short Term Aftershock Incompleteness (STAI)

In this section of the paper we conduct an investigation into STAI in the AE datasets. To do this, we follow the methodology described in Hainzl (2016), adjusting the bin size to an appropriate value for our experiments. As a proxy for earthquake magnitude, we divide amplitude values by 20, as per Sammonds and Ohnaka (1998).

This method estimates the b-value for filtered events with increasing event rate using the maximum likelihood method for event bins with at least 5 amplitudes using (Aki 1965; Marzocchi and Sandri 2003; Hainzl 2016):

$$\hat{b} = \frac{1}{\ln(10)(m - (M_c - 0.5\Delta m))} \tag{3.4}$$

where: m is magnitude.

$M_c = 2$ is the magnitude of completeness.

Δm is binning interval, set to 0.01 as per the precision of recorded amplitudes.

The magnitude of completeness is the magnitude above which we are certain we have complete recording of the data and is calculated by fitting the G-R Law to all the filtered data. It corresponds to the lowest magnitude to which the linear regime begins.

Figure 3.11 shows the G-R fitting for the AE data in both tests. Events were grouped by amplitude in bins of size 3 dB. The b-value is obtained by taking the gradient of the linear trendline, and was found to be b = 0.71 for Test 6b1 and b = 0.89 for Test 9b. We do not observe the non-linear regime existing at low magnitudes for earthquake data. In contrast to smaller scale experiments, where the G-R law is observed (e.g. Sammonds and Ohnaka 1998) the fit in our dataset has a greater uncertainty. This could be due to an increase in uncertainty of all parameters involved in all the calculations due to the large-scale nature of the experiment. The loose fit may also be a consequence of the heterogeneity of the source mechanisms, such as those reflecting breaking freeze-bonds compared with those generated during subsequent sliding. Due to the loose fit of the data, it is hard to obtain a precise b-value and the corresponding value for M_c. As a result, we take the $M_c = 2$, corresponding to the minimum amplitude datapoint.

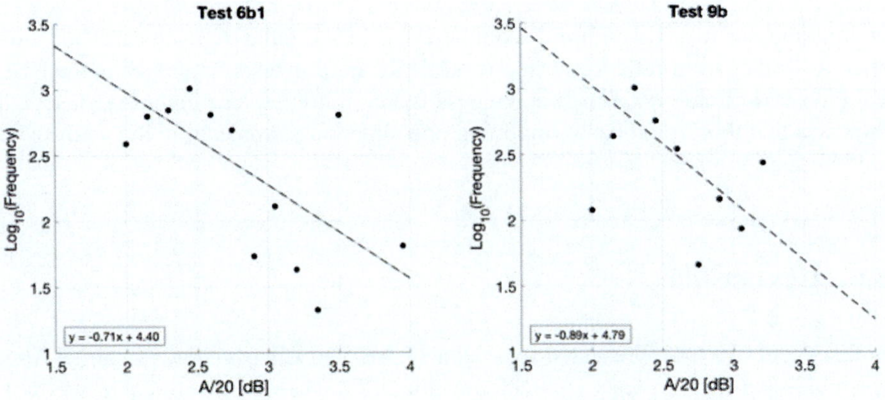

Fig. 3.11 Log(frequency)-magnitude data for both tests. Amplitudes were binned in 3 dB increments. Equations for G-R linear fittings are shown, which are used to determine b-values

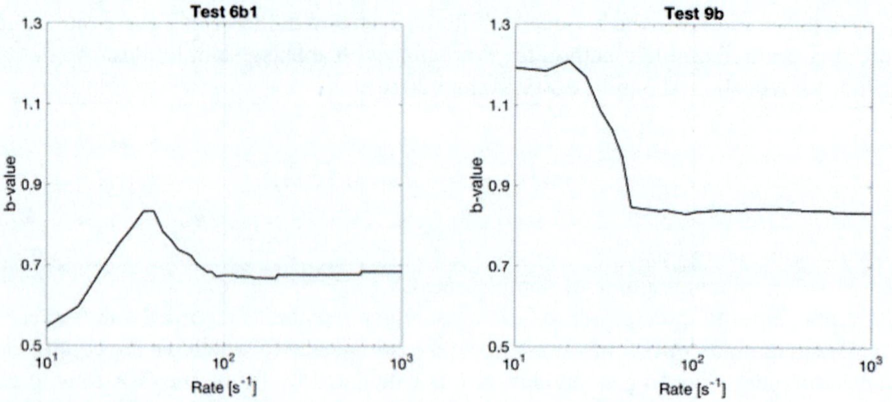

Fig. 3.12 Plots of b-value against hit rate for Test 6b1 and Test 9b. Both tests exhibit at decrease in b-value, which may indicate incompleteness

Figure 3.12 shows the rate-varying b-value obtained from data in both tests, estimated from Eq. 3.4, for rate bin-sizes between 10–1000 s^{-1}. These rate-bin sizes were deduced by observing the typical count rates occurring during our experiments, as well as those found in Marchenko et al. (2019), who measured AE data using the same apparatus described here. In both figures, there is a noticeable decrease in b-value, which may indicate incompleteness in the datasets, as per Hainzl (2016). Following this decrease, the b-value stays constant at around the values obtained in the Fig. 3.11. The onset of decreasing b-value occurs at a rate 40 s^{-1} for Test 6b1 and 20 s^{-1} for Test 9b.

The fact that we observe decreasing b-value in both tests indicates the potential for incompleteness in AE datasets in ice mechanics experiments. However, uncertainties in the calculations should be noted, particularly in the continuous calculation of b-value with rate from datasets which already give a loose fit to the G-R law. We aim to continue our investigation of incompleteness in AE datasets from other ice tank tests, as well as from concurrent smaller scale ice friction tests conducted in the UCL Ice Physics Laboratory. If incompleteness is a common feature in these datasets, it indicates that this would be an important phenomenon to consider in the analysis of any AE measurements in ice.

3.6 Conclusions

In this paper, we have presented two types of AE analysis from data obtained from ice-rubble-ice friction tests conducted at the HSVA LIMB. Specifically, we took data from two individual tests chosen on the basis of a clear spike in the value of μ at the initiation of sliding: Test 6b1 using large round rubble, and Test 9b using small angular rubble. This enabled a direct investigation into the influence of rubble geometry on the acoustic behaviour.

The AE data from both tests was filtered based on two criteria: (1) start and end times of the test, and (2) values for frequency at maximum amplitude (FMXA). For Test 6b1, further filtering based on hit duration and counts was conducted to eliminate noise in regions of high hit activity.

The spatial arrangement of AE transducers enabled source location of AE events to be conducted. P-wave velocity was deduced via calibration assuming an isotropic velocity model, with the average calculated to be $v_p = 3213$ m/s. P-wave arrival times were calculated using an RMS amplitude method. We found a greater number of events to be located at the sliding ice edges in Test 6b1, which we attribute to the rubble geometry and its influence on the breaking of freeze-bonds developed during the previous 10,000 s hold period.

The b-values were determined in both tests. We calculate $b = 0.71$ and $b = 0.89$ for Test6b1 and Test9b respectively. In both cases the data is only a loose fit to the linear G-R Law, probably due to a combination of heterogeneity of the source processes and uncertainties related with the large-scale nature of the experiment. Our b-values are low compared those calculated in Lishman et al. (2019, 2020) and indicate a comparatively high number of large events. We speculate that our lower values are a reflection of the different types of mechanical tests. In contrast to compression, indentation and flexure, where high levels of (small amplitude) microcracking are likely, in our experiment a significant proportion of recorded AE likely originated from larger-scale fracturing of freeze-bonds that formed between central block and the rubble during the previous 10,000 s hold period.

Finally, we conducted an investigation of STAI in the AE datasets by calculating the b-value at rates between 10–1000 s^{-1}. In both tests we see a decrease in the b-value, which may indicate incompleteness in the datasets, as per Hainzl (2016). Following this decrease, the b-values remained constant at levels similar to that obtained over the entire filtered datasets. However, due to loose fitting of the AE data to the G-R Law, a degree of uncertainty exists in this analysis. In future, we aim to continue this investigation of STAI in further AE datasets from both ice tank tests and smaller scale laboratory tests.

Acknowledgements The experiment described in this publication was supported by the European Community's Horizon 2020 Research and Innovation Programme through the grant to HYDRALAB-PLUS, Contract no. 654110. This project was led by Sally Scourfield, and we would like to thank her for permission to use this data. We would also like to thank the Hamburg Ship Model Basin (HSVA), especially the ice tank crew, for the hospitality, technical and scientific support and the professional execution of the test programme in the Research Infrastructure ARCTECLAB. Thanks to Ben Lishman for useful discussions and recommendations on both the experimental set-up and the analysis, and to Nic Brantut for providing the initial MATLAB code for AE source location. Finally, we would like to thank the two anonymous reviewers, whose comments improved our paper.

References

K. Aki, Maximum likelihood estimate of b in the formula logn= a-bm and its confidence limits. Bull. Earthq. Res. Inst. **43**, 237–239 (1965)

Applied Seismology Consultants, InSite Seismic Processor: user operation manual version 3.2. Author, Shrewsbury (2014)

N. Brantut, Time-resolved tomography using acoustic emissions in the laboratory, and application to sandstone compactions. Geophys. J. Int. **213**, 2177–2192 (2018)

D.M. Cole, J.P. Dempsey, In situ sea ice experiments in McMurdo Sound: cyclic loading, fracture, and acoustic emissions. J. Cold Regions Eng. **18**(4), 155–174 (2004)

D.M. Cole, J.P. Dempsey, Laboratory observations of acoustic emissions from antarctic first-year sea ice cores under cyclic loading. In: Proc. of the 18th Int. Conf. on Port and Ocean Eng. Under Arctic Conditions (POAC), Potsdam, USA (2006)

L.W. Gold, The cracking activity in ice during creep. Can. J. Phys. **38**(9), 1137–1148 (1960)

B. Gutenberg, C.F. Richter, Frequency of earthquakes in California. Bull. Seismologic. Soc. Am. **34**(4), 185–188 (1944)

S. Hainzl, Rate-dependent incompleteness of earthquake catalogs. Seismol. Res. Lett. **87**(2A), 337–344 (2016)

M. Ishimoto, K. Iida, Observations sur les seismes enregistres par le mi- crosismographe construit dernierement (1). Bull. Earthq. Res. Inst. Univ. Tokyo **17**, 443–478 (1939)

Y. Jiang, G. Wang, T. Kamai, Acoustic emission signature of mechanical failure: Insights from ring-shear friction experiments on granular materials. Geophys. Res. Lett. **44**(6), 2782–2791 (2017)

G. Kwiatek, K. Plenkers, M. Nakatani, Y. Yabe, G. Dresen et al., Frequency- magnitude characteristics down to magnitude-4.4 for induced seismicity recorded at Mponeng gold mine, South Africa. Bull. Seismol. Soc. Am. **100**(3), 1165–1173 (2010)

P.J. Langhorne, T.G. Haskell, Acoustic emission during fatigue experiments on first year sea ice. Cold Reg. Sci. Technol. **24**(3), 237–250 (1996)

A.J. Langley, Acoustic emission from the Arctic ice sheet. J. Acoust. Soc. Am. **85**(2), 692–701 (1989)

T. Lay, T.C. Wallace, Modern global seismology, vol. 58. Academic Press (1995)

D. Li, F. Du, Monitoring and evaluating the failure behavior of ice structure using the acoustic emission technique. Cold Reg. Sci. Technol. **129**, 51–59 (2016)

B. Lishman, A. Marchenko, M. Shortt, P. Sammonds, Acoustic emissions as a measure of damage in ice. In: Proc. of the 25th Int. Conf. on Port and Ocean Eng. Under Arctic Conditions (POAC), Delft, Netherlands (2019)

B. Lishman, A. Marchenko, P. Sammonds, A. Murdza, Acoustic emissions from in situ compression and indentation experiments on sea ice. Cold Regions Sci. Technol. **172**, 102987 (2020)

D. Lockner, The role of acoustic emission in the study of rock fracture. Int. J. Rock Mech. Mining Sci. Geomech. Abstracts **30**, 883–899 (1993)

I. Main, Statistical physics, seismogenesis, and seismic hazard. Rev. Geophys. **34**(4), 433–462 (1996)

K. Mair, C. Marone, R.P. Young, Rate dependence of acoustic emissions generated during shear of simulated fault gouge. Bull. Seismol. Soc. Am. **97**(6), 1841–1849 (2007)

A. Marchenko, A. Haase, A. Jensen, B. Lishman, J. Rabault, K.-U. Evers, M. Shortt, T. Thiel, Laboratory investigations of the bending rheology of floating saline ice, and physical mechanisms of wave damping, in the HSVA ice tank (2019). (in review) arXiv preprint arXiv:1901.05333

A. Marchenko, A. Grue, J. Karulin, E. Frederking, R. Lishman, B. Christy-akov, P. Karulina, M. Sodhi, D. Renshaw, C. Sakharov, A. Markov, V. Morozov, E. Shortt, M. Brown, J. Sliusarenko, D. Frey, Elastic moduli of sea ice and lake ice calculated from in-situ and laboratory experiments. In: 25th IAHR International Symposium on Ice, Trondheim, Norway (In review) (2020)

W. Marzocchi, L. Sandri, A review and new insights on the estimation of the b-valueand its uncertainty. Annals Geophys. (2003)

G. Michlmayr, D. Cohen, D. Or, Sources and characteristics of acoustic emissions from mechanically stressed geologic granular media—A review. Earth Sci. Rev. **112**(3–4), 97–114 (2012)
K. Mogi, Magnitude-frequency relation for elastic shocks accompanying fractures of various materials and some related problems in earthquakes (2nd paper) (1963)
M.A. Rist, S.A.F. Murrell, Ice triaxial deformation and fracture (1994)
P. Sammonds, P. Meredith, I. Main, Role of pore fluids in the generation of seismic precursors to shear fracture. Nature **359**(6392), 228–230 (1992)
P. Sammonds, M. Ohanaka, Evolution of microseismicity during frictional sliding. Geophys. Res. Lett. **25**(5), 699–702 (1998)
Scourfield, The influence of ice rubble on sea ice friction. Ph.D. thesis, University College London (2019)
C. Scholz, The frequency-magnitude relation of microfracturing in rock and its relation to earthquakes. Bull. Seismol. Soc. Am. **58**(1), 399–415 (1968)
N.K. Sinha, Acoustic emission and microcracking in ice. In: Proc. Joint Conference on Experimental Mechanics, Society of Experimental Stress Analysis/Japan Society for Mechanical Engineers, Honolulu/Maui, Hawaii, May, 1982, Part 11, pp. 767–772 (1982)
Sinha, Acoustic emission study on multi-year sea ice in an Arctic field laboratory. J. Acoust. Emission **4**(2/3), S290–S293 (1985)
W.F. St Lawrence, D.M. Cole, Acoustic emissions from polycrystalline ice. No. CRREL-82–21. Cold Regions Research and Engineering Lab. Hanover NH (1982)
D.L. Turcotte, *Fractals and Chaos in Geology and Geophysics* (Cambridge University Press, 1997)
C. Vogt, K. Laihem, C. Wiebusch, Speed of sound in bubble-free ice. J. Acoust. Soc. Am. **124**, 3613–3618 (2008)
J. Weiss, J.-R. Grasso, Acoustic emission in single crystals of ice. J. Phys. Chem. B **101**(32), 6113–6117 (1997)
Y. Xie, D.M. Farmer, Seismic-acoustic sensing of sea ice wave mechanical properties (1994)

Chapter 4
The Influence of Ice Rubble on Sea Ice Friction: Experimental Evidence on the Centimetre and Metre Scales

Sally Scourfield, Ben Lishman, and Peter Sammonds

Abstract Sea ice floes in the Arctic collide with each other, and this leads to the production of smaller pieces of broken ice, which we call rubble. Rubble is also produced when ice collides with offshore structures, and when ships pass through sea ice. Previous analyses of ice friction have considered the contact between two sliding ice surfaces. Here, we consider the effective friction between two ice surfaces separated by ice rubble. In particular, we present experimental results across a range of scales and environments. We show results from metre-scale experiments in the Barents Sea; from metre-scale experiments in the Hamburg Ship Model Basin (HSVA); and from centimetre-scale experiments in the Ice Physics laboratory at UCL. We show that the effective kinetic friction is consistent across these scales, and comparable to friction coefficients measured without rubble. Looking at static friction, we find that when floes are in static contact for a short time, the presence of rubble acts to reduce static friction. However, if floes and rubble remain in static contact for around 10^4 s (a few hours) then the presence of rubble promotes strengthening, and the floe-floe effective friction can be raised by the presence of rubble. This has implications for modelling Arctic Ocean dynamics and for assessing friction loads on ships making repeated passages through a channel.

4.1 Introduction

To model the dynamics of sea ice, we need to consider how sea ice floes interact through fracture and friction across a range of scales (Sammonds and Rist 2001). One

S. Scourfield · P. Sammonds
Institute for Risk and Disaster Reduction, University College London, Gower Street, London WC1E 6BT, UK
e-mail: sally.scourfield.13@alumni.ucl.ac.uk

P. Sammonds
e-mail: p.sammonds@ucl.ac.uk

B. Lishman (✉)
School of Engineering, London South Bank University, Borough Road, London SE10AA, UK
e-mail: ben.lishman@lsbu.ac.uk

J. Tuhkuri and A. Polojärvi (eds.), *IUTAM Symposium on Physics and Mechanics of Sea Ice*, IUTAM Bookseries 39, https://doi.org/10.1007/978-3-030-80439-8_4

part of this interaction is floe-floe in-plane sliding—i.e. large sheets of sea ice sliding past each other. To quantify resistance to this sliding, we need an understanding of friction. In some models of sea ice dynamics, ice friction has been set as a variable or tuning parameter (e.g. Hopkins and Thorndike 2006). This is, in part, because the friction of ice on ice is difficult to model. A complete model would account for the microphysics of melting, brine drainage, adhesion, lubrication and fracture, for example (Hatton et al. 2009). Even then, the results could vary with ice type (first-year or multi-year, for example) and with temperature. Recent work modelling ice friction, therefore, has looked for patterns in the sliding behaviour of ice, rather than focussing on a single friction coefficient.

Schulson (2018) shows data from twelve recent studies which support a model of velocity strengthening (i.e. friction increases with increasing sliding speed) at speeds below around 10^{-5}–10^{-4} ms^{-1}, and velocity weakening (friction decreases with increasing sliding speed) above this threshold. Schulson (2018) also separates experiments by roughness (specifically, by how the ice surfaces were prepared) and notes that friction appears to have a greater sensitivity to changes in roughness in the velocity weakening regime. Schulson (2018) goes on to suggest that the cause of the increased sensitivity at higher sliding speeds is due to sliding-induced fragmentation, where rougher surfaces produce larger fragments. The effect of this roughness or fragmentation is pronounced: at sliding speeds of 10^{-2} ms^{-1}, rougher ice has a friction coefficient around 0.6, while smoother ice has a friction coefficient around 0.06. This work is in contrast to Hatton et al. (2009), who investigated the effect of ice surfaces on ice friction, and found experimentally that "surface topography is controlled by the sliding process, not by … the cutting method initially used to produce the faults." Perhaps, as suggested by Schulson (2018), the important difference is not the eventual ice surface, but the possibility of ice fragments being dislodged, and acting as an intermediate layer between the two surfaces.

Lishman et al. (2011, 2013) show results from experiments on ice-ice sliding with varying sliding speeds, alongside slide-hold-slide (SHS) tests and acceleration-deceleration tests. These results show that ice friction has memory: the instantaneous friction coefficient depends not just on the instantaneous conditions, but also on what has gone before. By analogy with earlier experiments on sliding in rocks (Ruina 1983), we can say that friction depends on the *rate* of sliding and also on the *state* of the sliding interface.

In this work, we show results from slide-hold-slide tests, and tests with varying sliding speeds, conducted on ice-rubble-ice interfaces. These can then be compared directly with literature results from experiments which were conducted on ice-ice interfaces. In this way we can begin to test the hypothesis that ice fragmentation, and the presence of ice rubble, has a significant effect on sea ice friction.

One useful point of comparison is the literature on the effect of gouge on sliding in rock faults. A number of studies have compared rock-rock sliding and rock-gouge-rock sliding. Gouge material is the product of wear along sliding surfaces, and is naturally produced along faults when sliding occurs (Marone et al. 1990). Research has mainly focused on the presence and effects of gouge in rock systems, with the aim of understanding the dynamics of rock faults on tectonic scales. Studies on rock have shown that gouge generation at initially bare faults results in an increase in frictional resistance (Byerlee 1967; Scholz et al. 1972). There is evidence to suggest that the presence of gouge stabilises sliding (Byerlee and Summers 1976; Scholz et al. 1972; Marone and Scholz 1988) through shear localisation features which develop with increasing slip. Engelder et al. (1975) note that the transition from stick slip to stable sliding is accompanied by a transition from sliding occurring at the rock-gouge contact to sliding being accommodated by owing gouge. Increasing gouge thickness has the effect of increasing the normal pressure at which this transition occurs (Byerlee and Summers 1976).

Marone et al. (1990) conducted experiments using quartz sand at a range of velocities with and without an initial gouge layer. Velocity strengthening behaviour was observed when an initial gouge layer was present, but velocity weakening behaviour was observed when it was not. They also report that the magnitude of strengthening in the former case varies directly with gouge thickness and surface roughness, and inversely with normal stress.

Mair et al. (2002) compared the effect of varying angularity and particle size distribution (PSD) of gouge (using sand and glass spheres) on kinetic friction and sliding behaviour. They found that a narrow PSD of spherical gouge material resulted in unstable, stick–slip sliding, whereas angular and wide-PSD spherical gouge produced stable sliding. Fracture of spherical grains was suggested as the reason why sliding with this gouge type becomes stable with accumulated slip. Mair et al. (2002) also observed lower friction coefficients for spherical particles compared to angular ones ($\mu \approx 0.45$ compared to 0.6), and attributed this to a low friction translation mechanism such as grain rolling.

Dieterich (1972, 1981) noticed that the coefficient of static friction became highly time-dependent when gouge was present between the sliding faces, and an approximate logarithmic increase in gouge strength with hold time was observed in his experiments on rock. Subsequent studies have provided further evidence for the time- dependence of static friction between rock surfaces separated by gouge material (Scholz et al. 1972; Marone 1998).

Overall, then, we might expect that the presence of ice rubble might affect overall frictional resistance; the transition between velocity-strengthening and velocity-weakening regimes; the development of stick–slip cycles in sliding; and the time-dependence of static friction. Further, we might expect that the shape and size of the rubble (which may in turn change during sliding) will be important inputs. We are not aware of systematic studies investigating how ice rubble affects ice sliding. Fortt and Schulson (2007) do note that a gouge-like material is expelled from the sliding interface in their experiments on freshwater ice.

In this paper we present results from ice-rubble-ice sliding experiments in the laboratory, in the ice basin, and in the field. This range of experimental configurations allows us to make comparisons across scales.

4.2 Centimetre-Scale Laboratory Experiments

Double-direct shear experiments on the centimetre-scale were performed in the Ice Physics Laboratory at UCL using laboratory-grown saline columnar ice. The laboratory setting allowed for greater control of the set-up (for example, enabling the use of accurately measured ice blocks and rubble of a standardised size), and also for a greater control of the ambient temperature. Saline ice (water salinity 31ppt; bulk ice salinity 7-9ppt) was grown using an insulated cylinder, heated from below and cooled from above. The method is described, and example thin sections are shown, in Bailey et al. (2012).

Sliding experiments were conducted in a double-direct shear setup, as shown in Fig. 4.1. Three ice blocks, separated by two regions of ice rubble 12 mm wide, were arranged in the vertical plane inside a load frame which applied a normal force. The central, moving ice block is 150 mm (high) $\times$ 50 mm (wide) $\times$ 80 mm (deep), and the fixed ice blocks are 80 $\times$ 90 $\times$ 80 mm. Ice rubble was prepared by sieving broken lab-grown ice using two graduated sieves, so that the diameter of ice rubble grains were in the range 5.6–6.7 mm. This rubble size was chosen because it allowed three to four layers of rubble to fit in the gaps between middle and outer ice blocks, which is comparable to other experiments in this paper.

A shear force, applied by an actuator piston and detected by an in-line load cell (model 614 tension–compression Tedea Huntleigh, accuracy $\pm$ 0:1%), was applied to the middle block in order to push it through the two outer ice blocks, and its displacement was measured by a linear actuator (model PZ-34-A-150 Gefran, accuracy $\pm$ 0:075 mm). The force was evenly distributed across the middle block's top surface by a load spreading plate which was situated between the middle block and the load cell. A normal force was applied across the set-up via two side load panels, which controlled the load using a constant pressure hydraulic system. Perspex shims were placed underneath the outer ice blocks, and frozen onto their front and back (in the vertical plane) in order to confine the rubble to the rubble region. The load frame containing this set-up was contained in an environmental chamber in which the ambient air temperature was cooled using liquid nitrogen. This set-up was designed to be comparable to that used by Mair and Marone (1999) to investigate rock friction with a gouge.

Figure 4.2 shows a set of results from a slide-hold-slide test, conducted at -7 °C, using this setup. In this experiment, the middle ice block is moved at 10^{-4} ms^{-1} over distances of 6 mm, punctuated by static periods lasting 1, 10, and 100 s. Note that during hold periods, the actuator is moved away from contact with the ice, so that the middle block isn't moved when the actuator moves fractionally under hydraulic control. This "backing off" of the actuator accounts for the dips in pusher displacement seen around 150, 200, and 375 s in the bottom graph of Fig. 4.2.

These results show stick–slip sliding around a fairly constant average friction coefficient, with higher friction coefficients immediately after static holds. Note that we experienced some drop off of normal force due to leaks in the hydraulic system.

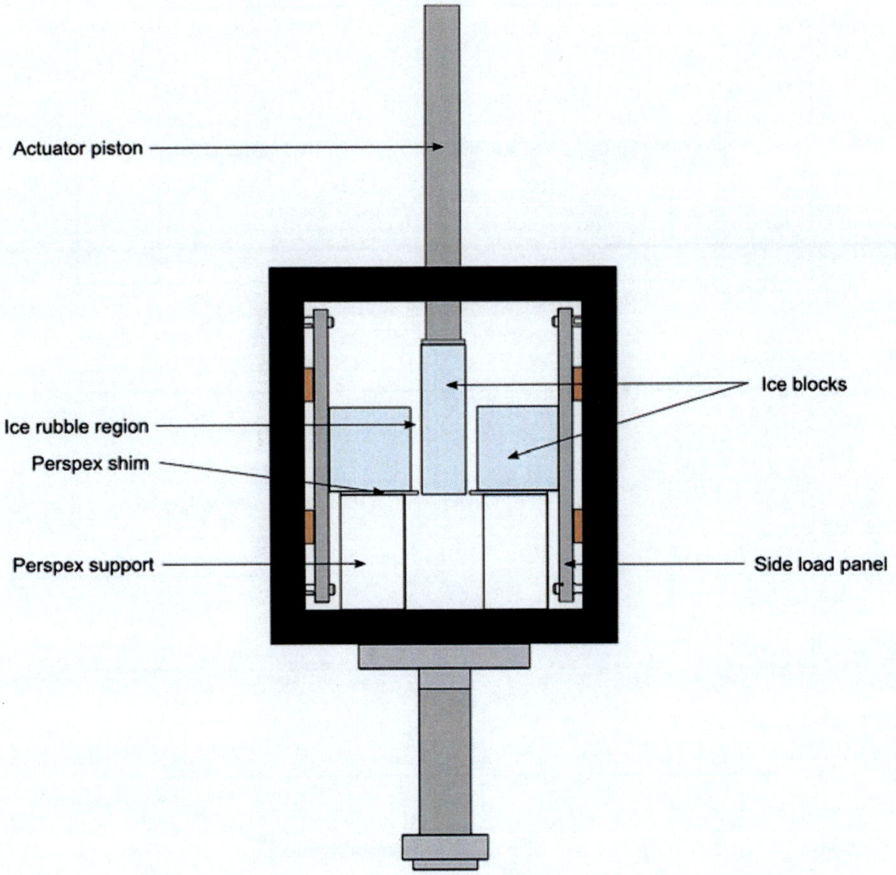

Fig. 4.1 The load frame in which the ice blocks and rubble were arranged. Plates on either side of this load frame applied a normal force, and the entire load frame was placed in an environmental chamber which allowed the ambient air temperature to be controlled. The central, moving ice block is 150 mm (high) x 50 mm (wide) x 80 mm (deep), and the fixed ice blocks are 80 × 90 × 80 mm

Friction coefficient μ is calculated as [shear force / (2 × normal force)], with the factor of two to account for the two separate sliding interfaces.

4.2.1 Laboratory Results

Two summary graphs of the experimental results from the laboratory experiments are shown in Figs. 4.3 and 4.4. In Fig. 4.3, the effect of hold time in slide-hold-slide experiments is shown. The peak friction coefficient required to reinitiate motion increases with the length of the hold period. The effect of temperature here doesn't lead to

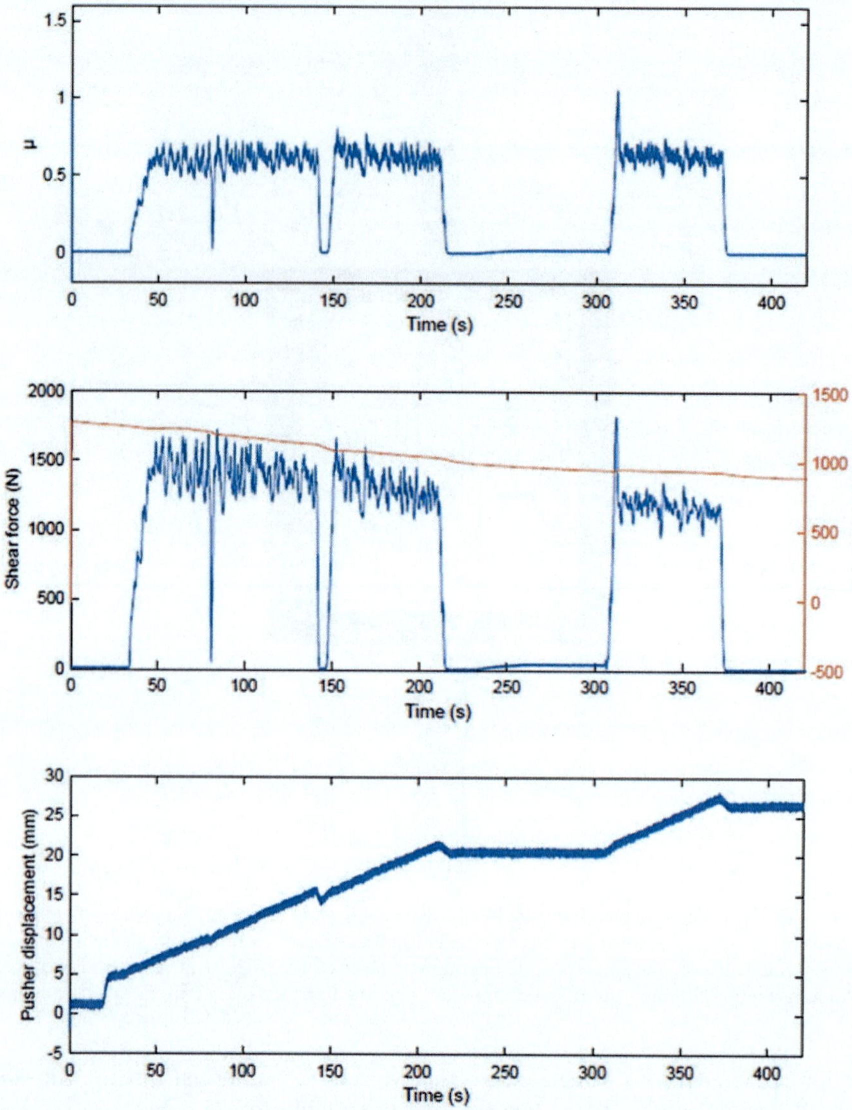

Fig. 4.2 Results from a slide-hold-slide experiment conducted at −7 °C in the laboratory apparatus shown in Fig. 4.1. The top figure shows calculated friction as a function of time; the middle figure shows the measured shear force and normal force; and the bottom figure shows the measured pusher displacement (programmed pusher displacement is similar but less noisy). Static holds of 1 s (at around 80 s), 10 s (at around 140 s) and 100 s (from 220 to 320 s) are shown. Increases in friction can be seen on resumption of motion after the 10 and 100 s static holds

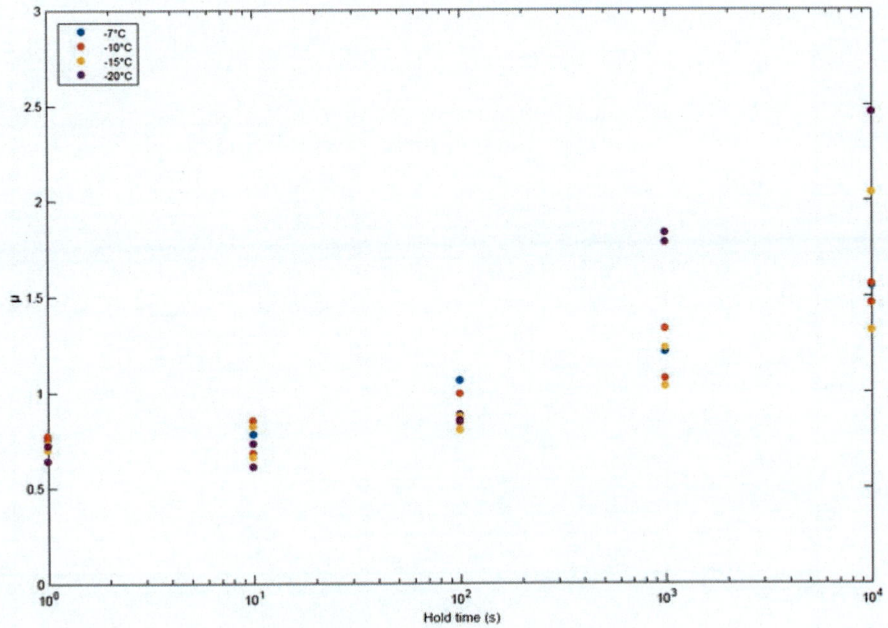

Fig. 4.3 The effect of static hold time on peak friction in laboratory experiments across a range of temperatures. Some experiments were repeated, and multiple data points are shown for these experiments: this gives a sense of experimental variation in our results

consistent trends across hold times. At longer hold times, lower temperatures lead to greater strengthening. The second graph shows the relationship between friction coefficient and sliding velocity, and shows evidence of both velocity-strengthening and velocity-weakening, with a transition somewhere around 10^{-3} ms^{-1}. Stick–slip behaviour is observed at 10^{-4}–10^{-3} ms^{-1} (shown by the vertical bars on the graph) and is less pronounced at 10^{-2} ms^{-1}.

4.3 Ice Basin Experiments

Double-direct shear experiments on the metre-scale were performed in the Large Ice Basin at HSVA using saline columnar ice grown in situ, following an approach described by Sammonds et al. (2019). The ice was formed from water of 6.8ppt salinity. Ice salinity was measured as 2ppt. The bulk density of the level ice was 920 kgm^{-3}. The ice was formed over the course of one week at an air temperature of −18 °C. All experiments were performed at an air temperature of −8 °C. During the experimental programme, ice thickness grew from around 20 to around 30 cm. Thin sections (Scourfield 2019) show a granular upper layer (grain size < 1 mm), around

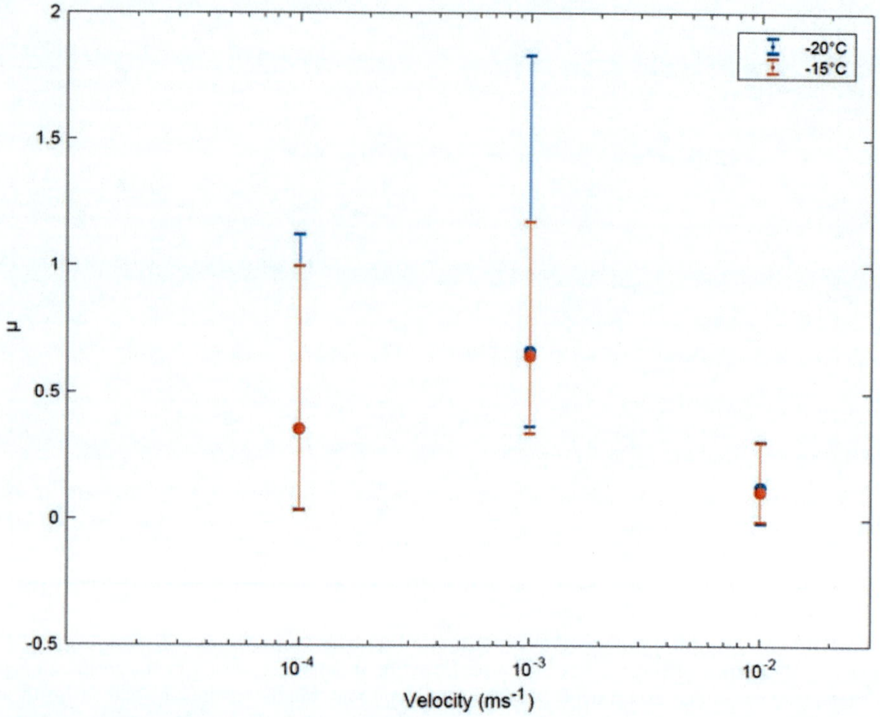

Fig. 4.4 The relationship between μ and sliding velocity for temperatures of −15 and −20 °C. Data points represent the mean kinetic friction during the period of steady sliding, and the maximum and minimum extent of vertical bars correspond to the highest and lowest kinetic friction experienced during this period. Higher vertical bars are associated with higher-amplitude stick–slip cycles

15 mm thick, with columnar ice below (grain size ≈1 cm). Further ice properties, including details of ice strength can be found in Scourfield (2019).

The experimental set-up consisted of a mobile central ice block (3.5 m long, 1.5 m wide) in a channel of open water. Either side of this were ice rubble regions (1.5 m long, 0.5 m wide), bound on one side by the middle block, and on the other by floating ice beams. These beams ensured that the rubble regions were bound by ice on both sides at all times during sliding.

Wooden side load frames housing pneumatic rams were situated on both sides of the set-up next to the floating ice beams, and applied a normal force across the ice rubble regions and middle block. The degree of extension or retraction of the side load frames was measured by laser distance sensors (Di-soric, type LHT 9–45 M 10 P3IU-B4, accuracy ± 15 mm), situated at either end of both frames; from these measurements, information about the contraction or dilation of the rubble regions could be extracted. A pusher plate attached to the facility's moveable main carriage provided a shear force to the middle block to push it through the channel and past the ice rubble regions, in double-direct shear fashion. Using the shear force applied by

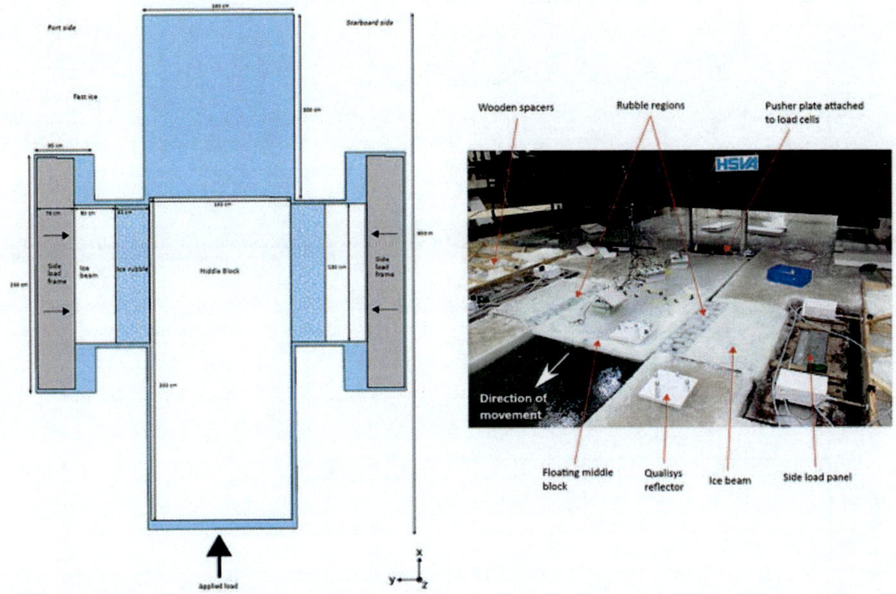

(a) (b)

Fig. 4.5 Experimental setup in HSVA

the pusher plate, and the normal force applied by the side load frames, the coefficient of friction, μ, may be calculated as the ratio of shear force to normal force as before.

To maintain a two-dimensional problem (to simplify modelling), the rubble region consisted of just one layer of floating rubble pieces. The effects of ice rubble angularity were explored by repeating steady sliding and SHS experiments with four different rubble types - small round rubble (i.e. discs, 9.5 cm diameter), large round rubble (16.5 cm diameter), small diamond-shaped rubble (10 cm short axis, 17 cm long axis) and large diamond-shaped rubble (18 cm short axis, 30 cm long axis). All rubble pieces were 7–10 cm thick (note that this means they are thinner than the surrounding level ice) (Fig. 4.5).

4.3.1 Ice Basin Results

Summary results from ice basin tests are shown in the same format as those from laboratory tests. Figure 4.6 shows a set of results from an individual experiment, in this case sliding at 3mms^{-1}. Figure 4.7 shows peak static friction as a function of hold time across all experiments. Again, peak friction increases with hold time. In these experiments, though, we see a marked transition in peak friction between 10^3 and 10^4 s hold times. Figure 4.8 shows average friction against sliding velocity for

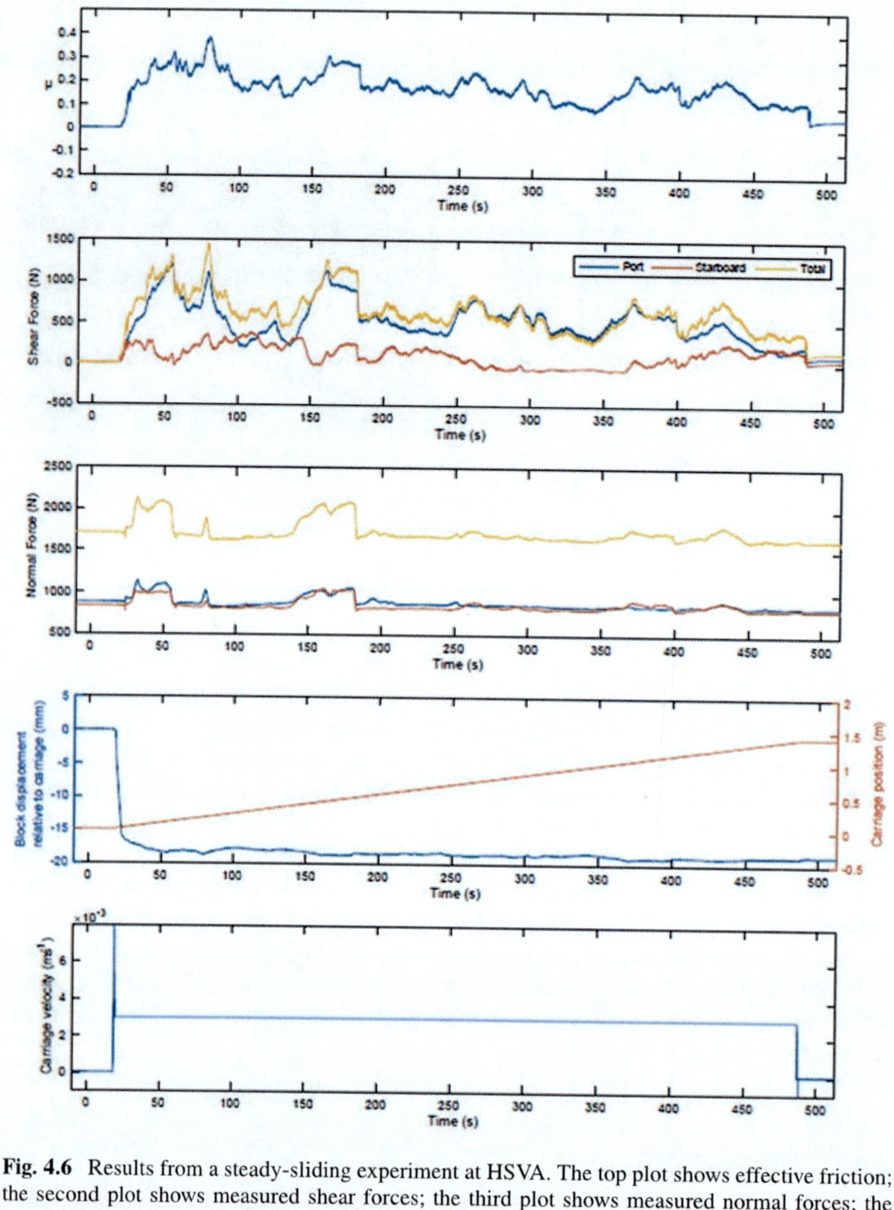

Fig. 4.6 Results from a steady-sliding experiment at HSVA. The top plot shows effective friction; the second plot shows measured shear forces; the third plot shows measured normal forces; the fourth plot shows the carriage position, and the relative displacement of the main sliding block relative to the carriage; and the final plot shows the carriage speed

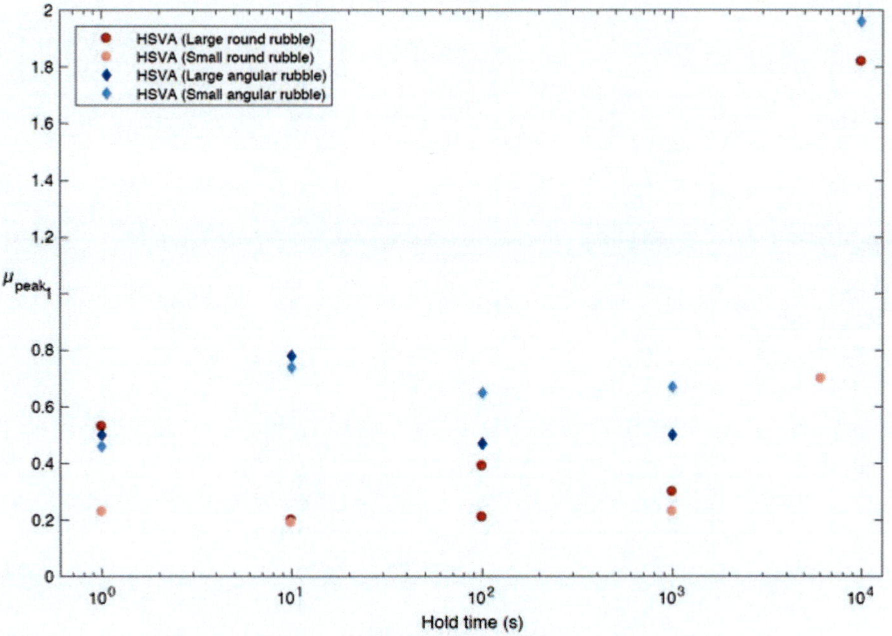

Fig. 4.7 The relationship between hold time and peak friction in ice basin experiments

the ice basin experiments. Within the range of sliding speeds we tested (3×10^{-3}–10^{-1} ms^{-1}), we see velocity-weakening behaviour, i.e. the average friction decreases with increasing sliding speed.

4.4 Field Experiments

Double direct shear experiments were conducted in the Van Mijenfjorden, near the mining town of Svea, Svalbard in March 2015, following an approach described by Scourfield et al. (2015). Air temperatures during the experiments were between −11 and −17 °C. The ice used in the experiments is natural sea ice, formed from water with measured salinity of 30ppt. The thickness of the level ice was around 60 cm. The experiments are conducted in a double shear configuration similar to that used in the lab and ice basin experiments. Ice rubble in these experiments is used by breaking up ice blocks using axes. The aim in this procedure was to produce a roughly fractal ice volume distribution. The largest pieces of ice were approximately 200 mm on each side. The ice rubble regions are approximately 300 mm deep. Side load was applied by a floating hydraulic load frame, and normal load was supplied by a mains-powered electrical actuator: the actuator speed for slide-hold-slide experiments was ~ 6×10^{-3} ms^{-1}. All the ice edges were saw-cut by hand. The experimental setup

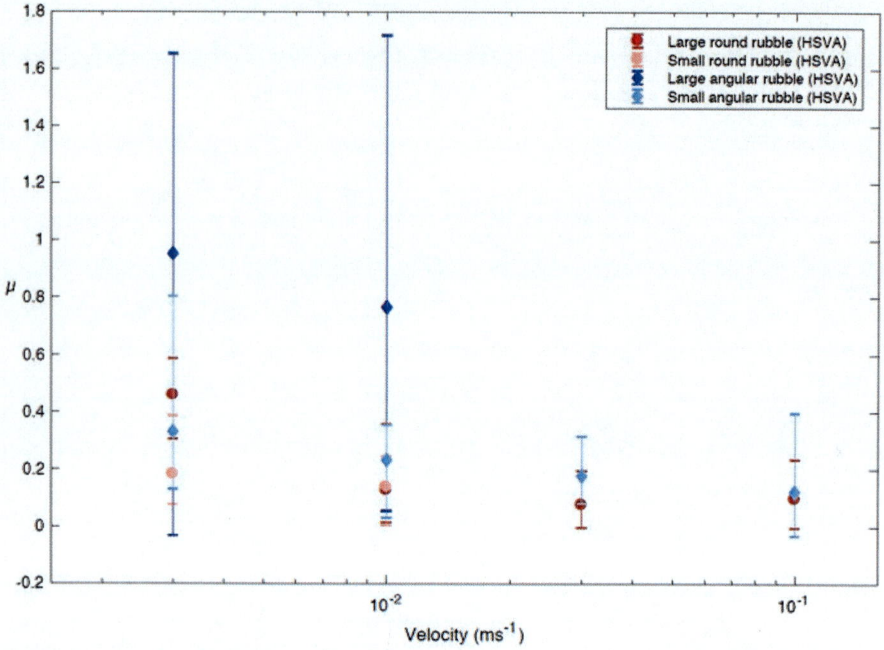

Fig. 4.8 The relationship between sliding velocity and friction in ice basin experiments. Data points represent the mean kinetic friction during the period of steady sliding, and the maximum and minimum extent of vertical bars correspond to the highest and lowest kinetic friction experienced during this period. Higher vertical bars are associated with higher-amplitude stick–slip cycles

is shown in Fig. 4.9, and a typical set of results for a slide-hold-slide experiment is shown in Fig. 4.10.

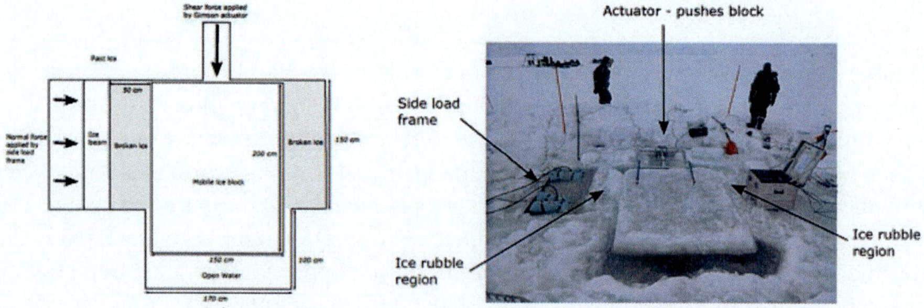

(a) (b)

Fig. 4.9 The experimental setup for field experiments

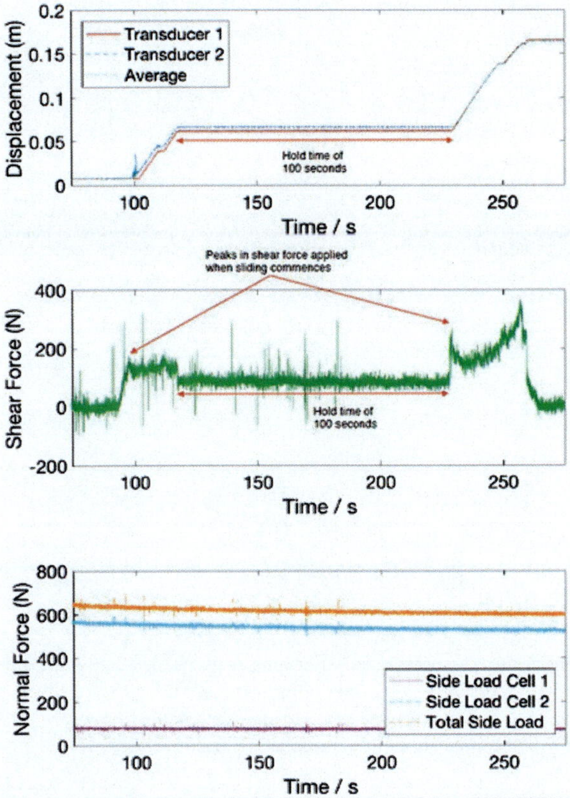

Fig. 4.10 Representative data for a SHS steady sliding test conducted in Svea, showing a period of sliding, followed by a 100 s hold time, followed by a second period of sliding

4.4.1 Field Results

Data from field experiments is shown in Figs. 4.11 and 4.12. These results are shown in the same format as results for lab and ice basin experiments: the first graph shows peak friction as a function of hold time, and the second graph shows average friction as a function of sliding velocity. We see peak friction increases with hold time, with a marked increase in peak friction between 10^3 and 10^4 s. One further experiment, with a hold time of 6.5×10^4 s, overnight, led to a shear force around 200 times greater than the normal force. This data point is not plotted as it distorts the axes, and also because it might be considered outside the scope of friction models (since the ice was fully consolidated.) Nevertheless, we note it here as evidence that the peak friction coefficient keeps on rising at longer hold times than those shown in Fig. 4.11.

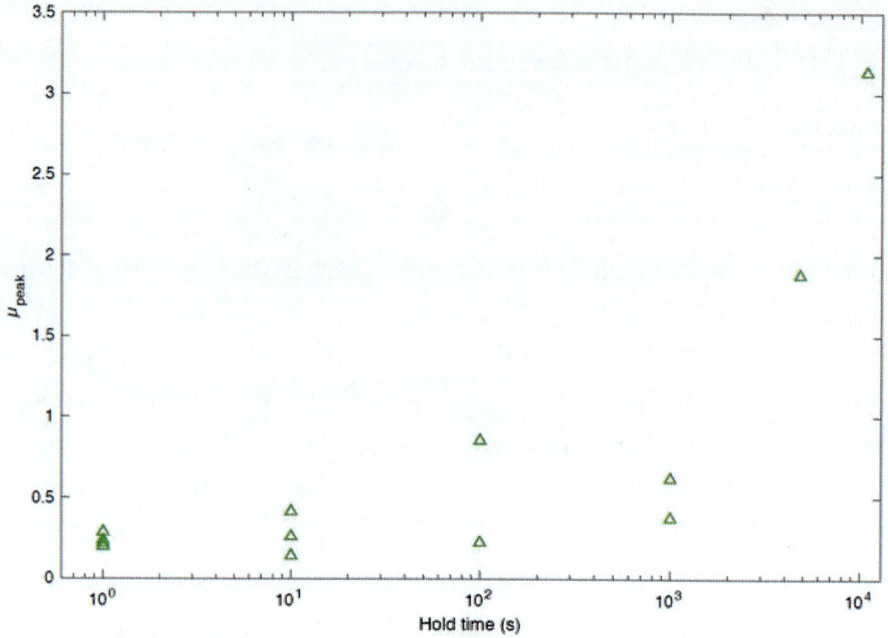

Fig. 4.11 The relationship between peak friction and hold time. Not shown here (since it leads to a distorted y-scale) is a data point from an 18-h (6.5×10^4 s) hold time, where the shear force was found to be a factor of nearly 200 greater than the normal force (i.e., $\mu_{peak} \approx 200$; see later discussion of whether this should be modelled as friction.)

4.5 Discussion

Experimental data have been presented across three scales, and three sets of experimental conditions. These sets of experiments are similar in configuration (double shear), and the slide-hold-slide experiments cover similar ranges of hold times (10^0–10^4 s). Sliding speeds vary from experiment to experiment, according to the capabilities of the actuators. Temperatures vary somewhat between experiments, within the range −20 to −7 °C. In each case we wished to produce results at around −10 °C, to allow comparisons to be made at constant temperature; in the ice basin, the temperatures were kept slightly warmer to limit ice growth, and in the field, it was not possible to control temperature. Salinity is similar in the lab and field experiments (8-9ppt ice salinity), and lower in the ice basin experiments (ice salinity around 3ppt), because of limitations in the amount of salt which could be added to the ice basin. Differences in the orientation of experiments (i.e., on the horizontal plane in the field and ice basin, but on the vertical plane in the laboratory) and the presence of water (i.e., field and ice basin experiments performed using ice floating on water, but dry in the laboratory) should also be noted. Acknowledging these sources of

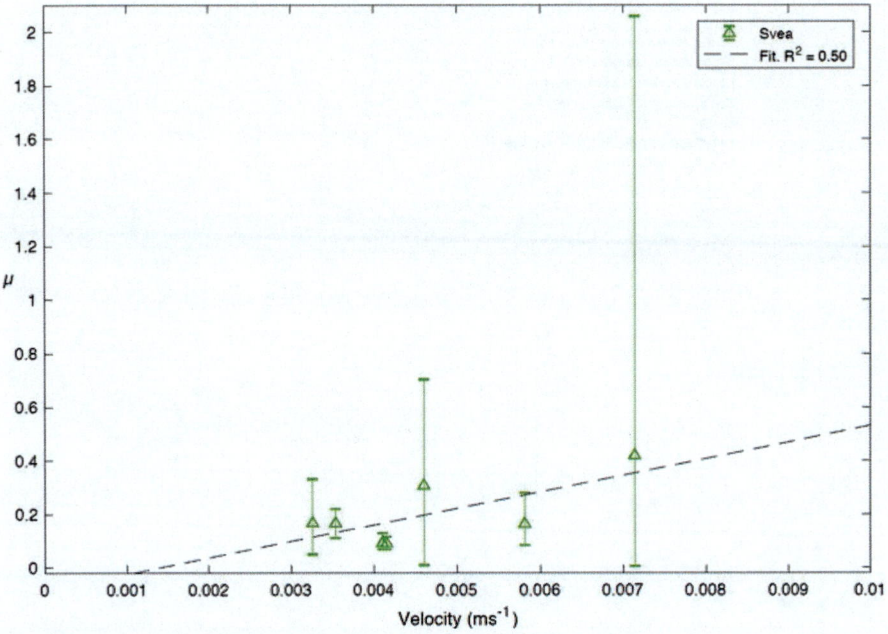

Fig. 4.12 The relationship between friction and sliding velocity. Data points represent the mean kinetic friction during the period of steady sliding, and the maximum and minimum extent of vertical bars correspond to the highest and lowest kinetic friction experienced during this period. Higher vertical bars are associated with higher-amplitude stick–slip cycles

variation between experiments, we still feel there is merit in comparing results from all experiments.

4.5.1 Time-Dependent Friction

As in the individual experimental results shown above, we first show the effects of varying hold times in slide-hold-slide experiments. In Fig. 4.13, alongside hold time results, we plot a theoretical model proposed by Schulson and Fortt (2013). This models explains static strengthening in terms of creep of asperities in contact.

In Fig. 4.13, we show the results from this ice-rubble-ice study (black data points) compared to results from previous studies of ice-ice sliding (Lishman et al. (2011), red data points; Schulson and Fortt (2013), blue data points).

Two outcomes can be seen in comparing the slide-hold-slide results from this paper to previous ice-ice results, and to the ice-ice model of Schulson and Fortt (2013). First, at low hold times ($<\approx 10^3$ s), the presence of ice rubble appears to reduce the static friction coefficient. The ice-ice experiments shown by red ‘x’ markers, and the ice-rubble-ice results shown by black ‘x’ markers show evidence that rubble

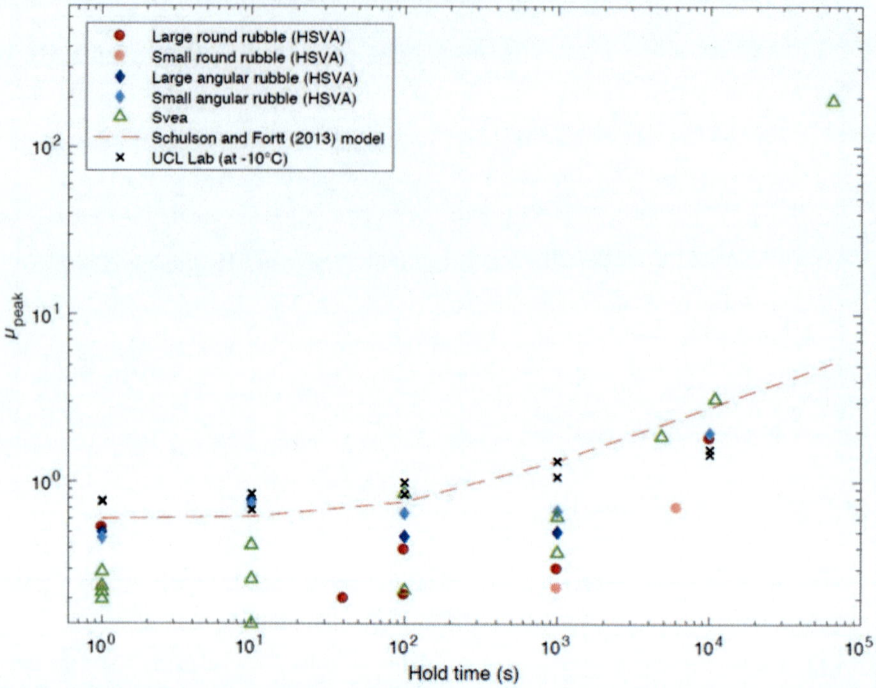

Fig. 4.13 Relationship between hold time and peak friction, compared across all experiments, and to ice-ice model proposed by Schulson and Fortt

decreases friction at hold times up to 10^3 s. Second, at higher hold times ($>\approx 10^4$ s), the presence of ice rubble appears to increase the static friction coefficient: the black markers in Fig. 4.14 curve upwards (on a logarithmic scale) at hold times greater than 10^3 s, in a way which isn't seen in the ice-ice data in Lishman et al. (2011) or Schulson and Fortt (2013). The behaviour is, however, a reasonable match for the ice-ice model (dashed red line in Fig. 4.13). In fact, the ice-rubble-ice data seems to fit the shape of the model more closely than ice-ice data does (cf. Schulson and Fortt (2013), Fig. 2).

The lab data in this study also shows a slight upwards curve (see Fig. 4.3), although this curve is less pronounced than for the field and ice tank data. The lab data also have higher friction values at low hold times, probably because the sliding between the hold times occurs at lower speeds. For this reason, the laboratory data are not shown in Fig. 4.14.

In summary, data from previous studies fall in straight lines on Fig. 4.14, while data from the laboratory (Fig. 4.3), ice tank (Fig. 4.7) and field (Fig. 4.11) in this study form upwards-trending curves.

The first outcome (that friction at low hold times may be lower for ice-rubble-ice sliding than for ice-ice sliding) makes some intuitive sense. The presence of ice rubble increases the number of pathways across which ice can slide. If one interface

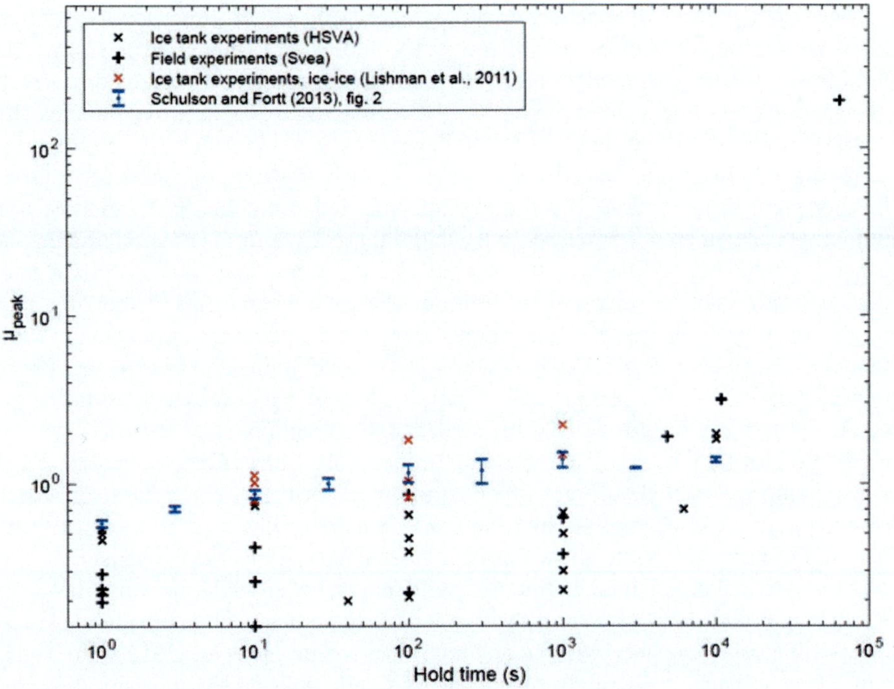

Fig. 4.14 Relationship between hold time and peak friction, comparing ice-ice sliding to ice-rubble-ice sliding. Black markers show results from this study. Red markers show results from Lishman et al. (2011). Blue markers show results from Schulson and Fortt (2013)

starts to stick, for any reason, then sliding can transfer to other interfaces. In some cases, sliding can also be replaced by rolling, which will also reduce static friction. Overall, cohesion is determined not by the strongest points of consolidation, as in ice-ice sliding, but by the weakest points—hence strengthening occurs more slowly. The most direct evidence for this weakening effect of rubble is in the comparison of ice-ice tests at HSVA (red markers on Fig. 4.14) with ice-rubble-ice tests at HSVA (black markers on Fig. 4.14).

The second outcome (that friction at higher hold times may be higher for ice-rubble-ice sliding than for ice-ice sliding) seems less intuitive, and the evidence presented here is less robust. The results in this study—specifically, those from the ice basin and from the field—show a discontinuity in friction at hold times of around 10^4 s (i.e., a few hours). Peak friction increases above the trend shown at lower hold times. This effect is not observed in data from previous studies on ice-ice sliding: in particular, there is no trace of it in experimental data from Lishman et al. (2011) or Schulson and Fortt (2013). This result suggests that ice-rubble-ice interfaces develop mechanical strength more quickly than ice-ice interfaces, when examined on a timescale of hours.

The exact process of mechanical strengthening, in either direct ice-ice friction or in ice-rubble-ice friction, is still not fully understood. Increased friction after hold times is observed in experiments with no melt, and can be explained by creep of asperities (see e.g., Dieterich 1979). However, in ice friction experiments the possibility also exists for freeze bonds to develop between sliding surfaces (Repetto-Llamazares et al. 2011), and for formation, through freezing, of a new ice matrix, linking previously separate ice blocks into one new solid block. In each of our hold time experiments, we believe that each of these three forms of strengthening—asperity creep, freeze bonds between surfaces, and freezing of interstitial water—occur, and affect the development of friction. The second two forms of strengthening might, in other contexts, be considered outside the realms described by friction models. In applications of ice friction, though, freeze bonding and freezing are likely to play important roles in practical situations, and so we consider it useful to try to incorporate them into ice friction models. This complexity of ice friction (that asperity-deformation type friction seems inseparable from effects of freezing and consolidation) means that forces involved in our experiments can grow over orders of magnitude, leading to plots like Figs. 4.13 and 4.14 where friction (or, at least, the ratio of normal force to shear force) is presented on a logarithmic axis.

Most data collected so far, in this study and others, cover hold times up to 10^4 s. At some long hold times, we expect that ice will have mechanically consolidated, and behave as a solid ice sheet, without any interface (see Bailey et al. (2012) for discussion of mechanical consolidation between vertically stacked ice sheets, a different but related problem). At this stage, we would not expect the strength of the ice to vary with further increases in hold times. Once mechanical consolidation has occurred, friction models (models which measure ratios of shear force or stress to normal force or stress) are not likely to be appropriate ways of understanding the shear strength of the ice sheet.

This behaviour is summarised in Fig. 4.15.

- In region 1 covering low hold times, rubble acts to weaken the sliding interface, by providing multiple sliding pathways and allowing rolling of rubble to replace sliding. The evidence for this weakening is presented in Fig. 4.13.
- In region 2, covering intermediate hold times (around 10^4–10^5 s), the ice-rubble-ice interface gains mechanical strength faster than the ice-ice interface. Tentative evidence for this transition is provided in this paper, in particular by the high-friction data points towards the right hand side of Figs. 4.7 and 4.11. Experimental evidence in this region is difficult to collect, since high friction leads to high forces, which can damage equipment. More data are required in this region.
- In region 3, at high hold times, we are not aware of any experimental data which are directly comparable to data for regions 1 and 2. We make two proposals about this region. First, we believe that as $t \to \infty$, the two lines should converge. Second, we propose that friction models are not appropriate ways of understanding or predicting the behaviour of the interface in this region.

Figure 4.16 shows a photo of the effects of resuming sliding after an 18-h hold during experiments in Svea. Motion does not occur by sliding on an interface parallel

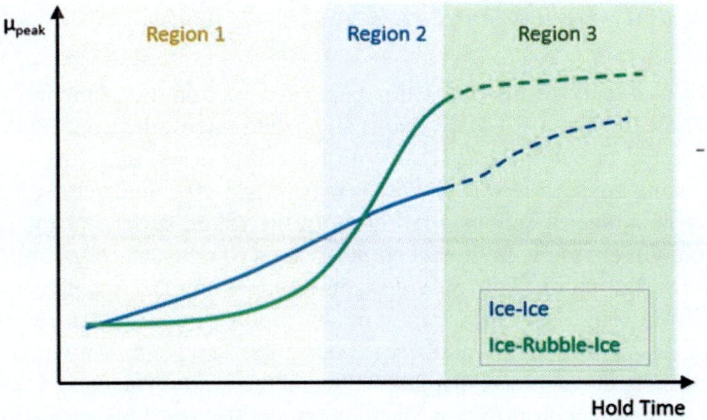

Fig. 4.15 Schematic diagram showing the effects of hold time on ice-ice interfaces (blue line) and on ice-rubble-ice interfaces (green line)

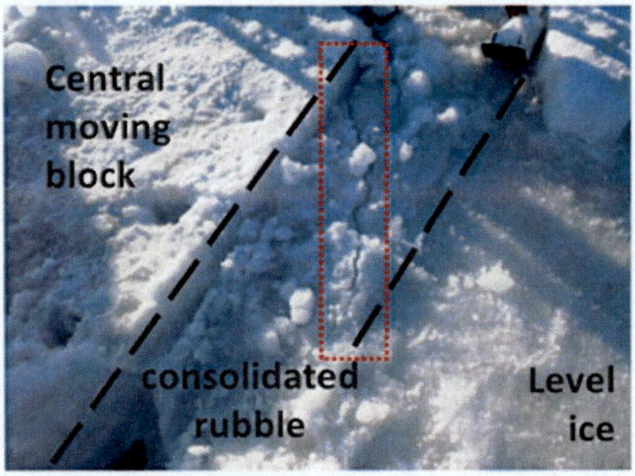

Fig. 4.16 Evidence of fracture (crack shown highlighted by red dotted box) within the consolidated rubble region as a result of the application of shear force after a hold time of 18 h

to the block and level ice edges, as in most sliding cases. Instead, a crack passes through the rubble region, at an angle of 30–45° to the direction of force, as might be expected in the failure of level ice. This data point gives some indication of where the transition between region 2 and region 3 occurs (after 18 h, the ice rubble behaves somewhat like level ice, i.e., region 3), and also supports the idea that in region 3, the ice rubble behaves like a solid sheet rather than like a pre-existing sliding interface.

4.5.2 Velocity-Dependent Friction

Figure 4.17 shows all results from this paper for friction as a function of sliding velocity. These results are broadly consistent with each other, and also with the results presented in Schulson (2018). With ice-rubble-ice sliding, the highest friction coefficients are seen at speeds between 10^{-3} and 10^{-2} ms^{-1}. Results for large angular rubble—blue diamonds—fall close to the rough-surface results presented by Schulson. Other results fall between Schulson's rough-surface grouping and the smooth-surface grouping. This provides some support for the hypothesis that fragmentation from rough-cut surface may be an important control on friction. Initial fragments from such surfaces are likely to be angular, and hence cause an increase in friction similar to the increased friction we see in this study. The data presented here do not show pronounced velocity-weakening above 10^{-4} ms^{-1}, which again suggests these ice-rubble-ice interfaces are more closely related to rough-cut ice interfaces than to the smooth surfaces.

We note that there is significant scatter within the velocity-dependent data we present, and that the different studies overlap but don't all follow the same trends. For example, the field data from Svea show velocity strengthening up to 10^{-2} ms^{-1},

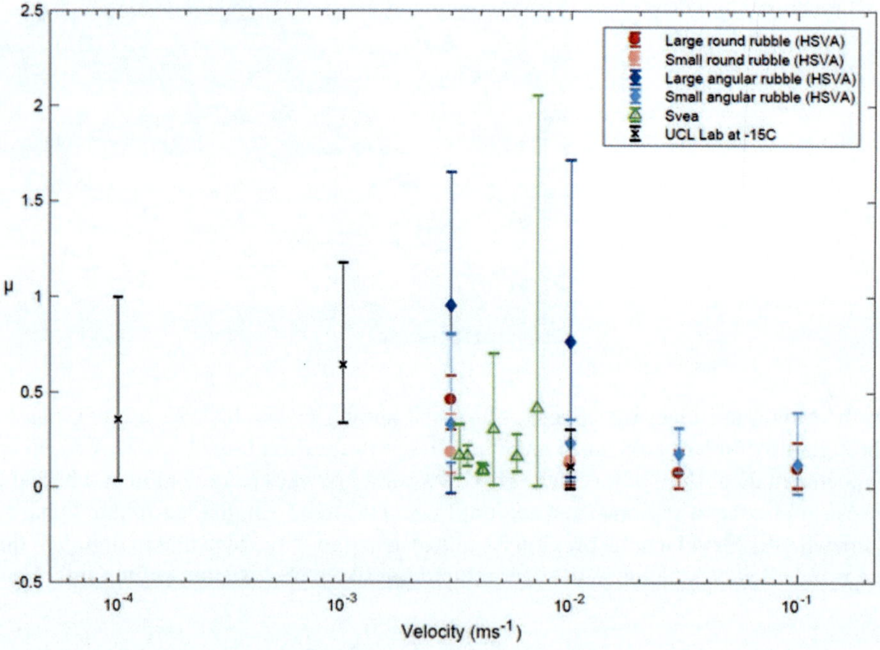

Fig. 4.17 Relationship between sliding velocity and friction, compared across all experiments. Data points represent the mean kinetic friction during the period of steady sliding, and the maximum and minimum extent of error bars correspond to the highest and lowest kinetic friction experienced during this period

whereas the ice tank data from HSVA show velocity weakening in this region. It's also worth noting that we can only present very limited data at low sliding speeds (due to limitations on the actuator and carriage controls in Svea and HSVA respectively).

4.5.3 Suggestions for Further Work

Several aspects of ice-rubble-ice friction are not discussed in this work. For simplicity, the rubble has been made using a similar process within each family of experiments, but further experiments might investigate the effects of rubble size, rubble size distribution, and porosity. Slush—tiny particles of floating ice—seems to play an important part in promoting consolidation, and should be investigated. Data at low sliding speeds are limited in the present paper, and more data here would give a clearer picture of the transition between velocity-strengthening and velocity-weakening regions.

4.5.4 Implications for Understanding Sea Ice Dynamics

Over time, and with continued sliding, we expect ice to be abraded and wear away. This happened in all the experiments described in this paper. Examples from the ice basin and the field are shown in Fig. 4.18. In both cases, sharp corners are rounded down during experiments. This means that the rubble pieces tend to become more

(a) (b)

Fig. 4.18 Ice before and after experiments in (**a**) the ice basin, and (**b**) the field. In both cases, freshly made ice rubble is shown on the left. The right-hand container in each image shows ice which has been through a sliding experiment spherical. If angular rubble leads to higher friction than rounded rubble, and rubble tends to become less angular during frictional sliding, then we expect that average friction might decrease during prolonged frictional sliding of natural ice-rubble-ice interfaces.

Improved understanding of ice-rubble-ice interfaces, as described above, may help to increase our understanding of how natural sea ice slides. When sea ice floes break up, rubble forms. Further, the ice edges which form are likely to be rough, which means further fragmentation and rubble production is likely, as described above. The presence of rubble will mean that short term memory—static friction from short hold times—is likely to be reduced, compared to ice-ice interfaces. Mechanical consolidation of ice-rubble-ice interfaces may occur more quickly than for ice-ice interfaces. Over prolonged sliding, the ice rubble will abrade and become more spherical, and this will lead to reductions in the sliding friction. Direct observations of the evolution of ice-ice interfaces and ice-rubble-ice interfaces in nature (either on small scale through ship-based or land-based observation, or on large scale through aerial or satellite observation) will help to test the predictions in this paper and provide a further scale across which data can be compared.

4.6 Conclusions

We show results from experiments on the sliding of ice surfaces separated by ice rubble, across three different scales and environments. On the metre-scale, the presence of ice rubble lowers the coefficient of static friction (compared to ice-ice sliding) at hold times $< \approx 10^3$ s. This is because the rubble offers multiple pathways for movement, so that the static friction is dependent on the weakest cohesion, rather than the strongest. At higher hold times, experiments in the ice basin and the field suggest that rubble leads to a transition to rapid strengthening, so that ice-rubble-ice interfaces can be stronger than ice-ice interfaces for a given hold time. This would suggest that vertical planar ice-ice interfaces reach mechanical consolidation more slowly than aggregations of ice rubble. At very high hold times ($t > \approx 10^5$ s) friction becomes a less useful model for understanding the strength of the interface. During constant sliding experiments, ice-rubble-ice experiments show comparable results to ice-ice sliding experiments. Results with diamond-shaped rubble blocks are similar to results with sliding of rough-cut ice-ice interfaces, suggesting that fragmentation and rubble interference may play a part in the sliding of rough ice-ice interfaces. This has implications for the sliding of natural sea ice, which fractures into rough ice surfaces: in turn, these ice interfaces will fracture, leading to ice-rubble-ice sliding.

Acknowledgements We would like to thank Aleksey Marchenko and the University Centre in Svalbard (UNIS) for the opportunity to join the Lance RV cruise, and for guidance and the substantial logistical support during fieldwork in Svea. We also acknowledge and thank the Research Council of Norway for funding field work through the SFI SAMCoT. We would like to thank Mark Shortt, Aleksey Marchenko, Ellie Bailey and Sammie Buzzard for assistance with the experiments at HSVA and Neil Hughes, Steve Boon and John Bowles for assistance with laboratory experiments at UCL. We would like to thank Kaj Riska (TOTAL SA) for advice and continual support.

The work described in this publication was supported by the European Community's Horizon2020 Research and Innovation Programme through the grant to HYDRALA-PLUS, Contract no. 654110. The authors would like to thank the Hamburg Ship Model Basin (HSVA), especially the ice basin crew, for the hospitality, technical and scientific support and the professional execution of the test programme in the Research Infrastructure ARCTECLAB.

SS was supported by a UCL Impact Studentship funded by the Institute for Risk and Disaster Reduction and TOTAL S.A.

References

E. Bailey, P.R. Sammonds, D.L. Feltham, The consolidation and bond strength of rafted sea ice. Cold Reg. Sci. Technol. **83**, 37–48 (2012)

J.D. Byerlee, Frictional characteristics of granite under high confining pressure. J. Geophys. Res. **72**(14), 3639–3648 (1967). https://doi.org/10.1029/JZ072i014p03639

J. Byerlee, R. Summers, A note on the effect of fault gouge thickness on fault stability. Int. J. Rock Mech. Mining Sci. Geomech. Abstr. **13**(1), 35–36 (1976)

J.H. Dieterich, Time-dependent friction in rocks. J. Geophys. Res. **77**(20), 3690–3697 (1972)

J.H. Dieterich, Modeling of rock friction: 1. Experimental results and constitutive equations. J. Geophys. Res. Solid Earth **84**(B5), 2161–2168 (1979)

J.H. Dieterich, Constitutive properties of faults with simulated gouge. Mech. Behav. Crust. Rocks **24**, 103 (1981) doi: https://doi.org/10.1029/GM024p0103

J.T., Engelder, J.M., Logan, & J. Handin, The sliding characteristics of sandstone on quartz fault-gouge. Pure Appl. Geophys. **113**(1), 69–86 (1975)

A.L. Fortt, E.M. Schulson, The resistance to sliding along Coulombic shear faults in ice. Acta Mater. **55**(7), 2253–2264 (2007)

D.C. Hatton, P.R. Sammonds, D.L. Feltham, Ice internal friction: standard theoretical perspectives on friction codified, adapted for the unusual rheology of ice, and unified. Philos. Magaz. **89**, 2771–2799 (2009). https://doi.org/10.1080/14786430903113769

M.A. Hopkins, & A.S. Thorndike, Floe formation in Arctic sea ice. J. Geophys. Res. Oceans **111**(C11) (2006)

B. Lishman, P. Sammonds, D. Feltham, A rate and state friction law for saline ice. J. Geophys. Res. Oceans 116(C5) (2011)

B. Lishman, P.R. Sammonds, D.L. Feltham, Critical slip and time dependence in sea ice friction. Cold Reg. Sci. Technol. **90**, 9–13 (2013)

K. Mair, K.M. Frye, C. Marone, Influence of grain characteristics on the friction of granular shear zones. J. Geophys. Res. Solid Earth, **107**(B10), ECV-4 (2002)

K. Mair, C. Marone, Friction of simulated fault gouge for a wide range of velocities and normal stresses. J. Geophys. Res. Solid Earth **104**(B12), 28899–28914 (1999)

C. Marone, C.B. Raleigh, C.H. Scholz, Frictional behavior and constitutive modeling of simulated fault gouge. J. Geophys. Res. Solid Earth **95**(B5), 7007–7025 (1990)

C. Marone, C.H. Scholz, The depth of seismic faulting and the upper transition from stable to unstable slip regimes. Geophys. Res. Lett. **15**(6), 621–624 (1988)

C. Marone, The effect of loading rate on static friction and the rate of fault healing during the earthquake cycle. Nature **391**(6662), 69 (1998)

A.H.V. Repetto-Llamazares, K.V. Høyland, K-U. Evers Experimental studies on shear failure of freeze-bonds in saline ice: Part I. Set-up, failure mode and freeze-bond strength. Cold Reg. Sci. Technol. **65**(3), 286–297 (2011)

A. Ruina, Slip instability and state variable friction laws. J. Geophys. Res. **88**, 10359–10370 (1983). https://doi.org/10.1029/JB088iB12p10359

P.R. Sammonds, M.A. Rist, Sea ice fracture and friction, in *Scaling laws in ice mechanics and ice dynamics*, eds by J.P. Dempsey, H.H. Shen (Amsterdam, The Netherlands: Kluwer, 2001), pp. 183–194

P. Sammonds, S. Scourfield, B. Lishman, M. Shortt, E. Bailey, A. Marchenko, Sea ice dynamics: the role of broken ice in multi-scale deformation, in *Proceedings of the HYDRALAB+ Joint User Meeting* (Bucharest, Romania, 2019) pp. 166–171

C. Scholz, P. Molnar, T. Johnson, Detailed studies of frictional sliding of granite and implications for the earthquake mechanism. J. Geophys. Res. **77**(32), 6392–6406 (1972)

E.M. Schulson, Friction of sea ice. Philos. Trans. R. Soc. A Math. Phys. Eng. Sci. **376**(2129), 20170336 (2018)

E.M. Schulson, A.L. Fortt, Static strengthening of frictional surfaces of ice. Acta Mater. **61**(5), 1616–1623 (2013)

S. Scourfield, P. Sammonds, B. Lishman, A. Marchenko, The effect of rubble on ice-sliding, in *Proceedings of the 23rd Int. Conf. Port and Ocean Engineering under Arctic Conditions* (Trondheim, Norway, 2015)

S. Scourfield, The influence of ice rubble on sea ice friction. Ph.D. Thesis, University College London, 2019

Chapter 5
Ice Fracture

Wenjun Lu

Abstract This paper presents an overview of ice fracture at engineering scales. Although ice fractures are ubiquitous in the Arctic and it spans a wide range of scales, e.g., from geophysical to engineering scales, fracture mechanics is not always adopted to describe ice fractures. This is largely due to that a fracture mechanics – based framework is lacking regarding Arctic offshore structural design; and characterising the ice fracture by the strength theory is often simpler and more conservative. Limiting our ice fracture studies at the engineering scale (i.e., Arctic offshore structural design and Arctic marine operation), a fracture mechanics—based framework is proposed in this paper to describe the two most typical ice failure modes (i.e., out-of-plane bending and in-plane splitting) during ice and sloping structure interactions. In this regard, two numerical methods (i.e., a hybrid approach and a purely analytical fracture-based approach) are introduced and related simulation examples and validations are presented in this paper. While characterising the splitting failure mode, a review on the strength theory—based approach and Linear Elastic Fracture Mechanic (LEFM)—based approach is conducted. This reveals the size effect in ice fracture and the introduction of the fictitious crack model (or cohesive zone model, CZM) to characterise the fracture of sea ice from geophysical scale down to grain scale. Because of the following two reasons: (1) most of the calculation methods in the fracture mechanics—based framework requires the fracture toughness of sea ice as an important input; and (2) based on the CZM, we see vividly a size effect in the fracture of sea ice and it deviates from the LEFM scaling at the laboratory scale, we carried out three field ice fracture experiments campaigns in a consecutive of three years (i.e., 2016–2018). Albeit only preliminary unprocessed results are available, we observed the importance of including the creep behaviour of sea ice in deriving the fracture properties of sea ice. This was also highlighted in previous studies through the development of the so-called Viscoelastic Fictitious Crack Model (VFCM). In addition, based on the most recent theoretical development in the CZM, we reviewed

W. Lu (✉)
Norwegian University of Science and Technology (NTNU), Trondheim, Norway
e-mail: wenjun.lu@ntnu.no

The Norwegian Academy of Science and Letters, Oslo, Norway

J. Tuhkuri and A. Polojärvi (eds.), *IUTAM Symposium on Physics and Mechanics of Sea Ice*, IUTAM Bookseries 39, https://doi.org/10.1007/978-3-030-80439-8_5

our field tests' set-ups; and we also presente some initial results in relation to the loading rate effect.

Keywords Ice fracture · Ice–structure interaction · Arctic marine operation · Analytical fracture · Size effect

5.1 Introduction

Ice fractures are ubiquitous in polar regions and directly influence the global climate and human activities, e.g., in the high North. The fracture of sea ice spans a wide range of spatial and temporal scales, i.e., from a geophysical scale in the orders of hundreds of kilo meters to the human engineering scale in the order of several hundred meters. To date, there exists no a generally-accepted universal model to describe sea ice's fracturing process at all relevant scales. For example, fracture mechanics is not always utilised to describe ice fractures. This paper starts from the general description of ice fractures at different scales, to a review of how ice fracture is studied in history; then focus is given to two engineering practices with relatively smaller ice fracture scales, namely, the Arctic marine structural design and marine operations. Examples of how fracture mechanics can be applied in these scales are described. Afterwards, the importance of conducting field experiment to retrieve the scale-invariant fracture properties of sea ice is described with initial measurements presented.

5.2 Multi-Scale Ice Fractures

Depending on the involved spatial and temporal scales, we can categorise sea ice fractures into those taking place at the geophysical scale (see Fig. 5.1), at the operational scale (see Fig. 5.2a), and man-made structure scale (see Fig. 5.2b). A similar scale categorisation into those 'geophysical scale, floe scale and structural scale' have been

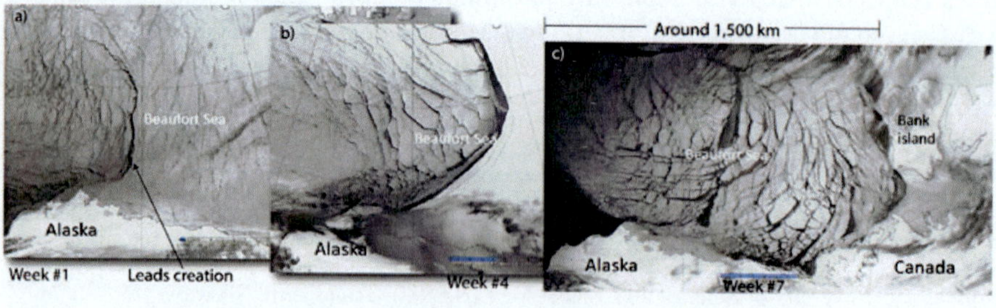

Fig. 5.1 Geophysical-scale sea ice fracture in the Beaufort Sea due to a storm during February–March 2017 (from NASA Earth Observatory)

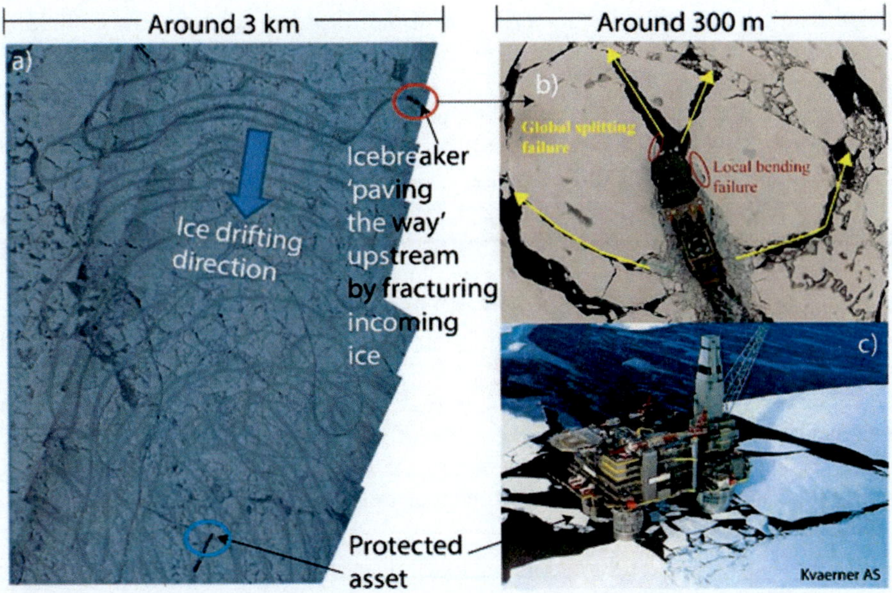

Fig. 5.2 **a** A typical Arctic marine operation involving fracturing dangerous ice features to protect an asset (e.g., an ice going vessel, LNG tanker, or an Arctic offshore platform) downstream; **b** Arctic structures designed to be able to withstand impacts from ice floes

proposed by Dempsey (2000). Here, we replace the 'floe scale' into the 'operational scale'; this is to highlight the involvement of human engineering practices in these different scales.

At different scales, a hierarchical model with different dominant ice mechanical processes has been proposed (Overland et al. 1995). At the large geophysical scale, an 'averaging' process is often adopted and it treats the ice sheet within the entire, e.g., Arctic basin, as a continuum (i.e., discrete fractures are not considered explicitly); and the continuum-based theories are often adopted to characterise sea ice fracturing (i.e., leads opening and ice ridging) (Hibler III 2001). This continuum-based theoretical paradigm leads to one of the key scaling problems which is that the measured ice strength shows two order of magnitude differences at different scales (Leppäranta 2011). To address these scale issues, we believe and foresee that a theoretical paradigm shifts from continuum-based theories to fracture-mechanics (i.e., to include each fracture explicitly in the model) based theories are necessary and possible (thanks to the advancement in computational mechanics) in characterising sea ice fractures at the geophysical scale. However, before that, sufficient knowledge on the ice fractures at different scales should be established.

Figure 5.2a shows a typical Arctic marine operation, i.e., ice management, in which, an icebreaker is working upstream to break the incoming large ice floes into smaller ones for the downstream protected vessel. The back-and-forth tracks of the icebreaker forms parallel channels in the scale of several kilometres in length. In between these channels, ice fracturing with the formation of long cracks linking neighbouring channels are often taking place to effectively reduce the incoming ice floes' size. However, such ice fracturing process is rarely treated from a fracture mechanics point of view. The prevailing engineering practice adopts a 'kinematic' approach, in which, the ice fracturing's occurrences and consequences are treated empirically, e.g., see simulators developed in literatures (Hamilton et al. 2011a, b).

In Fig. 5.2b and c, ice fracture takes place almost at the same size of the structure, i.e., in the order of several hundreds of meters. Ice failure is one of the limiting mechanisms that determine the design ice load. As shall be introduced later, there are many ice failure modes during ice—structure interactions. However, most of the ice failure modes are treated based on the strength theory (Sanderson 1988) instead of fracture mechanics.

In summary, while treating ice fracture at different scales in Fig. 5.2, fracture mechanics is not always utilised. This is partly due to the difficulties in applying conventional/classic fracture mechanics in ice engineering related problems and partly due to the fact that the ice fracture mechanics topic is not yet well established. As Timco and Weeks (2010) wrote:

> It is not entirely clear how the fracture toughness[1] should be used in ice engineering. When the concept was first applied to address the issue of ice forces on offshore structures, it was felt that this approach would help to explain why measured full-scale loads were lower than those predicted by simple physics-based models. This was never fully realized and the use of the fracture mechanics concepts fell out of fashion for a while.

In this paper, we shall focus on reviewing and introducing the application of fracture mechanics at engineering scales. Afterwards, it will be clear that the sea ice's fracture properties are crucial in these applications as one of the important input values. In light of this, field experiments need to be designed and performed. The tests and preliminary results are described.

5.3 Arctic Marine Structural Design

When designing an Arctic marine structure, the standard (ISO19906 2019) lists several important items and their associated parameters that should be taken into account (see the diagram in Fig. 5.3). In this diagram, the item 'failure mode' is highlighted. From a solid mechanics point of view, most of the failure modes, except for 'creep' are all taking place in a form of ice fracture. The failure mode largely influences the ice action and consequential structural design. For example, a sloping structure is preferred over a vertical structure in the Arctic as the sloping structure promotes the bending failure mode which alleviates the ice load as opposed to a vertical structure, on which, the ice crushing failure mode is more prevalent.

In fact, even for the same structural type, most of the failure modes in Fig. 5.3 co-exit and compete with each other (Lu et al. 2016a). For example, the failure mode

[1] Fracture toughness is one of the central concepts within fracture mechanics.

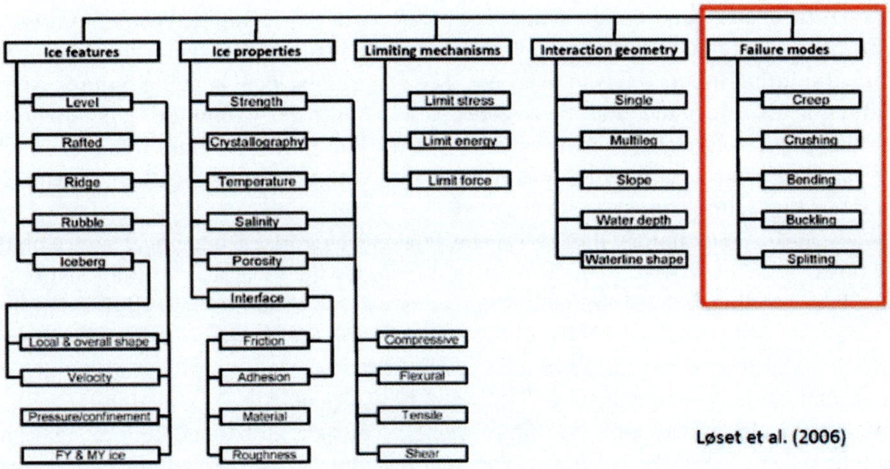

Fig. 5.3 Important items and associated parameters that should be considered during Arctic offshore structural design, by Løset et al. (2006)

competition is documented by a video camera system installed on the bow of the KV Svalbard in a voyage to the Northern Greenland Sea in March (Lubbad et al. 2012). During this period, first-year sea ice forms a relatively large and uniform icefield. The icebreaker primarily travelled within a large ice floe with continuous local bending failure (refer to Fig. 5.4a). This dominant flexural failure mode is featured by the initial contact with ice that is crushed (i.e., crushing failure) and the subsequent

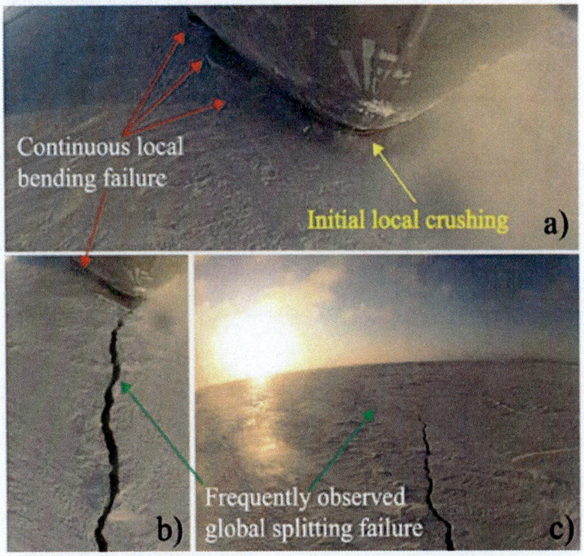

Fig. 5.4 Interaction process and failure modes competition during the transit of an icebreaker in ice

formation of cusp- and wedge-shaped ice blocks (bending failures). When conditions permit (e.g., a relatively small ice floe or an ice floe with minimal confinement), the global splitting failure tends to overtake. The continuous local bending failures were subsequently alleviated, and the structure tended to travel within the 'lead' created by the global splitting process (Fig. 5.4b and c). The splitting failure was presumed to act as a load-releasing mechanism compared with the continuous local crushing and bending failures.

These different failure modes can be further categorised into local and global failures (Lu et al. 2016a). The local failure includes the local ice crushing and out-of-plane bending failure; the global failure is used to describe the splitting failure mode (see Figs. 5.2b and 5.4b, c). The reason for such classification is because that local failure is to a large extend controlled by crack initiation and therefore the application of strength theory (more detail in Sect. 5.3.3) is justified. On the other hand, the global splitting failure mode involves a global crack's propagation and therefore, a fracture mechanics treatment is considered more appropriate (more detail in Sect. 5.3.2).

5.3.1 *Splitting Failure Mode by the Strength Theory*

In spite of the frequently observed long splitting cracks in sea ice, fracture mechanics was not always adopted to model this splitting failure mode. When dealing with the splitting crack induced by a bridge pier, Korzhavin (1971) assumed a continuous shear and tensile stress along a splitting crack within sea ice. This approach was re-presented in the book by Michel (1978) as in Fig. 5.5. Two representative cracks, $S_1$and S extending to the far-side and near-side borders are considered separately. Taking the near-side crack S as an example, its force balance along a the crack and the ice—structure contact can be written as in Eq. (5.1),

$$T(\cos\theta + \mu \sin\theta) = hS\tau_0, \tag{5.1}$$

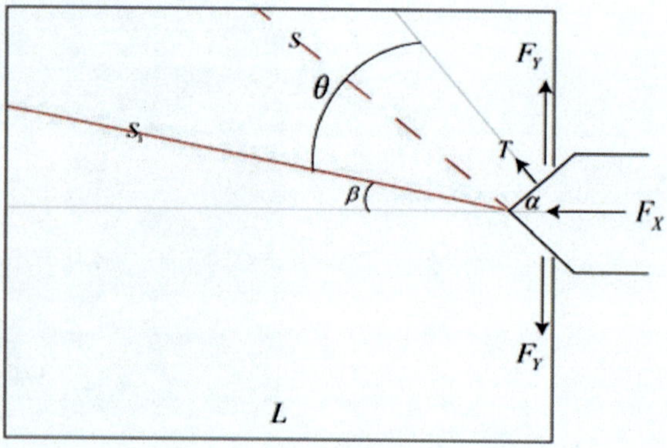

Fig. 5.5 Splitting of an ice floe, crack and force diagram according to Korzhavin (1971), reproduced from Michel (1978)

in which,

- h is ice thickness, [m];
- τ_0 is the shear strength of the sea ice, [kPa];
- T is a normal contact force on the structure's wedge surface (see Fig. 5.5) and is related to the horizontal force $F_X = 2T(\sin\alpha + \mu\cos\alpha)$;
- S is the crack path length which extends to the near-side of the ice floe; it is related to the ice floe width as $S = (B/2)/\sin\beta$. When the crack S_1 reaches the far-side border, we have $S_1 = L\cos\beta$;
- μ is the ice—structure friction coefficient.

Rearranging Eq. (5.1), neglecting the ice – structure friction and noting that $\theta = 90^o - (\alpha + \beta)$, we obtain Eq. (5.2). The minimum value of F_X is obtained while $\beta = 90^o - \alpha$, which leads to the splitting force for type S crack as in Eq. (5.3). Similarly, for type S_1 crack, the force is derived as Eq. (5.4) while setting $\beta = 0$.

$$F_X = \frac{\tau_0 h B \sin\alpha}{\sin\beta\sin(\alpha + \beta)} \tag{5.2}$$

$$F_X = \tau_0 h B \tan\alpha \tag{5.3}$$

$$F_X = 2\tau_0 h L \sin\alpha \tag{5.4}$$

There is a similar derivation for the tensile type splitting crack. All these derivations are rooted in the strength theory assumption, i.e., the entire body is considered as a continuum and there is a continuous shear band or tensile band activated simultaneously within the ice floe. As Michel (1978) stated, this type of approach is mainly applicable to smaller ice floes. Later, similar strength theory, i.e., the plastic limit theory, was adopted by Ralston (1981) to analyse the splitting of an ice floe. Here, following the same approach by Bažant (2005), we can define a nominal strength σ_N of a 'structure' as in Eq. (5.5), in which, F_{peak} is the peak force that leads to the failure of a structure; and L and h represent the size of the 'structure' in two different dimensions, i.e., length and thickness. Following the formula format in Eq. (5.5), the above Eqs. (5.3) and (5.4) can be organised in the format in Eq. (5.6).

$$\sigma_N = \frac{F_{peak}}{Lh} \tag{5.5}$$

$$\begin{aligned} \sigma_N^1 &= \frac{F_X}{Lh} = \tau_0 \frac{B}{L}\tan\alpha = f_1(\text{material property, geometric shape}) \\ \sigma_N^2 &= \frac{F_X}{Lh} = 2\tau_0\sin\alpha = f_2(\text{material property, geometric shape}) \end{aligned} \tag{5.6}$$

Equation (5.6) shows that the nominal strength of a structure is a function of material property (τ_0) and geometric shape (B/L and α). Structural size (L or B) does not enter the strength formulation. Thus, following the strength theory, given ice floes of the same geometric similarity (B/L) interacting with the same structure (α), the nominal strength is not varying with the size of the ice floe, or equivalently, the splitting force is linearly scaling with the ice floe size (L or B).

5.3.2 *Splitting Failure Mode by Fracture Mechanics*

Comparing to the strength theory, the application of fracture mechanics is an improvement in treating the splitting failure mode. At first, it was the Linear Elastic Fracture Mechanics (LEFM) that was applied. Palmer et al. (1983) discussed the role fracture mechanics in calculating different failure modes and introduced Eq. (5.7) to calculate the splitting force in a semi-infinite ice sheet,

$$F_Y = \frac{K_{IC}}{2.59} h\sqrt{\pi A}, \tag{5.7}$$

in which,

- F_Y is the splitting direction force and can be assumed to be related to the force component F_X depending on the contact geometry (reference to (Lu et al. 2015b)), for example, $F_X = 2\tan\alpha F_Y$ in Fig. 5.5;
- K_{IC} is the fracture toughness of sea ice and is a material property, [kPa$\sqrt{\text{m}}$];
- A is the length of the crack in a semi-infinite ice sheet, [m].

Bhat (1988, 1991) presented the formulas to calculate the splitting a square and disk shape ice floes based on Finite Element Analysis (FEA). Later, Dempsey et al. (1993) presented analytical solutions to the splitting of a square ice floe based on the weight function method. The weight function was later extended to rectangular ice floes of arbitrary width to length ratio (Dempsey and Mu 2014). For comparison purpose, we arrange and present the formulas to calculate the splitting force F_Y without the influence of F_X in the previously mentioned literatures (Figs. 5.6, 5.7 and 5.8).

$$F_Y = 0.19hK_{IC}\sqrt{L} \tag{5.8}$$

Equation (5.8) is for a square ice floe and calculate the peak splitting force (solid squares) in Fig. 5.9 of Bhat (1988).

$$F_Y = 0.17hK_{IC}\sqrt{L} \tag{5.9}$$

Equation (5.9) is for a disk shape ice floe and calculate the peak splitting force (lower curve) in Fig. 5.5 of Bhat et al. (1991).

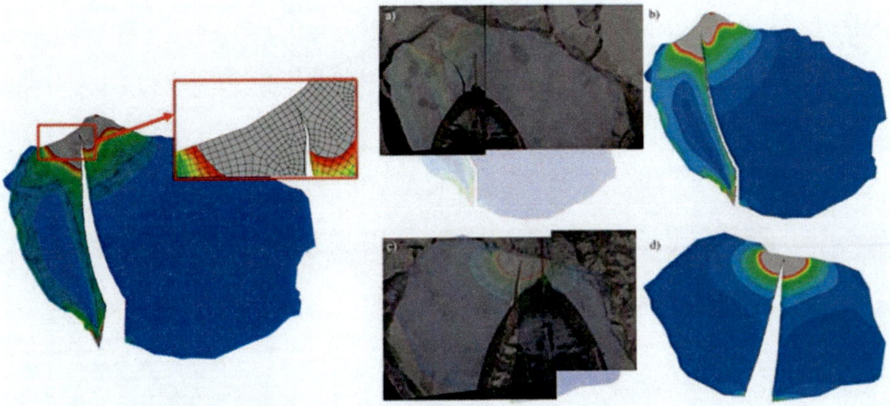

Fig. 5.6 An XFEM-based approach to simulate the splitting failure of an arbitrary shaped ice floe during its interaction with icebreaker Frej (Lu et al. 2018a)

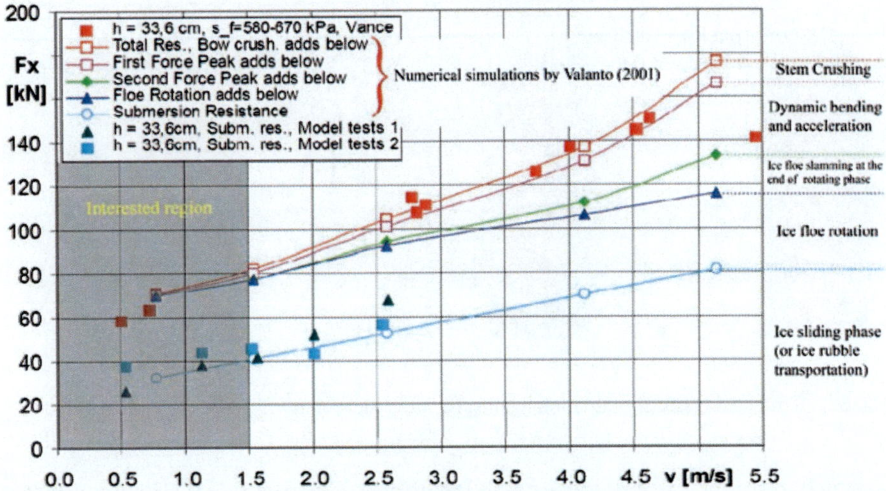

Fig. 5.7 Different physical processes and their corresponding force components contributing to the total ice resistance for icebreakers in level ice (originally from (Valanto 2001) and adapted from (Lu 2014)). The shaded interested region is mainly intended for fixed and floating structure in ice

$$F_Y = \max\left(\frac{(1-a)^{3/2}}{\sum_{i=0}^{7}\alpha_i^P a^i} \cdot \sqrt{\pi a}\right) \cdot h K_{IC} \sqrt{L}, \tag{5.10}$$

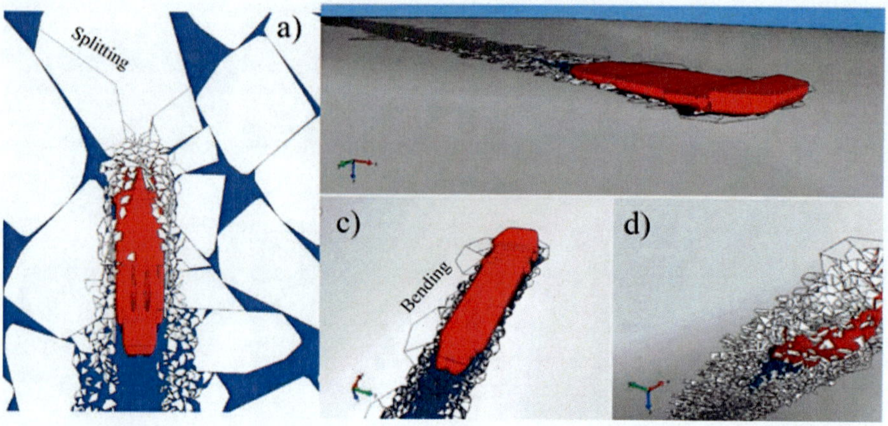

Fig. 5.8 Example of ice—structure interaction simulation by SAMS employing the fully analytical approach for ice fracture calculations

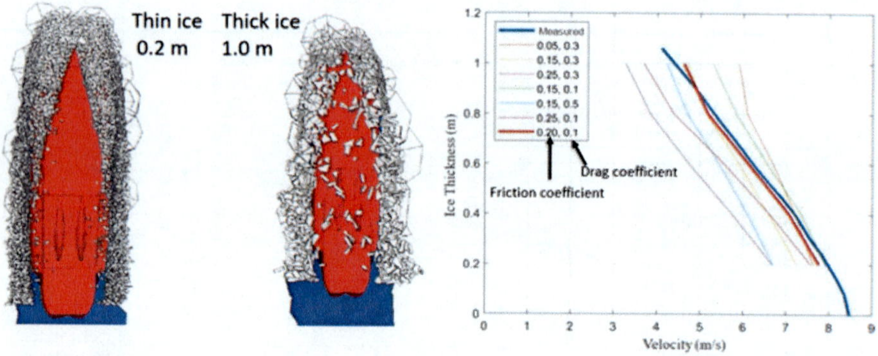

Fig. 5.9 Simulation results' validation against the field measured *h*-*v* curve of icebreaker Oden

in which, $a = A/L$ is a unitless crack length varying from 0 to 1. Equation (5.10) can be used to calculate the peak splitting force for a rectangular ice floe with a wide range of width to length ratios that are of engineering interests. Different width to length ratios are reflected by the parameter α_i^P, which is presented in Table 4 of Dempsey and Mu (2014).

Following the same definition of nominal strength σ_N in Eq. (5.5), the previous Eqs. (5.7) to (5.10) can be re-written as in Eq. (5.11).

$$\sigma_N^3 = \frac{F_Y}{Ah} = K_{IC}\frac{\sqrt{\pi}}{2.59}A^{-0.5} = f_3(\text{material property, geometric shape,size}^{-0.5})$$
$$\sigma_N^4 = \frac{F_Y}{Lh} = 0.19K_{IC}L^{-0.5} = f_4(\text{material property, geometric shape,size}^{-0.5})$$
$$\sigma_N^5 = \frac{F_Y}{Lh} = 0.17K_{IC}L^{-0.5} = f_5(\text{material property, geometric shape,size}^{-0.5})$$
$$\sigma_N^6 = \frac{F_Y}{Lh} = \max(\frac{(1-a)^{3/2}}{\sum_{i=0}^{7}\alpha_i^P a^i} \cdot \sqrt{\pi a}) \cdot K_{IC}L^{-0.5}$$
$$= f_6(\text{material property, geometric shape,size}^{-0.5}) \tag{5.11}$$

Equation (5.11) shows that according to the LEFM, the nominal strength of an ice floe for the splitting failure is scaled with size$^{-0.5}$, whereas size is not a part of the scaling in the strength theory in Eq. (5.6).

5.3.3 *Bending Failure Mode*

Aside from the global splitting failure mode, the local bending failure mode is often observed during ice—sloping structure interactions (see Figs. 5.2 and 5.3). The bending failure mode is a type of out-of-plane failure of an ice plate. For an ice plate under a concentrated vertical load, radial cracks firstly emanate from the loading area. After the radial cracks reaching almost twice the characteristic length ℓ (Sodhi 1996), keep increasing the vertical load will leads to the formation of circumferential crack. The initiation of the circumferential crack determine the final out-of-plane failure of ice plate (Kerr 1976; Lu et al. 2015c). It is possible to take a fracture mechanics approach to solve the out-of-plane failure of an ice plate (e.g., see the analyses by Dempsey et al. (1995) and by Bažant and Li (1993)). However, as the final failure is governed by the initiation of the circumferential cracks, a strength theory is considered sufficient for most engineering applications. The trade-off here is that if we utilise the strength theory to calculate the out-of-plane failure of an ice sheet, we get only the maximum failure force. The consequent crack propagation and eventual fragmentation of ice rubbles are simplified. In order to capture the entire failure process, ice fracture mechanics must be applied to achieve detailed energy balance during the material damage/softening/fracturing process. Notably, the cohesive fracture mechanics has been adopted to tentatively simulate the bending failure of a three-dimensional plate (Lu et al. 2012, 2014); and to the successful applications of two-dimensional beam bending fracture and fragmentations in the combined FEM-DEM scheme (Paavilainen et al. 2009; Paavilainen and Tuhkuri 2012, 2013). Most of the previous researches regarding the out-of-plane bending failure are based on the strength theory (e.g., see the review by (Kerr 1976) and work carried out by Nevel (1958, 1961, 1965, 1972) and Sodhi (1995, 1996, 1997)). Among these studies, Nevel's (1961) analytical solution are often utilized to characterize the out-of-plane bending failure for ice and structure interactions (Lubbad and Løset 2011; Li et al. 2019; Lubbad et al. 2018a). More recently, numerical methods including the combined FEM-DEM (Lilja et al. 2017, 2019) and the lattice model (Berg 2015) have been demonstrated to capture the out-of-plane behavior (i.e., no fracture involved) of a rectangular ice plate rather well.

5.3.4 *Ice Fracture Simulation: The Hybrid and Analytical Approaches*

To simulate the fracture of sea ice during ice—structure interactions, various methods can be utilised depending on the application purpose. For Arctic structural design, simulations with high accuracy are often expected. In this regard, computational fracture mechanics is preferred. There is a wealth of computational methods (Ingraffea and Wawrzynek 2004) that can be utilised to 'try to' solve ice—structure interactions. In this section, two alternative approaches are introduced.

Instead of engaging in a fully computational fracture mechanics—based approach, it is possible to only simulate the two most frequent fracture modes, namely, the out-of-plane bending failure mode and the in-plane splitting failure mode. For the out-of-plane bending failure mode, a pure fracture mechanics treatment starting from the radial crack initiation to propagation, until the formation of circumferential crack is rather challenging. Since the out-of-plane bending failure is mainly controlled by the 'initiation of circumferential crack' (see Sect. 5.3.3), a strength theory based analytical solution would serve the purpose. For the splitting failure mode, it is heavily influenced by the geometry, size, and confinement of the ice floe in contact, an analytical treatment requires further simplifications. Given the choice of only simulating the two most frequent failure modes, i.e., bending and splitting failures, depending on if choosing an analytical approach or numerical approach to simulate the splitting failure, two different methods are proposed in this paper:

1. A hybrid approach, in which, the bending failure is characterised by a set of analytical formulas whereas the splitting failure is simulated numerically.
2. Fully analytical approach, in which, both the bending and splitting failure modes are characterised by a set of analytical formulas.

For the first method, the eXtended Finite Element Method (XFEM) has been adopted in the hybrid approach to simulate the splitting of an arbitrary shaped ice floe (see Fig. 5.6). This has been applied for two interaction cases during the icebreaker Frej's transit through a broken ice field (Lubbad et al. 2016). The simulated impact force was compared with the onboard Inertia Measurement Units' measured data and calculations. Satisfactory comparisons were obtained (Lu et al. 2018a). This method, comparing to the fully computational fracture mechanics (e.g., (Lu et al. 2014)), is rather effective as it only deals with one type of failure mode numerically.

For the fully analytical fracture approach, the motivation stems from that there are at least three different ice force components that are present during ice—structure/ship interactions. These are the ice breaking, ice rotating/submergence, and ice sliding components (Lindqvist 1989). Among these force components, the ice sliding component covers almost half of the total ice resistance whereas the ice breaking (or fracture) component covers only a small portion of the total force (see Fig. 5.7). As long as a 'good enough' calculation on the ice breaking (fracture) part is captured and a more realistic ice sliding simulation is achieved, the final ice resistance prediction is still trustworthy. Given this background, the analytical formulas for both the bending and splitting failure modes are implemented into the Simulator for Arctic Marine Structures (SAMS). More information regarding the theoretical background of SAMS can be found in related literatures (Lubbad et al. 2018a, b). Here, among many validation cases (Tsarau et al. 2018), we present the validation case concerning the icebreaker Oden's transit in various ice conditions. A simulation snapshot containing both the local bending failure and splitting failure modes (in a broken ice field) are presented in Fig. 5.8a. Simulation views from different perspectives containing both ice breakings (fracture) and ice sliding processes are presented in Fig. 5.8b, c for the level ice condition.

Oden's performance in level ice (i.e., the *h*-*v* curves) was simulated by SAMS (Raza et al. 2019) and compare to measurements (Johansson and Liljestrom 1989) in Fig. 5.9. To construct the *h*-*v* curves, different ice thickness scenarios must be simulated. Figure 5.9 shows that the average ice breaking length's dependence on the ice thickness is well reflected in the simulation. In addition, Fig. 5.9 presented the quantitative comparison of simulated and measured *h*-*v* curves. Despite there are more physical processes, such as ice rubble accumulation, hydrodynamics, buoyancy et al. that are involved in the simulations, the satisfactory comparisons in both the visual and quantitative results in Figs. 5.8 and 5.9 indicate the appropriate treatment of one of the important physical process, i.e., the simulation of ice fractures with a set of analytical algorithms for ice—structure interactions.

5.4 Arctic Marine Operations

The analytical treatment of ice fracture in simulating ice—structure interactions are computationally efficient. This opens the possibility of simulating Arctic marine operations (e.g., see Fig. 5.2a), which span a much larger spatial (i.e., tens of kilometres) and temporal scales (up to several tens of hours). Two recent such simulation involving ice fractures applications are presented here.

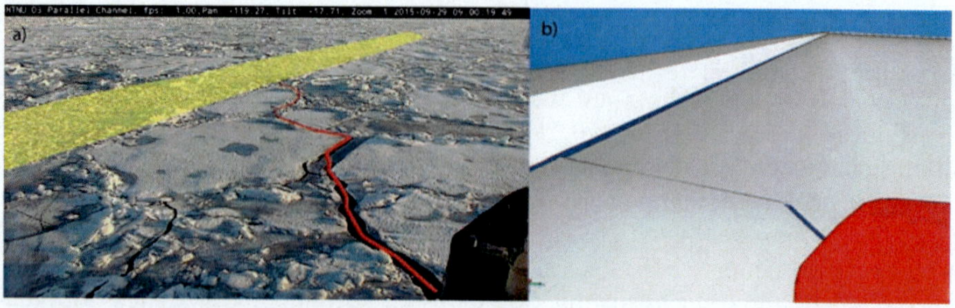

Fig. 5.10 **a** An observed long crack formation and extending towards a neighbouring parallel channel; **b** simulated long crack formation and extending towards the nearby free boundary

5.4.1 Parallel Channels' Fracture Simulations

With the recent development of analytical formulas concerning crack kinking in an edge-cracked rectangular plate (Lu et al. 2018b, c), long crack formations between two parallel channels during an ice management operation are simulated by SAMS and presented in Fig. 5.10.

Figure 5.10a is the visual results based on a camera system installed on board of Oden from a dedicated field experiment (Lu et al. 2016b); Fig. 5.10b is a SAMS simulation snapshot from a similar perspective. Due to the ice condition differences, the visual results are not exactly compared. However, the physical essence that a long crack tends to kink to the nearby free boundary is reflected in the simulation.

5.4.2 Ice Management (IM) Efficiency

Another application of utilising the analytical ice fracture algorithm within Arctic marine operation is the quantification of ice manage (IM) efficiencies. Up to now, the state-of-the-art method to evaluate and plan an ice management strategy is largely based on kinematic models (Hamilton et al. 2011a, b); and the operations are largely experience based. Mechanics concerning ice – structure interaction, let alone ice fracture mechanics, is not directly considered in these models. With the advancement in developing analytical formulas to characterise different ice fracture modes, it is now at our disposal to develop a 'mechanically-based ice management (IM) simulation tool'. Comparing with the kinematic approach, the mechanically based IM simulation takes into account major mechanical processes (e.g., ice resistance, ice—structure contact, and ice fractures) and quantify the IM efficiency based on more tangible concepts such as ice force and ice resistance on the protected assets.

Figure 5.11 illustrates the large-scale simulation (1 km by 2 km) of four different IM strategies. Not only does the floe size reduction or floe size distribution in the downstream after the IM operation are explicitly simulated, the ice resistance

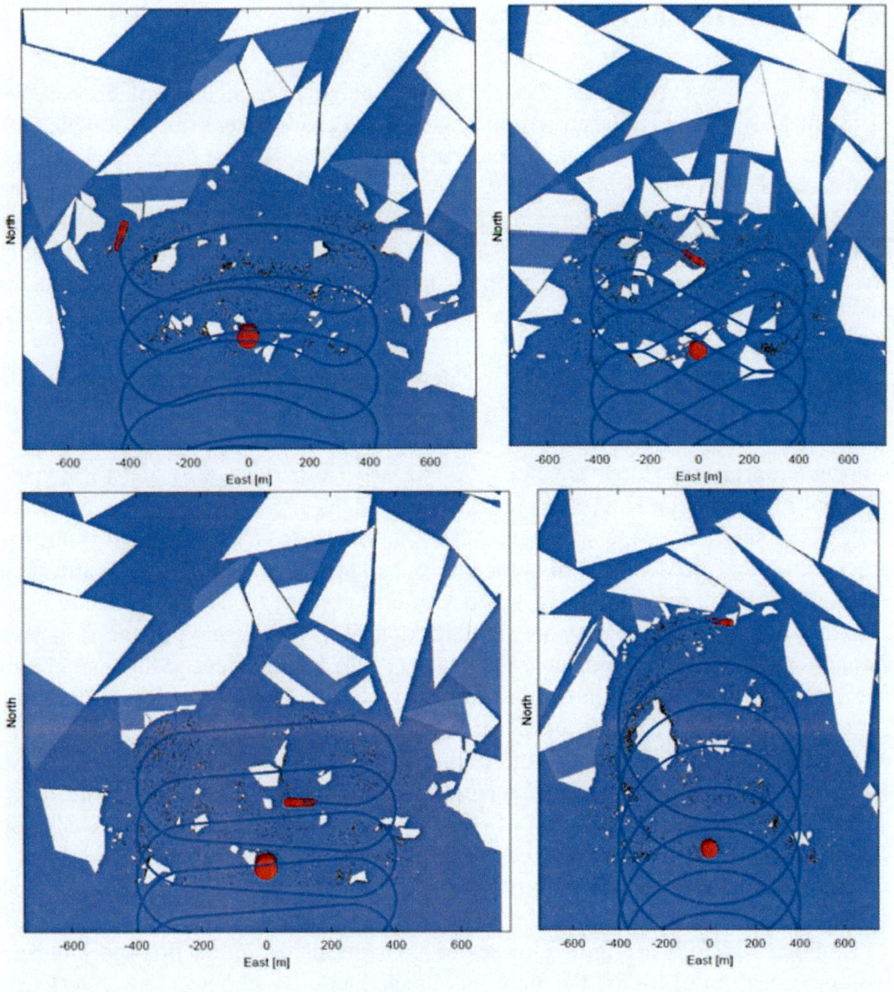

Fig. 5.11 IM efficiency evaluation by simulating four different IM strategies in a same drifting ice environment

encountered by the operating icebreaker and the protected assets are also available for evaluations. Largely because of the analytical fracture algorithms' efficiency, the mechanically-based operational simulations are rather effective and offer more insight into the efficiency of various IM strategies and position us towards the pursuit of 'Green Ice Managements'.

5.5 Size Effect in Ice Fracture

The previous two sections described ice fracture and their related calculations in two different scales, namely, the structural scale and operational scale, both of which are larger than several tens of meters. The size effect in ice fracture has been discussed and debated heatedly over the years. Previous researches mainly concern the ice fracture properties' testing requirements, i.e., the ice test sample's size. More review on the experiment will be presented in Sect. 5.6. For engineering applications (mainly Arctic structural design), size effect in the fracture mechanics is often considered as one of the explanations to the observed Pressure-Area relationship (Sanderson 1988; Løset et al. 2006).

In Sects. 5.3.1 and 5.3.2, methods to calculate the splitting failure mode have been introduced. It is mentioned that when the ice floe is rather small, a strength theory (Sect. 5.3.1) can be used; whereas a fracture mechanics treatment (Sect. 5.3.2) is more appropriate for large ice floes. As we see from Eqs. (5.6) and (5.11), the nominal strength of an ice 'structure' is not scaled the same with size, i.e., for small size, $\sigma_N \propto \text{size}^0$, for large size $\sigma_N \propto \text{size}^{-0.5}$. This is to say, for a tensile splitting crack initiating from length 0 m to the entire ice floe, there is not a simple universal scaling law (Dempsey et al. 2018). One method to bridge the strength scaling is by adopting the fictious crack model (Hillerborg 1991), also termed as the Cohesive Zone Model (CZM). Mulmule and Dempsey (1998) first introduced this model into the ice community and extended this model to the so-called 'Viscoelastic Fictitious Crack Model (VFCM)' to decode the large-scale experiments performed by Dempsey et al. (1999a, b). Comparing to the fictitious crack model (Hillerborg 1991), the VFCM additionally introduced a creep compliance function to characterise the time-dependent visco-elastic behaviour of sea ice upon loading (Mulmule and Dempsey 1998). It turns out that the ice creeps significantly. Without the creep compliance function, the measured crack mouth opening displacement is largely underestimated by the 'elastic' fictious crack model (Mulmule and Dempsey 1997).

Limiting our study at engineering scales, considering the interaction speed during ice—structure interactions, the fictitious crack model (without creep effect) was re-implemented for the splitting failure of an edge cracked square plate with the results presented in Fig. 5.12. It shows that the peak splitting forces at the field scale (or engineering scale larger than several tens of meters), both the LEFM, the fictitious crack model (or the CZM), and the field measurements agree well with each other. This signifies that, from a force level's perspective, the application of LEFM in treating ice fractures at the engineering and operational scales (Sects. 5.3.1 and 5.3.2) is justified. The reason that we stress *'from a force level's perspective'* is because that sea ice creeps rather significantly (as will be demonstrated later in Sect. 5.6). Interestingly, different experimental and theoretical studies (including our studies) show that for sea ice's tensile fractures, the creep has a much larger influence on the deformation while a much less significant influence on the force level, at which ice fractures. For ice—structure interaction studies, it is often the peak force at which ice fails that is of primary interests. Therefore, the application

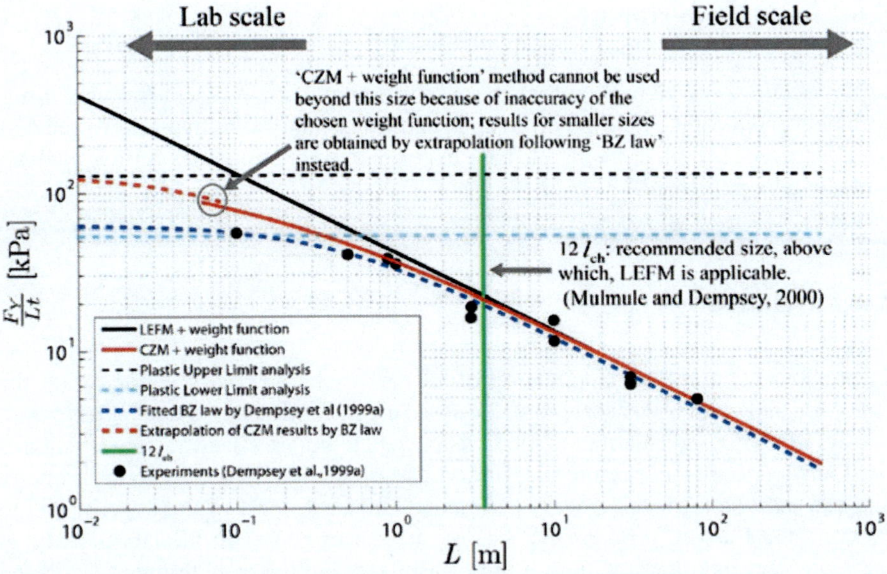

Fig. 5.12 Normalised splitting force for an edge cracked square ice plate with an initial crack length of 30% of the plate size L

of LEFM (i.e., excluding the simulation of creep/viscoelasticity of sea ice) can yield good enough ice force levels at the engineering and operation scales. In these scales, the 'strength' of a cracked ice plate is scaled with its size$^{-0.5}$. This is also supported by the analyses results of Mulmule and Dempsey (2000) stating that LEFM becomes applicable when the cracked body size is larger than 12 l_{ch} (i.e., green vertical curve in Fig. 5.12). l_{ch} is the characteristic length of sea ice and is defined in Eq. (5.12),

$$\ell_{ch} = \frac{K_{IC}^2}{\sigma_f^2}, \tag{5.12}$$

in which, K_{IC} and σ_f are the fracture toughness and flexural strength of sea ice.

On the other hand, in the lab scale, both the linear fictitious crack model (CZM), LEFM, and strength theories (i.e., plastic upper and lower limit methods) are not sufficient to explain the measured data. In this scale, the size of the Fracture Process Zone (FPZ) is comparable to the size of the cracked specimen; in addition, the creep effect, depending on the loading rate, starts to have more significant influence of the experimental results. In this regard the VFCM (Mulmule and Dempsey 1998, 1999) is theoretically more appropriate to be applied in this scale to characterise the ice fracture.

5.6 Field Experiments

In this section, we mainly present the motivation and some recent findings in the technical details of the field experiment. In addition, some preliminary measured data are illustrated. Detailed results of these experiments are still under further analysis and yet to be published.

5.6.1 *Motivation*

In previous sections, the application of LEFM to characterise ice fracture at the structural and operational scales are discussed together with ice fracture's size effect. All the LEFM based calculation methods presented before (see Eq. (5.11)) requires the knowledge of the fracture toughness K_{IC} of sea ice. It is a material property and can only be retrieved from experiments. However, most of the experiments to measure the fracture properties of sea ice were carried out in the lab, namely, at lab scale; and the interpretation of the measured value (often in terms of force and displacement) into the fracture toughness is following the LEFM theory. However, as illustrated by Fig. 5.12, the LEFM theory seems to have deviated from reality at the lab scale. A summary of the experiments and measured values can be found in Schulson and Duval (2009). Dempsey (1991) made a thorough review of the previous sea ice fracture toughness experiments and conclude that most of the lab experiments are 'sub-sized'. According to Fig. 5.12 and Mulmule and Dempsey (2000), a size that is larger than 12 ℓ_{ch} (around 3 m) for the edge cracked ice plate is needed for the LEFM theory to apply. This size requirement is larger than most, if not all, the laboratory experiments. Another important issue for the ice fracture experiments is 'loading rate', which influences ice's creeping process significantly (Schulson and Duval 2009). To resolve both the size and loading rate issues, a series of field ice fracture experiments were designed (Lu et al. 2015a) and executed in the year 2016, 2017 and 2018.

5.6.2 *Fracture Process Zone (FPZ) Size*

The experimental design is illustrated in Fig. 5.13 using the sample size 10 m by 5 m as an example. 7 group of displacement sensors are installed along the centre line of the test sample. Among these displacement sensors, 3 sets of sensors (i.e., LP803-1) are installed on the initial crack; 4 sets of sensors (LD sensors) are installed ahead of the initial crack to measure the fracture process zone.

By the time the experiments were planned and executed, there exists no knowledge on the size of the fracture process zone for the rectangular plate. Later, Wang et al. (2019) calculated that the upper limit of the fracture process zone size for a fully

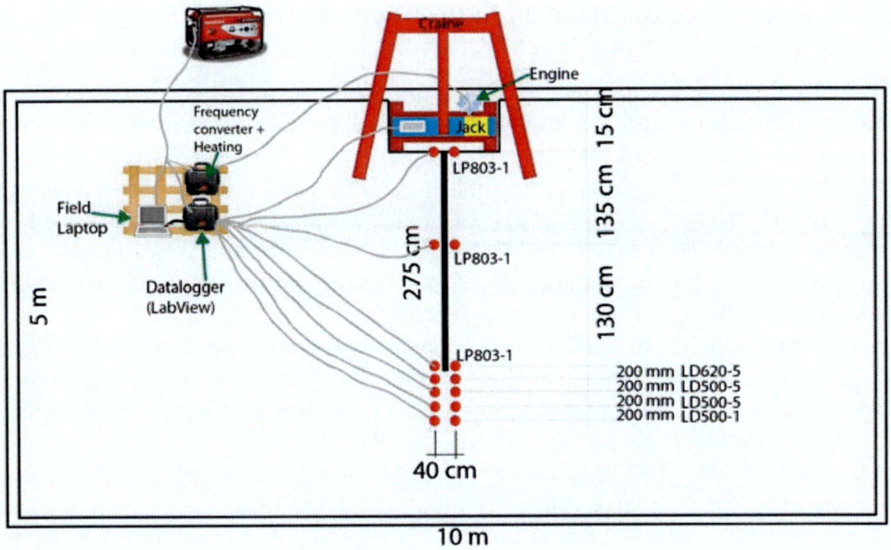

Fig. 5.13 Instrumentation of the prepared sample (using the 10 m by 5 m ice sample as an example)

grown crack in a semi-infinite plate is Eq. (5.13) for rectangular softening material (originally from (Schapery 1975)) and Eq. (5.14) for linear softening material.

$$R = \frac{\pi}{8}\ell_{ch} \tag{5.13}$$

$$R = 0.465\frac{\pi}{2}\ell_{ch} = 0.232\pi\ell_{ch} \tag{5.14}$$

Supplying available values of fracture toughness (i.e., 100 kPa$\sqrt{\text{m}}$ to 250 kPa$\sqrt{\text{m}}$) and a typical tensile strength for sea ice, i.e., 500 kPa, the characteristic length ℓ_{ch} is calculated in the range of 40 mm to 250 mm. Using Eqs. (5.13) and (5.14), we can estimate that the upper limit of FPZ size as in Eq. (5.15).

$$R = \begin{cases} 98 - 182\,\text{mm} & \text{if} \quad K_{IC} = 250\,\text{kPa}\sqrt{\text{m}} \\ 15 - 30\,\text{mm} & \text{if} \quad K_{IC} = \ 100\,\text{kPa}\sqrt{\text{m}} \end{cases} \tag{5.15}$$

Comparing to the calculation in Eq. (5.15) with the displacement sensor (LD sensors) arrangements in Fig. 5.13, it shows that a large enough LD sensors' measuring range (i.e., from −100 mm to 700 mm with the crack tip at 0 mm) encompassing the expected Fracture Process Zone (see Eq. (5.15)) is achieved.

5.6.3 Size, Loading Rate and Orientations

Over the three field experiments campaigns, different test specimen sizes were varied in 2016; three different loading rates were applied in 2017; and the specimen orientations were varied in 2018.

For each of the tests, 7 displacement histories and 1 crack mouth loading history are available. In order to decode these measurements and derive the fracture properties of sea ice, the VFCM is to be re-implemented. A same procedure as by Mulmule and Dempsey (1998) shall be adopted to derive the fracture properties of sea ice (i.e., the cohesive law) through matching the measured data.

The test specimen size that has been adopted during the tests vary from 5 to 20 m. As the raw measured peak force data indicates, the influences from test specimen size in the field are following the size effect of $\sigma_N \propto \text{size}^{-0.5}$(the LEFM trend). The results are illustrated in Fig. 5.14 together with the LEFM fitting, which scales strictly with $L^{-0.5}$. The fitting is not exact; however, good enough to indicate that our test sample size is large enough. All our tested specimen sizes are larger than the recommended size, i.e., 3 m as shown by the green line in Fig. 5.12 according to Mulmule and Dempsey (2000).

With regard to the loading rate's effect, a direct presentation of the measured raw data, i.e., load versus the crack mouth opening displacement (i.e., the top most LP803-1 in Fig. 5.13) is presented in Fig. 5.15. It shows that the peak splitting force is reversely influenced by the loading rate. In addition, the maximum crack opening displacement at the peak force is largely dependent on the loading rate. A slower loading rate leads to a larger crack opening displacement. This signifies the importance of including the creep behaviour into the material model of sea ice.

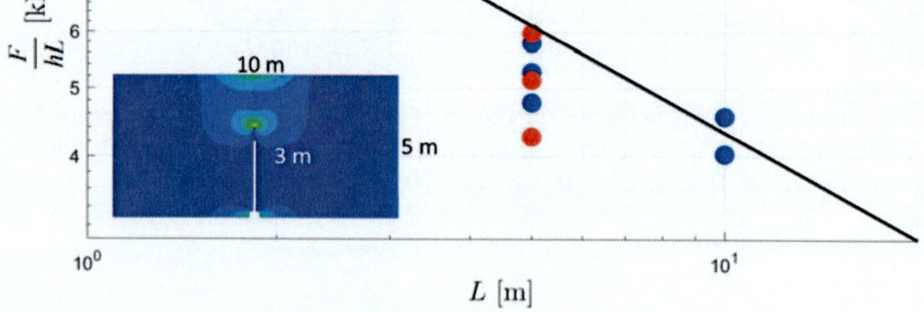

Fig. 5.14 Measure peak splitting force versus the test specimen size

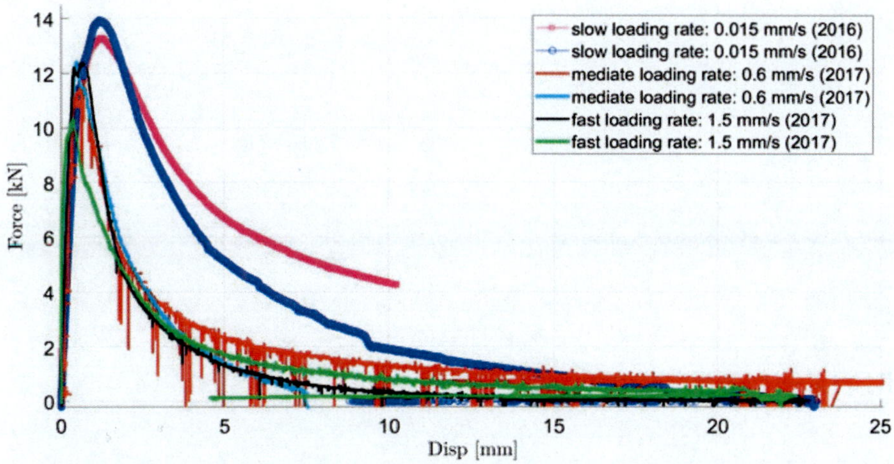

Fig. 5.15 Splitting force versus crack opening displacement at various loading rates

Concerning the orientation of the test specimen, there is a predominant current direction (yellow arrow in Fig. 5.16) at the test size. As the current direction shall influence the c-axis of the ice formation; and the fracture properties are dependent on the c-axis alignment (DeFranco and Dempsey 1994), we varied the test specimens' orientation during the test campaign in 2018 (see Fig. 5.17).

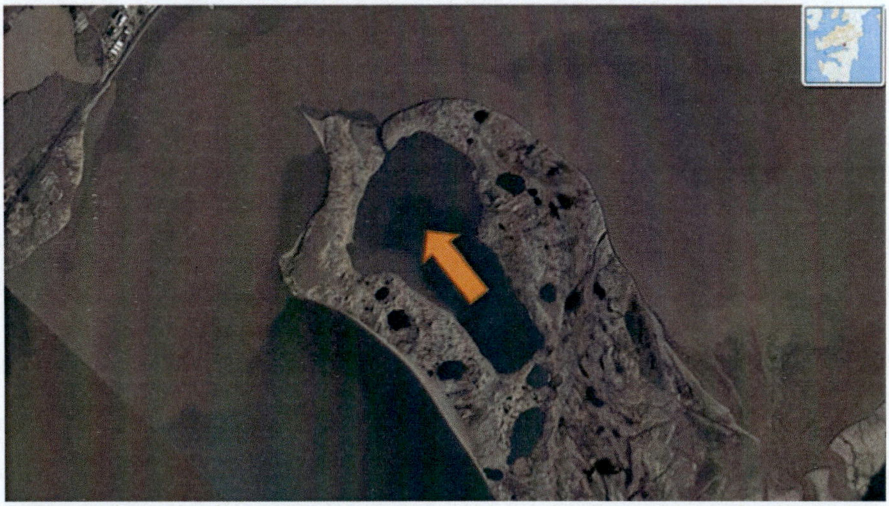

Fig. 5.16 Current direction and the test site during summer time (from https://toposvalbard.npolar.no/)

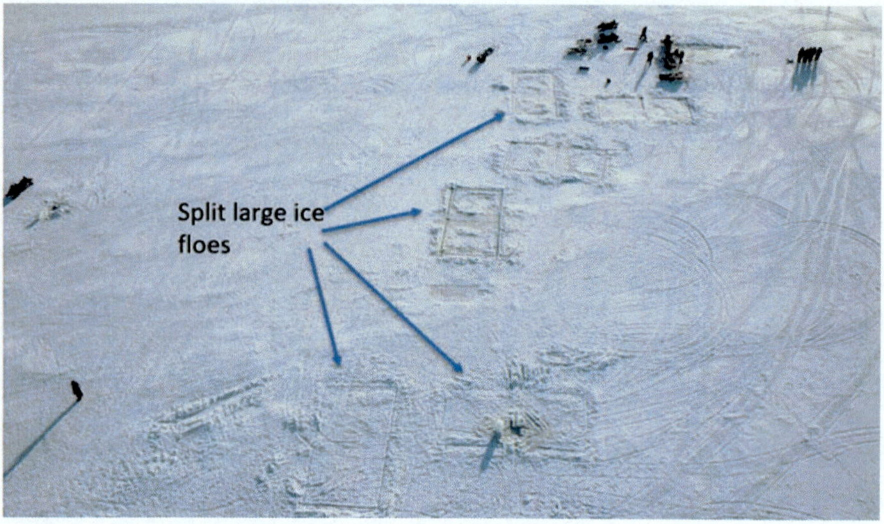

Fig. 5.17 Orientation variation during the test campaign in 2018; test specimens' size is 5 m by 2.5 m

By the time when this paper is written, the preliminary data does not indicate any influences from the orientation of the test specimen. More detailed microscopic structure of the test specimen is needed.

5.7 Conclusions

This paper gives an overview of how ice fracture is considered, and its related fracture mechanics can be applied at engineering scales in relation to Arctic offshore structure design and Arctic marine operations. Surprisingly, fracture mechanics related theories are not always utilised to deal with ice fracture in history. This is largely due to the facts that fracture mechanics is more difficult; and the strength theory is relatively simpler and often offers conservative results. However, when the fracture scale becomes larger, i.e., larger than several tens of metres, a fracture mechanics treatment becomes necessary as the strength theory is too conservative in calculating ice forces on a structure. In this regard, this paper summarises the theoretical backgrounds and analytical solutions to two types of failure modes that are often observed during ice and sloping structure interactions. These are the out-of-plane bending and in-plane splitting failure modes. Their numerical treatments are further elaborated with the introduction of the hybrid and purely analytical approaches. simulation and validation examples are presented.

While presenting the theoretical background and the calculation methods, the size effect in ice fracture is summarised in conjunction with the introduction of the fictitious crack model. From ice force's perspective, it is concluded that a Linear Fracture Mechanics (LEFM) treatment of ice fracture at engineering scales (i.e., larger than tens of meters) is sufficient. However, at lab scale, the creep behaviour and the size of the fracture process zone should be considered in decoding the measured displacement and force data.

As most of the fracture mechanics applications to sea ice engineering requires the knowledge on the fracture properties of sea ice, we presented our on-going work in relation to the field experiments. The preliminary results indicate that:

- Ice creeps significantly in the process of loading and fracturing.
- In accordance to previous studies, it appears that the conducted test specimen sizes are large enough and the instrumented area ahead of the crack tip is large enough to capture the entire fracture process zone's deformation.
- The preliminary test data does show an orientation effect.

Acknowledgements This paper presents many aspects of ice fracture mechanics that the author has been engaged in the past several years. The author would like to thank Professor Sveinung Løset, Associated Professor Aleksey Shestov, Professor Jukka Tuhkuri, Professor John Dempsey, and Miss Iman Gharamti for participating the field ice fracture experiments. In addition, the author would like to thank Associate Professor Raed Lubbad and Dr. Marnix van den Berg in their contribution of establishing the software platform of SAMS within which the analytical fracture algorithms are implemented. SAMS simulation results are presented in both the Arctic offshore structural design and Arctic marine operational validation cases. Last but not least, the author would like to thank the Norwegian Research Council through the research centre of SAMCoT CRI for financial support in carrying out the experiment. The author also like to thank VISTA—a basic research program in collaboration between The Norwegian Academy of Science and Letters, and Equinor (former Statoil), for financial support in writing this paper.

References

Z.P. Bažant, *Scaling of structural strength* (Butterworth-Heinemann, 2005)

Z.P. Bažant, Y.N. Li, Fracture analysis of penetration through floating sea ice plate and size effect. ASME Appl. Mech. Division-Publications--AMD **163**, 131–131 (1993)

V.D.M. Berg, A 3-D random lattice model of Sea Ice. *Arctic Technology Conference*. St John's, Newfoundland and Labrador, Canada (2015)

S.U. Bhat, Analysis for splitting of ice floes during summer impact. Cold Reg. Sci. Technol. **15**, 53–63 (1988)

S.U. Bhat, S.K. Choi, T. Wierzbicki, D.G. Karr, Failure analysis of impacting ice floes. J. Offshore Mech. Arct. Eng. **113**, 171 (1991)

S.J. Defranco, J.P. Dempsey, Crack propagation and fracture resistance in saline ice. J. Glaciol. **40**, 451–462 (1994)

J.P. Dempsey, The fracture toughness of ice. *Ice-Structure Interaction*, 109–145 (1991)

J.P. Dempsey, Research trends in ice mechanics. Int. J. Solids Struct. **37**, 131–153 (2000)

J.P. Dempsey, S.J. Defranco, D. Blanchet, A. Prodanovic, Splitting of ice floes, in *The 12th International Conference on Port and Ocean Engineering under Arctic Conditions* (POAC 93), 17–20 August, Hamburg, 17–22 (1993)

J.P. Dempsey, R.M. Adamson, S.V. Mulmule, Scale effects on the in-situ tensile strength and fracture of ice. Part II: First-year sea ice at Resolute NWT. International Journal of Fracture **95**, 347–366 (1999a)

J.P. Dempsey, S.J. Defranco, R.M. Adamson, S.V. Mulmule, Scale effects on the in-situ tensile strength and fracture of ice Part I: Large grained freshwater ice at Spray Lakes Reservoir, Alberta. Int. J. Fract. **95**, 325–345 (1999b)

J. Dempsey, D. Cole, S. Wang, E. Sciences, Tensile fracture of a single crack in first-year sea ice. J. Philosophic Trans R Soc A Math Phys **376**, 20170346 (2018)

J.P. Dempsey, Z. Mu, Weight function for an edge-cracked rectangular plate. Eng. Fract. Mech. **132**, 93–103 (2014)

J.P. Dempsey, L.I. Slepyan, I.I. Shekhtman, Radial cracking with closure. Int. J. Fract. **73**, 233–261 (1995)

J. Hamilton, C. Holub, J. Blunt, D. Mitchell, T. Kokkinis, Ice management for support of arctic floating operations, in *OTC Arctic Technology Conference* (2011a)

J. Hamilton, C. Holub, J. Blunt, Simulation of ice management fleet operations using two decades of Beaufort Sea ice drift and thickness time histories, in *Proceedings of International Society of Offshore and Polar Engineers (ISOPE), Maui, Hawaii.* (2011b)

W. Hibler III, Sea ice fracturing on the large scale. Eng. Fract. Mech. **68**, 2013–2043 (2001)

A. Hillerborg, Application of the fictitious crack model to different types of materials. *Current Trends in Concrete Fracture Research* (Springer, 1991)

A.R. Ingraffea, P.A. Wawrzynek, Computational fracture mechanics a survey of the field, in *European Congress on Computational Methods in Applied Science and Engineering* (2004)

ISO19906, *Petroleum and natural gas industries—Arctic offshore structures,International Standard, International Standardization organization* (Geneva, Switzerland, 2019)

B. Johansson, G. Liljestrom, Oden-A state-of-the-art icebreaker. *POAC 89* (1989)

A.D. Kerr, The bearing capacity of floating ice plates subjected to static or quasi-static loads. J. Glaciol. **17**, 229–268 (1976)

K.N. Korzhavin, Action of ice on engineering structures. Cold Regions Research and Engineering Lab Hanover NH (1971)

M. Leppäranta, *The Drift of Sea Ice* (Springer, Helsinki, 2011)

F. Li, M. Kotilainen, F. Goerlandt, P. Kujala, An extended ice failure model to improve the fidelity of icebreaking pattern in numerical simulation of ship performance in level ice. Ocean Eng. **176**, 169–183 (2019)

V.-P. Lilja, A. Polojärvi, J. Tuhkuri, J. Paavilainen, A three-dimensional FEM-DEM model of an ice sheet, in *International Conference on Port and Ocean Engineering under Arctic Conditions*, 2017. POAC.

V.-P. Lilja, A. Polojärvi, J. Tuhkuri, J. Paavilainen, A free, square, point-loaded ice sheet: afinite element-discrete element approach. Marine Struct. **68**, 102644 (2019)

G. Lindqvist, A straightforward method for calculation of ice resistance of ships, in *Proceedings of POAC*, 722–735 (1989)

S. Løset, K. Shkhinek, O. Gudmestad, K. Høyland, *Actions from Ice on Arctic Offshore and Coastal Structures* (Krasnodar, St Petersburg, Russia, 2006)

W. Lu, *Floe Ice—Sloping Structure Interactions* (Norwegian University of Science and Technology, Doctor of Philosophy Doctoral, 2014)

W. Lu, R. Lubbad, S. Løset, K.V. Høyland, Cohesive zone method based simulations of ice wedge bending: a comparative study of element erosion, CEM, DEM and XFEM, in *The 21st IAHR International Symposium on Ice*, ed by Li and Lu, June 11–15, 2012 Dalian, China, pp. 920–938 (2012)

W. Lu, R. Lubbad, S. Løset, Simulating ice-sloping structure interactions with the cohesive element method. J. OMAE, 136 (2014)

W. Lu, S. Løset, A. Shestove, R. Lubbad, Design of a field test for measuring the fracture toughness of sea ice, in *The 23rd International Conference on Port and Ocean Engineering under Arctic Conditions*, ed by Kim, E., Lu, W., Høyland, June 14–18, 2015, Trondheim, Norway (2015a)

W. Lu, R. Lubbad, S. Løset, In-plane fracture of an ice floe: A theoretical study on the splitting failure mode. Cold Regions Sci. Technol. (CRST) **110**, 77–101 (2015b)
W. Lu, R. Lubbad, S. Løset, Out-of-plane failure of an ice floe: Radial-crack-initiation-controlled fracture. CRST **119**, 183–203 (2015c)
W. Lu, R. Lubbad, S. Løset, M. Kashafutdinov, Fracture of an ice floe: local out-of-plane flexural failures versus global in-plane splitting failure. CRST **123**, 1–13 (2016a)
W. Lu, R. Lubbad, S. Løset, R. Skjetne, Parallel channel tests during ice management operations in the Arctic Ocean, in *Arctic Technology Conference 2016* (St. John's, Newfoundland and Labrador, 2016b)
W. Lu, H.-M Heyn, R. Lubbad, S. Løset, A large scale simulation of floe-ice fractures and validation against full-scale scenario. Int. J. Naval Architect. Ocean Eng. (2018a)
W. Lu, R. Lubbad, S. Løset, Parallel channels' fracturing mechanism during ice management operations. Part II: Experiment. Cold Reg. Sci. Technol. (2018b)
W. Lu, R. Lubbad, A. Shestov, S. Løset, Parallel channels' fracturing mechanism during ice management operations. Part I: theory. Cold Reg. Sci. Technol. (2018c)
R. Lubbad, S. Løset, A numerical model for real-time simulation of ship-ice interaction. Cold Reg. Sci. Technol. **65**, 111–127 (2011)
R. Lubbad, S. Løset, U. Hedman, C. Holub, D. Matskevitch, Oden Arctic Technology Research Cruise 2015, in *Proc. of the Arctic Technology Conference (ATC)*. St. Johns's, Newfoundland and Labardor, Canada (2016)
R. Lubbad, S. Løset, W. Lu, *Oden Arctic Technology Research Cruise 2012: Data Report Part I, SAMCoT Report* (Norwegian University of Science and Technology, Trondheim, 2012)
R. Lubbad, S. Løset, W. Lu, A. Tsarau, M. Van Den Berg, An overview of the Oden Arctic Technology Research Cruise 2015 (OATRC2015) and numerical simulations performed with SAMS driven by data collected during the cruise. Cold Reg. Sci. Technol. (2018a)
R. Lubbad, S. Løset, W. Lu, A. Tsarau, M. Van Den Berg, Simulator for Arctic Marine Structures (SAMS), in ASME 2018 37th International Conference on Ocean, Offshore and Arctic Engineering. American Society of Mechanical Engineers, V008T07A020-V008T07A020 (2018b)
B. Michel *Ice mechanics* (Quebec, Less Presses de L'Universite Laval, 1978)
S.V. Mulmule, J.P. Dempsey, Stress-separation curves for saline ice using fictitious crack model. J. Eng. Mech. **123**, 870 (1997)
S.V. Mulmule, J.P. Dempsey, Viscoelastic fictitious crack model for fracture of sea ice, in *Mechanics of Time-dependent Materials* (1998)
S.V. Mulmule, J.P. Dempsey, Scale effects on sea ice fracture. Mech. Cohesive Frict. Mater. **4**, 505–524 (1999)
S.V. Mulmule, J.P. Dempsey, LEFM size requirements for the fracture testing of sea ice. Int. J. Fract. **102**, 85–98 (2000)
D.E. Nevel, The theory of a narrow infinite wedge on an elastic foundation. U. S. Army Snow Ice and Permafrost Research Establishment, Corps of Engineering (1958)
D.E. Nevel, The narrow free infinite wedge on an elastic foundation. U. S. Army Snow Ice and Permafrost Research Establishment, Corps of Engineering (1961)
D.E. Nevel, A semi-infinite plate on an elastic foundation. U. S. Army Snow Ice and Permafrost Research Establishment, Corps of Engineering (1965)
D.E. Nevel, The ultimate failure of a floating ice sheet. International Association for Hydraulic Research, Ice Symposium, 17–22 (1972)
J.E. Overland, B.A. Walter, T.B. Curtin, P. Turet, Hierarchy and sea ice mechanics: a case study from the Beaufort Sea. J. Geophys. Res. (1995)
J. Paavilainen, J. Tuhkuri, Parameter effects on simulated ice rubbling forces on a wide sloping structure. Cold Reg. Sci. Technol. **81**, 1–10 (2012)
J. Paavilainen, J. Tuhkuri, Pressure distributions and force chains during simulated ice rubbling against sloped structures. Cold Reg. Sci. Technol. **85**, 157–174 (2013)

J. Paavilainen, J. Tuhkuri, A. Polojärvi, 2D combined finite-discrete element method to model multi-fracture of beam structures. Eng. Comput. **26**, 578–598 (2009)

A.C. Palmer, D.J. Goodman, M.F. Ashby, A.G. Evans, J.W. Hutchinson, A.R.S. Ponter, Fracture and its role in determining ice forces on offshore structures. Ann. Glaciol. **4**, 216–221 (1983)

T.D. Ralston, Plastic limit analysis of ice splitting failure, in Proc. 6th POAC. Quebec **1**, 205–215 (1981)

N. Raza, V.D.M. Berg, W. Lu, R. Lubbad, Analysis of Oden icebreaker performance in level ice using simulator for arctic marine structures (SAMS), in *Proceedings - International Conference on Port and Ocean Engineering under Arctic Conditions*, Delft, The Netherlands (2019)

T.J.O. Sanderson, Ice mechanics and risks to offshore structures (1988)

R.A. Schapery, A theory of crack initiation and growth in viscoelastic media II. Approximate methods of analysis. Int. J. Fract. **11**, 369–388 (1975)

E.M. Schulson, P. Duval, *Creep and Fracture of Ice* (Cambridge University Press, Cambridge, 2009)

D.S. Sodhi, Breakthrough loads of floating ice sheets. J. Cold Regions Eng. **9**, 4 (1995)

D.S. Sodhi, Deflection analysis of radially cracked floating ice sheets. OMAE, 97–102 (1996)

D.S. Sodhi, Vertical penetration of floating ice sheets. Int. J. Solids Struct. **35**, 4275–4294 (1997)

G.W. Timco, W.F. Weeks, A review of the engineering properties of sea ice. Cold Reg. Sci. Technol. **60**, 107–129 (2010)

A. Tsarau, M. Van Den Berg, W. Lu, R. Lubbad, S. Loset, Modelling results with a new simulator for arctic marine structures-SAMS. ASME 2018 37th International Conference on Ocean, Offshore and Arctic Engineering, 2018. American Society of Mechanical Engineers, V008T07A035-V008T07A035 (2018)

P. Valanto, The resistance of ships in level ice. SNAME **109**, 53–83 (2001)

S. Wang, I.E. Gharamti, J.P. Dempsey, Cohesive cracking in negative geometries. Eng. Fracture Mech. **221**, 106681 (2019)

Part II
Ice Loads

Chapter 6
Ice Action on Ship Hull: What Do We Know and What Do We Miss?

Kaj Riska

Abstract The paper describes different methods to clarify ice action on ship hull; what different methods have given and what information on ice action is still missing. First what can be obtained by observations is described. Observations, be they visual or about consequences of ice action like hull damage, give an idea of the loading process based on the breaking pattern in level ice or impact with a discrete ice floe. Also an understanding about the load patch being relatively narrow vertically and long horizontally emerges. An estimate of the load magnitude can be obtained by estimating what load gives the observed damages. Theoretical studies include a model for collision with level ice or smaller ice floes (Popov model) and also a model for collision with massive multi-year ice floes. Theoretical studies have also analysed the creation of the breaking pattern of level ice. All these studies give information about the load patch and total maximum force. In the last chapter some ship measurements are described. These have mostly endorsed the observed ice load magnitude and also the small load height of the load patch. The aim of the paper is to give an insight of how the ice action on ship hull can be described, even if only a selected number of studies is included.

6.1 Introduction

Much knowledge about the magnitude of ice action on ship hull, especially the magnitude of ice action to be used in design, has been gleaned by observing the damage ice has caused on the shell structure of the ship hull. Estimating the excitation from response or just describing ice action qualitatively requires that several assumptions on the factors influencing ice action must be made. If the load patch size—load patch is the area on which ice is acting in any distinct ship-ice interaction event—is assumed to be known and the contact pressure on the patch is assumed uniform, then measuring the dent depth in the shell gives an estimate of the maximum average pressure. Some knowledge on the ice action can also be obtained from observing

K. Riska (✉)
Helsinki, Finland

J. Tuhkuri and A. Polojärvi (eds.), *IUTAM Symposium on Physics and Mechanics of Sea Ice*, IUTAM Bookseries 39, https://doi.org/10.1007/978-3-030-80439-8_6

the ice action onboard ships. Here the problem is that only the ice top surface—which usually is covered with snow—can be seen. Supporting observations with an insight in basic mechanics yield anyhow valuable information about the ice breaking process.

Direct measurement of ice action, especially the ice pressure distribution on ship hull is still beyond the practical technical possibility, even if some panels and special pressure gauges have been used in very short-term measurements. Gaining knowledge on ice action on ship hull requires collection and synthesis of various clues obtained both empirically and theoretically. This kind of heuristic approach is not necessarily the most efficient and does not necessarily lead to a unique theoretical methodology in describing the ice action, but in practical terms it may lead to an understanding what is known and also importantly what is less well known. The aim of this article is to give an introduction what is known about ice action on ship hull in a heuristic way by describing first what can be obtained from observations, then what is gained from theoretical modelling and finally what data are obtained from measurements. The description is not always necessarily causal as the interpretation of e.g. observations has sometimes benefitted from theoretical studies (and vice versa).

This article is based on a presentation in the IUTAM Symposium on Physics and Mechanics of Sea Ice given by the author at the Aalto University, June 3–7, 2019 (presentation given on June 4). These presentations were published at the site https://www.aalto.fi/en/department-of-mechanical-engineering/iutam-symposium-on-physics-and-mechanics-of-sea-ice-june-3-7, and the figures in this article are taken from the author's presentation—only if there is a specific need, the original source is mentioned.

6.2 Description of Ice Action Based on Observations

A usual benchmark case of ice action is a ship progressing in level ice. Observations with a common wedge shaped hull form—see Fig. 6.1—suggest that ice fails in bending down. Observations suggest further that the bending cracks create a pattern of consecutive semi-circular floes. This forms when an ice floe that has broken off from the ice sheet starts rotating pushed by the ship, This pattern is called the breaking pattern. Each floe is formed when the ship hull contacts the sharp cusp formed by a junction between two semi-circular floe edges, and crushes the edge until a bending crack forms. The crack path to create the floe is termed 'circular' even if its exact shape is unknown; it just surely looks circular.

Sailing in level ice is considered a benchmark case as only one ice parameter—apart from ice deformation characteristics—is needed; ice thickness h_i. Collision with a distinct ice feature like multi-year ice floe or an iceberg could also be a benchmark case. In a collision the ice edge and ice floe geometry have a large influence and the number of required parameters increase much. In first year level ice the bending forces are dominating while in a collision case inertial forces are more important.

Fig. 6.1 The breaking pattern created on level ice by the bow of IB Urho (photo: Kaj Riska)

Already this simple analysis of what is seen from a ship proceeding in level ice yields several pieces of knowledge about ice action on ship hull. These can be listed as follows:

- Ice action forms a repetitive crushing-bending cycle;
- Each ice action event consists of a distinct contact between ice edge and ship hull where first ice edge is crushed followed by a bending failure. A clear distinct load patch can be inferred;
- Contact force due to ice can be assumed to first increase as the contact area increases and finally reach the bending strength of the ice cover;
- Bending strength of the ice cover determines the maximum force in each contact event. As the contact event is of short duration, hydrodynamic support force must be taken into account in determining the bending strength of ice cover;
- Load patch shape depends on the ice edge geometry. As the cusps are quite blunt, the load patch can be inferred to be much wider horizontally than in vertical direction;
- Ice load patch moves along the ship hull due to ship motion relative to ice cover,
- In thick level ice the ice cover bending strength may be that large that it is not reached before the relative velocity between the ship and ice is zero. In this case ship stops or changes direction

This description contains already the basics of ice action on ship hull in level ice. There are some implicit terms in the description; What is meant by 'crushing'? What is meant by the 'bending strength of ice cover'? Actually, to explain these invokes the main unknown aspects about ice action on ship hull.

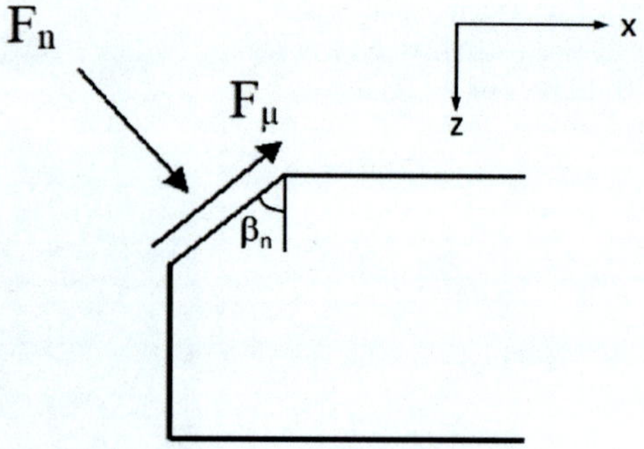

Fig. 6.2 Force components when ship hull indents an ice edge (x is longitudinal and z vertical coordinate) (Riska 2018)

Ice action (force) on ship hull is commonly investigated based on the bearing capacity of ice. For this purpose, force components acting on the ice edge are to be defined. Figure 6.2 shows the force components on a vertical plane containing the normal of the ship hull. The frame angle versus vertical is usually denoted as β_n on this normal plane. The contact force can be divided into normal and tangential (friction) components. If these are resolved into horizontal (*x*-direction, positive away from the ship) and vertical (*z*-direction, positive downwards), these are

$$\begin{aligned} F_x(t) &= F_n(t)cos\beta_n + F_\mu(t)sin\beta_n \\ F_z(t) &= F_n(t)sin\beta_n - F_\mu(t)cos\beta_n \end{aligned}, \tag{6.1}$$

where the time dependency of the force is highlighted. If then frictional force is described by a coefficient of friction μ i.e. $F_\mu = \mu F_n$, these equations simplify to

$$\begin{aligned} F_x(t) &= F_n(t)(cos\beta_n + \mu sin\beta_n) \\ F_z(t) &= F_n(t)(sin\beta_n - \mu cos\beta_n) \end{aligned}. \tag{6.2}$$

The coefficient of friction μ depends at least on the sliding speed and contact pressure but in ship ice load applications it is commonly treated as constant with a value of roughly 0.1.

The bending crack occurs when $F_z(t)$ exceeds the dynamic bending strength of the ice cover, F_D. At that moment $t = t_B$ the normal force reaches its maximum

$$F_n^{max} = \frac{F_D}{sin\beta_n - \mu cos\beta_n} \tag{6.3}$$

Equation (6.3) shows that friction increases the required normal force to break ice and further the horizontal component is increased due to two factors; increase in $F_n{}^{max}$ and increase of the trigonometric term. Further, if $tan\beta_n = \mu$, the vertical force component disappears. This occurs at the frame angle of roughly 6° as the coefficient of friction between ice and ship hull is typically around 0.1.

Equation (6.3) gives a first estimate of the ship ice load F_n^{max} if the bending strength of ice sheet is considered static and results from bearing capacity of ice cover are used. The vertical force required to break ice cover statically is of form (see e.g. Kerr 1976)

$$F_D = C\sigma_f h_i^2, \tag{6.4}$$

where σ_f is flexural strength of the ice cover. Ice flexural strength is considered as a material property even if here the flexural strength refers to the flexural strength of the whole ice sheet. This is different from a small scale ice bending strength as natural ice cover may contain layers of different ice and usually also has a temperature gradient. The constant C depends on the ice cover geometry and is typically 1,...,2. It must be noted that if the crushing-bending event is of short duration, the situation is dynamic and the bending strength of ice cover can be an order of magnitude larger than the static strength, see Tan et al. (2014).

If the ship hits a thick ice floe embedded in broken ice as Fig. 6.3 shows, the forces involved stem from inertia and crushing. This kind of situation occurs when multi-year ice floes exist in deformed first year ice field. Also frequently navigated ship channels full of brash ice can contain large rounded ice floes. When the relative motion between the ship hull and the ice floe is zero, the crushing at the ice edge stops. If the floe is thinner, also bending and bending failure can occur. The load patch shape in the collision case depends on the ice floe edge geometry but typically in the collision cases the load patch is rectangular or rounded with a length-height ratio close to one.

Fig. 6.3 Collision of IB Healy with a multi-year ice floe embedded in deformed first year ice field in the Baffin Bay (photo: K. Riska)

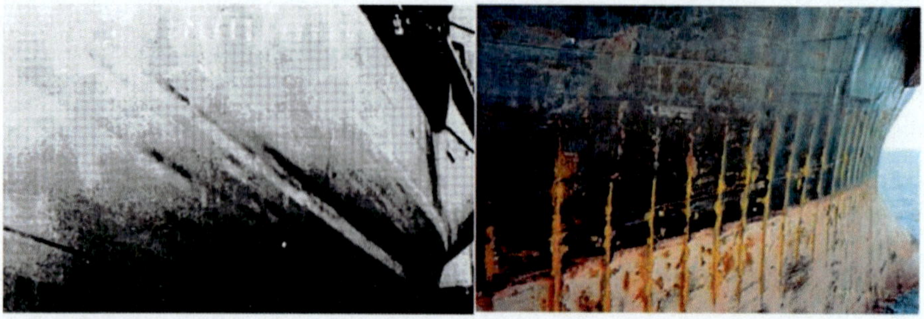

Fig. 6.4 Typical ice damage on the ship shell, left longitudinally and right vertically framed structure (Kujala 1991)

Observations of the damage caused by ice suggest a description of ice action on ship hull (see Kujala 1991; Hänninen 2003). Figure 6.4 shows typical ice damages that has been caused by first year ice. If the ship shell is longitudinally framed, the ice damage is a long dent between the longitudinals; the reason why in the Baltic most often only the plating was damaged in longitudinally framed ships, is that earlier (for example in the 1971 Finnish-Swedish ice class rules) there was incorrect idea about the load patch height resulting in somewhat underdesigned plating. If the ship shell is transversely framed, typical ice damage is denting of plating between frames. The fact that the frames are rarely plastically deformed, shows that strength hierarchy in shell structure is correct. The conclusions from observation of ice damages support those from observing ice action, described above.

Observations of ice action on ship hull and its consequences in terms of ice damage lead to a basic parameterization of ice action with a rectangular load patch, see Fig. 6.5. This description of parameterization is based somewhat on hindsight; especially how the theoretical methods have dealt with modelling of the ice action. It still focuses the attention to certain parameters while obscuring other issues like the interplay between the effect of relative velocity between ice and ship and the so called pressure-area relationship—more about this below. The ice load patch is

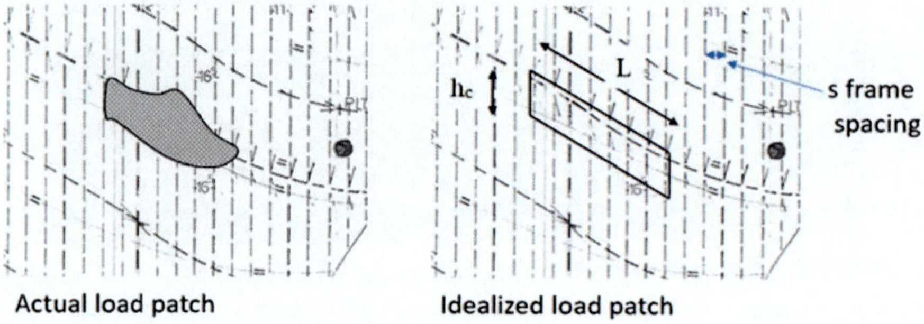

Fig. 6.5 Ice load patch idealization with load height h_c and load length L (Riska 2018)

idealized to a rectangle of load (contact) height h_c and width L. The average ice pressure on the load patch is p_{av}. This gives the total force acting on the load patch as

$$F_n(t) = p_{av} h_c L. \tag{6.5}$$

Each of these parameters evolve during acontact event. The utility of these four parameters lies partly in the practicality of measuring them or in using them for design of shell scantlings. The total force can be determined based on ice cover strength or floe mass by making an assumption on edge geometry and ice pressure. Frames or plating in transversely framed structures are not sensitive to load width L (if this is larger than two frame spacings) and the frame response is not sensitive to load height if the assumed force level is correct. This last observation has led to use of the line load in design. This is defined as

$$q = p_{av} h_c. \tag{6.6}$$

Transverse frames can be designed using the value of qs where s is the frame spacing. The design line load value can be obtained from ice damage surveys, and value of about 2 MN/m has been indicated. This is mostly based on back calculating what force causes the observed damages, but supported by frame force measurements which are described below.

In summary, the observations suggest that the maximum ice force is associated with the dynamic bending fracture of the ice sheet, that the load patch height is small and the line load value is known as about 2 MN/m. The effect of speed on the ice action is unclear, an early study (Johansson 1967) suggested that instead of speed,ice force depends on a factor $k = \sqrt{P\Delta}$ (P ship propulsion power and Δ ship displacement), see Fig. 6.6. Even if the Finnish-Swedish ice class rules are based on this dependency, it can be said that the linear fit in Fig. 6.6 is not wholly convincing.

Apart from the line load value, the load height h_c is needed for design. As discussed above, for transverse frames the value of load height is not important provided that the line load value is correct. For design of plating and especially longitudinal frames it is important to have correct load height. In early design considerations the load height was assumed to be about the ice thickness (800 mm) but later studies use a lower height, see Fig. 6.7. The question of load height has become important in designing larger, AFRAMAX size tankers which often are longitudinally framed. For construction efficiency the spacing of longitudinal frames tends to be large (close to one meter) leading to thick plating (up to 50 mm) when the load height is only a fraction of the frame spacing used.

Some discussion on the observations done about the load patch height is in place. Observations and measurements have shown that in some cases the load patch height is very small (Joensuu et al. 1989; Riska et al. 1990), only less than 1 cm. At the same time local pressure peaks are high and the load patch forms a line-like feature. These observations have been endorsed by using tactile sensors, see for example Sodhi et al. (1998, 2001). These measurements suggest also that the high local pressures

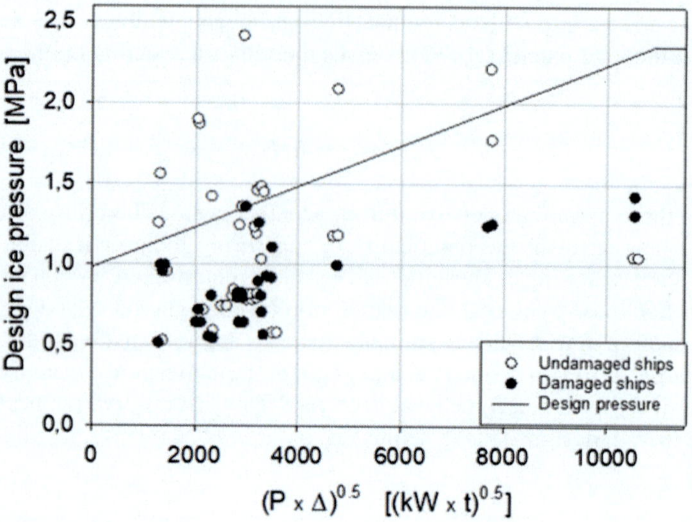

Fig. 6.6 Effect of ship displacement and propulsion power to design ice pressure (Modified from Johansson 1967)

Fig. 6.7 Understanding about the ice load height at the time of the first ice damage survey and when the smaller height first time was incorporated into the Finnish-Swedish Ice Class Rules (Riska 2018)

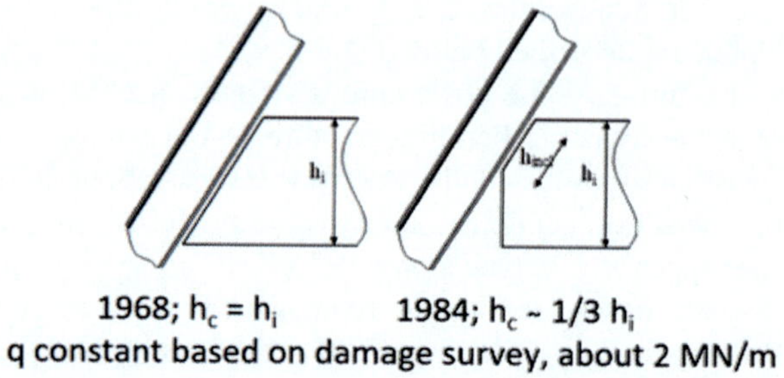

and line-like load patch occurs in high indentation speeds (more than, say, 3 cm/s) and in low temperatures. In low indentation speeds and high homologous temperatures the contact is different with load patch height covering the whole ice thickness (or so called apparent contact area). At the same time the local ice pressures are low, close to the compressive strength of ice. The line load value along the ice edge is, however, higher in the latter, ductile case. This makes the application of the small contact height observation difficult to apply in design. In many ship-ice collisions, the indentation speed is in the beginning of contact close to ship speed (depending on the direction of shell normal) but at the end of the contact the relative speed between ice and ship is small or zero. This suggests that ice failure changes from brittle in the beginning of contact to ductile in the end of contact.

The final matter which is not clarified by observations alone is the value and distribution of ice pressure.

6.3 Theoretical Ice Action Studies

The aim of theoretical studies of ice action on ship hull has mostly been to calculate the maximum normal force acting on the ship shell structure, and at the same time to determine the load patch. Knowledge from direct measurements on the average contact ice pressure, especially in form of the pressure-area curve, has been added to the theoretical modelling, as will be seen. Theoretical studies have usually defined an idealized ice-structure interaction scenario and then equations of motion (EoM) have been derived, and solved (integrated). This way also a simulation model for the whole ship-ice contact is obtained and the solution of this has led to simulating i.e. determining the force and ship motion time histories in a ship-ice collision event.

Ship-ice interaction modelling must take into account all the deformation and motion components; these can generally be considered as degrees of freedom (DoF) of the problem. The definition of these deformations for each problem at hand is a crucial stage of the problem solving. The relative importance of these is not always self evident so a too eagar simplification in the modelling may lead astray. These deformation and displacement components may include:

- Rigid body motions of ice floe or ship (maximum 6 + 6 DoF)
- Bending deformation of ice (1 DoF)
- Elastic deformation of ice or ship shell at the contact point (2 DoF)
- Crushing deformation of ice edge (1 DoF)

Thus there might be 16° of freedom in a general ship-ice interaction case. The solution proceeds so that the EoM's for each DoF are derived. These equations may be coupled and nonlinear. Tackling all the 16 EoM's is definitely a large task and the case is usually reduced to smaller number of degrees of freedom and simpler equations.

The solution cannot be, however, obtained solely based on the EoM's as there is always one unknown more than the number of equations (EoM's); the contact force $F_n(t)$. The simulation of multi body assemblies has been carried out by several somewhat different Discrete Element Methods (DEM's). The treatment of the local ice (or ship) deformation as well as the frictional force poses difficulties for these methods, see e.g. Hopkins (1992), Rabatel (2015), Tuhkuri and Polojärvi (2018). Below a straightforward method for the solution of ship-ice interaction is described (which in essence is a two body problem).

The one additional equation for the solution of the interaction is obtained from the requirement of the bodies being in contact with each other. This kinematic condition has been presented in a simple form for example in Goldsmith (1960). The kinematic condition, when ice and ship are in contact, can be expressed by setting the displacement components for ice and for ship along the normal of the contact surface equal. If the ship displacements are less than those of ice, then there is no contact, no penetration into ice and the contact force is zero. The ship displacements along the normal cannot be larger than those of ice. The situation can be complicated by the contact surface not being planar and also rotation between bodies occurring, resulting in rotation of the contact plane.

The need for a kinematic condition is sometimes not apparent. In a very simple case where ship side impacts an ice edge, two ice deformations can be defined; the crushing displacement $u_{cr}(t)$ and displacement due to elastic deformation of ice $u_e(t)$—here by 'crushing' is meant all local failure of ice at the contact. Then the total ice displacement is $u_{ice}(t) = u_{cr}(t) + u_e(t)$. At some point of the contact event it may happen that the total ice displacement starts to decrease. Using the kinematic condition is the best way to resolve this kind of case. If it can be assumed that ice is crushed throughout the impact event (ice displacement increases monotonically), then the contact force $F_n(t)$ can simply be related to ice crushing pressure by setting $F_n(t) = p_{cr} \cdot A(t)$ where p_{cr} is the average ice pressure required to cause local (crushing) failure and $A(t)$ the contact area on which the pressure acts, this area can be expressed as a function of the crushing (indentation) depth into ice.

In order to illustrate the use of the kinematic condition, a collision of an ice floe with the ship bow is described. The situation is considered on a plane and ship is considered to have large inertia i.e. to move with a constant speed v_s. The ice floe length is $2R$ and ice thickness h_i. The normal force acting at the contact is $F(t)$. The situation is illustrated in Fig. 6.8.

There are four displacement components when the bending of the floe is ignored; ship rigid body motions u_x, u_z and φ as well as the crushing depth into ice $\underline{u_{cr}}$ with the contact force being the fifth variable. The equations of motion for the first three (rigid body motions of the floe) are straightforward to form (hydro static buoyancy force may be ignored as a small force in comparison with the contact force but the hydrodynamic forces may be significant). The fourth equation for crushing arises from the so called pressure-area relationship of which more when describing measurement results. This relates the average pressure during crushing to the load patch area $A(u_{cr})$ depending on the crushing depth and of course on ice edge geometry:

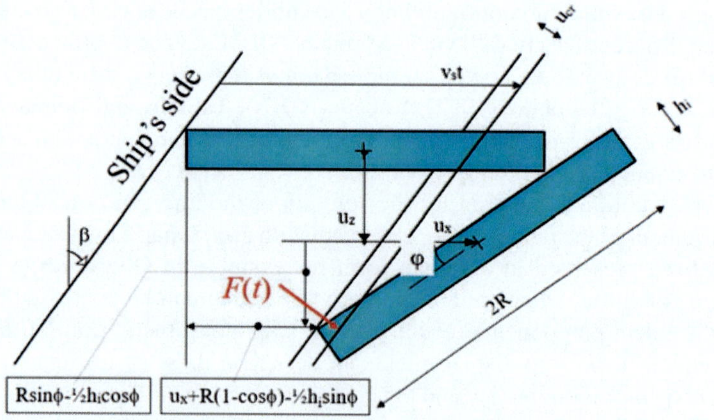

Fig. 6.8 Modelling the ice motions in the collision between ship hull and an ice floe (Riska 2018)

$$F(t) = A(u_{cr}(t)) \cdot p_{cr}(A(u_{cr}(t))) \tag{6.7}$$

The pressure-area relationship is commonly considered to be a power relationship with an exponent n ($p_{cr} \sim A^n$, the empirical value of n is about –0.5, with a range of –0.1 to –0.7). As in this planar case the contact area depends linearly on the crushing depth, the force depends on the crushing depth to an exponent of $1 + n$. The fifth equation needed is the kinematic condition that can be expressed in this case as

$$v_s t = tan\beta\left[u_z(t) + Rcos\varphi(t) + \frac{h_i}{2}(1 - cos\varphi(t))\right] + u_x(t) + R(1 - cos\varphi(t)) - \frac{h_i}{2}sin\varphi(t) + \frac{u_{cr}(t)}{cos\beta}. \tag{6.8}$$

The solution of the equations of motion together with the kinematic condition requires numerical time step solution.

It is not the aim here to describe the solution of this specific collision case but view what knowledge the problem formulation in this way brings. This formulation suggests—with some benefit of hindsight:

- If the ship bow is wedge shaped and the ship-ice collision is (as it is commonly called) oblique, the load is moving on the ship which adds complication in solving the ship equations of motion;
- Moving load also makes the definition of the local ship structure deformation difficult;
- Closed form solution is obtained by simplifying the situation much (see below);
- In formulating the EoM's for ice the hydrodynamic reaction force must be taken into account (see here for example Keijdener et al. 2018; Keijdener 2019 or Tan et al. 2014). For the ship equations of motion, the standard added mass approach is adequate;
- It is not necessarily clear what motion components are important and which can be ignored (one example of this below) and
- Ship and ice rigid body motions and ice flexure must be treated as dynamic but the displacement due to the local ice and ship elastic deformation can be treated statically.

The first—and at present the only general—theoretical ship-ice interaction model was formulated in the Golden Era (1950–70) of ship related ice research in Russia, Popov et al. (1967). The model treats the ship oblique impact on an ice edge, where ice can either be level ice (bending and crushing displacement of ice included) or an ice floe where the rigid body motions of ice were taken into account. It is, however, straightforward to combine these two cases. The main thrust of the modelling is to formulate a one degree of freedom equation of motion along the normal of the contact area. This is done using the following assumptions:

- The hydrostatic reaction forces are ignored for rigid body motions of ice and ship (the rigid body EoM's contain only acceleration and force terms);

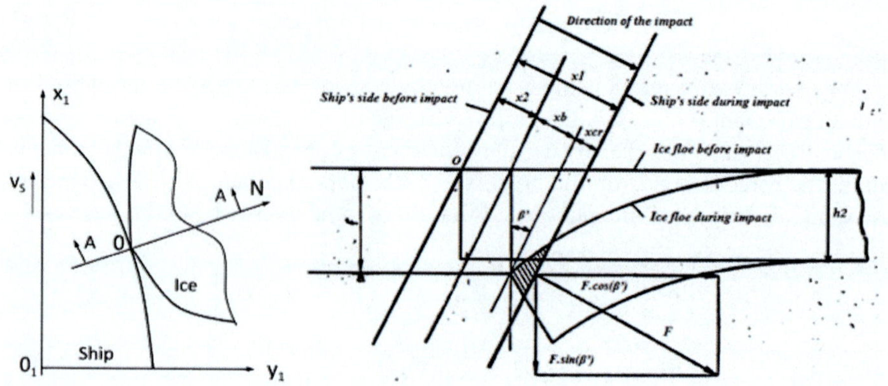

Fig. 6.9 The case modelled by the Popov model (left) and the displacements taken into account on the A-A plane (right), originals from or modified from Popov et al. (1967)

- Elastic local deformation of ice or ship shell is ignored;
- The movement of the load along the ship hull is ignored;
- The contact force is determined assuming ice to fail in crushing;
- Maximum force occurs when the relative motion between ship and ice is zero or when ice fails in bending

The last assumption means that the contact force is an increasing function on time and no kinematic condition is required. If ice does not fail in bending before, the contact is assumed to terminate when the contact force starts to decrease after reaching its maximum with the maximum indentation into ice. The situation modelled is shown in Fig. 6.9.

The Popov model is the only general albeit simplified model for ship-ice interaction at present as it takes into account all the six degrees of freedom for ship motions as well as the ice floe rigid body motions (assumed symmetric and thus only 3 DoF's) and both level ice and ice floes. Other models described below tackle restricted cases vis-à-vis ship motions. The simulation of the breaking pattern, described also below, tackles essentially the same question as the Popov model but does not derive an equation for the indentation.

In the original Popov model, the contact pressure was assumed constant; and it was obtained utilizing a concept of specific crushing energy, energy required to crush a unit volume of ice, obtained from drop ball tests. Later modifications for the Popov model used a pressure distribution obtained by assuming the crushed ice to form a layer between intact ice and ship, a liquid layer that is squeezed out from the contact. The pressure was calculated using the thin film Reynolds equations (Kurdjumov and Kheisin 1976). Other modification was to use the empirical pressure-area relationship (Daley 1999). The benefit of the Popov solution is that a closed form solution for the indentation $u_{cr}(t)$ is obtained even for the case of a finite floe with bending deformation (see for example Jumeau 2017).

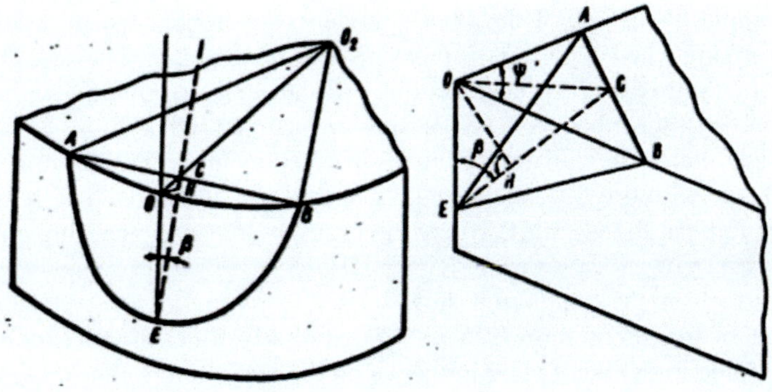

Fig. 6.10 Two ice edge geometries that were used in the original work of Popov (Popov et al. 1967)

What causes uncertainty in applying the Popov model for determining the ship design ice force is that ice edge geometry influences much the maximum force. Popov et al. (1967) used mainly two different edge geometries, see Fig. 6.10. The circular edge is used in formulating the Russian Maritime Register of Shipping ice class rules with an assumed radius of 25 m (mentioned for example in the backgound material for Polar Classes). A sharp edge is suggested by observations on the breaking pattern whereas rounded edge gives a wide load patch suggested by ice damage observations. This question of edge geometry is not likely to be settled by any theoretical means as any geometry could be selected as the study by Tunik (1984) shows. Ice load measurements may suggest preference to some geometry—but more likely the edge geometry will remain a source of a lot of scatter in measurements and uncertainty in calculations.

In summary, the outcomes from the Popov modelling are:

- A clear simple model for indentation into ice with a closed form solution;
- Maximum contact force and load patch obtained;
- Speed dependency of the maximum force obtained.

The uncertainties or restrictive assumptions in modelling are:

- The maximum force and load patch shape depend much on ice edge geometry;
- The maximum force in the case when ice cover bending is included is assumed to correspond to the static bending strength of ice cover. This underestimates largely the force;
- The frictional force is ignored. This directly underestimates the normal force $F_n(t)$ by roughly 10% but the effect on the motions or on the maximum force $F^{max}{}_n$ is more difficult to estimate.

The problem with modelling the collision with friction is that the direction of the frictional force is not clear as it opposes the sliding and the path of the contact on the ship hull is difficult to evaluate. Popov model is also essentially one dimensional so coupling the frictional force into the model is not straightforward.

An example of the use of the Popov model is given next which highlights the questions arising when applying the pressure-area relationship. The case analysed is a collision of a ship (the ship used here is the bulk carrier Eira) colliding with a circular ice floe (iceberg), see Fig. 6.11. The pressure-area curve used is $p_{av} = C \cdot {A_c}^n$, where A_c is the load patch area. Three different area exponents, n = 0.0, –0.1 and –0.4, were used. The resulting load patch is circular as Fig. 6.12 shows. The contact point is assumed to be at the waterline where the beam is B/2 (in typical ship coordinates y = B/4).

The results from the calculation show, for each pressure-area curve, a clear effect of speed on the maximum force (see Fig. 13). Similarly, for each area exponent n, the floe radius (floe mass) has a clear influence on the maximum force. However, if the response of only one frame is investigated, a different effect of speed emerges. One specific frame is sensitive to pressure only in the area marked dark blue in Fig. 6.12.

MV Eira

Lpp	148.00 m
B	24.60 m
T	9.03 m
dwt	19625 t
MCR	7860 kW

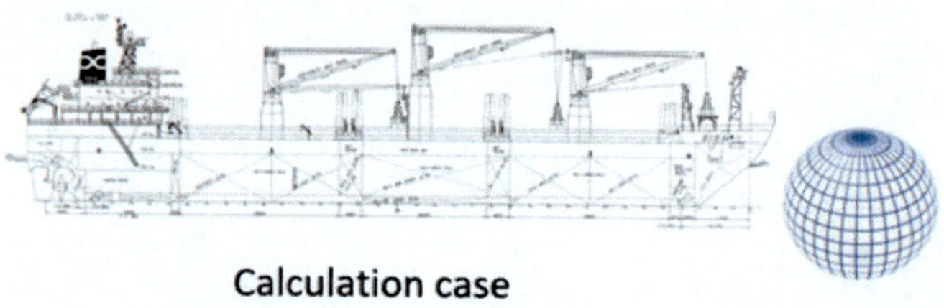

Fig. 6.11 Calculation case of a collision with a spherical ice floe and the bulk carrier Eira (Riska 2018)

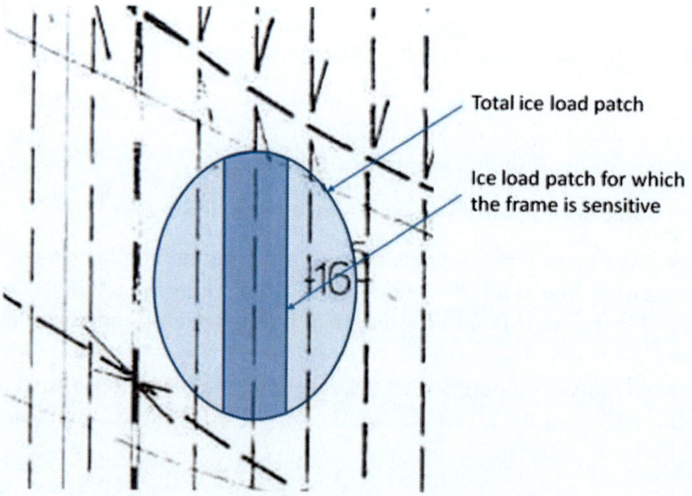

Fig. 6.12 Load patch in the collision case with a spherical floe depicted on the shell expansion (Riska 2018)

This area is load height times frame spacing, $A = h_c \cdot s$. Using the approximate equations from the Finnish-Swedish Ice Class Rules (FSICR 2017), the maximum bending stress on the frame (normalized by design values from FSICR for this ship) is obtained as

$$\frac{\sigma_{MY}}{\sigma_{FS}} = \frac{p_{MY}(v) \cdot h_c \cdot (1 - 0.5h_c) \cdot m_t}{4p_{FS} \cdot h_{c,FS} \cdot L}, \tag{6.9}$$

where the subscript 'FS' refers to design values for this specific ship (ice class IA Super), m_t is a boundary condition factor obtained also from the rules, L frame span. The results for the maximum stress versus ship speed with different area exponents n are given in Fig. 6.14. As the average pressure decreases when the contact area increases, the frame response which is sensitive only for part of the whole contact area is quite insensitive to ship speed. This lack of effect of ship speed on frame response has been observed also in full scale measurements (see for example Kotilainen et al. 2017, St. John et al. 1984).

The second case where theoretical modelling has been used to obtain the ice force and at the same time structural response is ramming large multi-year ice floes. The situation is shown in Fig. 6.15. The modelling is described in Riska (1987). The displacements that were included in the modelling were (the motions were in the *xz*-plane):

- The ship rigid body motions; surge, heave and pitch (3 DoF);
- Ice indentation due to crushing and displacement due to ice local elastic deformation (2 DoF);
- Ship hull elastic bending displacement at the bow (1 DoF).

The ice floe was assumed to be that large that its motions could be neglected. In a later analysis also the floe was considered finite (Riska 1991). The ice force included the frictional component. Ship bending response was solved using modal

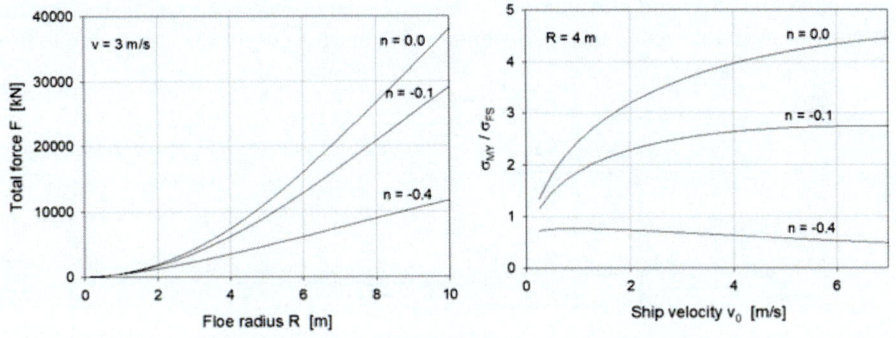

Fig. 6.14 Total force in the collision with a spherical ice floe and the ship Eira versus the floe size with three different area exponents (left) and the normalized frame response versus ship speed in the same collisions (right) (Riska 2018)

Fig. 6.15 CANMAR Kigoriak ramming massive multi-year ice floe during ice trials in 1981

superposition of the bending (including shear response) natural modes with the heave and pitch being the first and second mode. The motion components are shown in Fig. 6.16.

The resulting equations are nonlinear as the contact pressure was assumed to be dependent on the contact area which in this ramming case in proportional to $u_{cr}{}^{2}$. The integration of EoM's was done time stepwise. The calculation results were compared with laboratory test results and also with full scale trials with the bulk carrier MV Arctic. Comparison between calculated and measured ramming force and ship response results is quite reasonable as seen in Fig. 6.17. It is interesting that even if the ice and ship elastic deformations were small, at most some centimetres while the ship rigid body motions and ice indentation were several metres, the elastic deformations made a large difference for the ship response. This is due to the

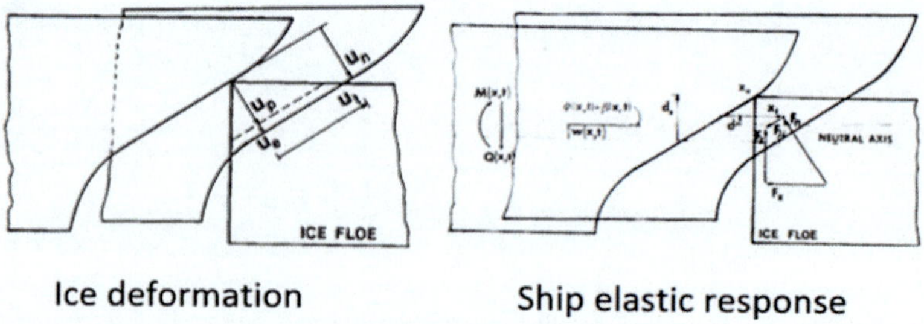

Fig. 6.16 Displacements taken into account in simulating ramming (Riska 1987)

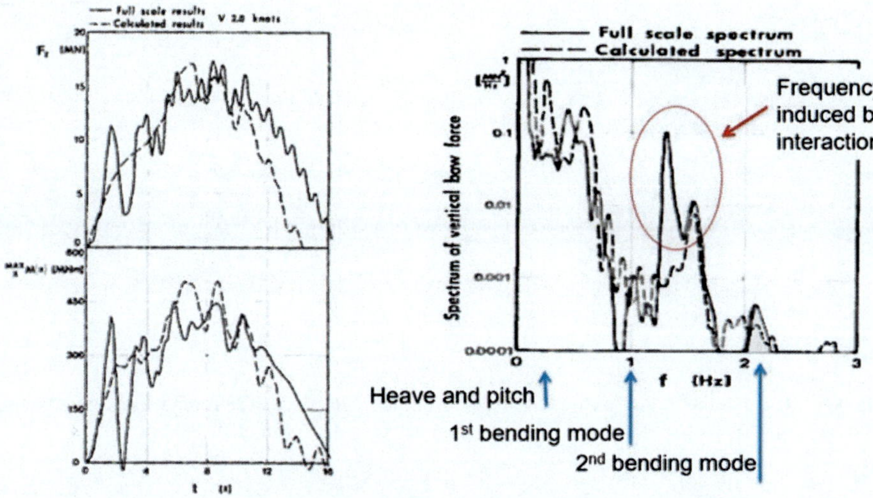

Fig. 6.17 Results for the bulk carrier MV Arctic; calculated and measured bow force and bending moment (left) and the spectrum of the calculated and measured bow force (right) (Riska 1987)

intermittent ship bow motion away from the ice edge (a rebound due to elasticity of the ship and ice), this is caused by the elastic reaction force and shows in the contact force as a decreasing trend at some stages during the contact. The ice elasticity creates also another natural frequency to the ship-ice system corresponding to the bow being supported by an elastic support. This frequency is between the first and second bending mode as Fig. 6.17 shows.

The ship local ice action has been studied by simulating the breaking pattern in order to obtain realistic load patch (Su et al. 2011; Tan et al. 2014). In the simulation ice was assumed to fail into circular floes and the force as well as the floe size determined by the bending crack were taken from measurements, see Fig. 6.18. The simulation method emphasizes the variability in the breaking pattern of level ice and the apparent randomness in force. The method has given quite good results when the ship performance in terms of the $h_i - v$ curve as well as turning circle has been simulated.

Here it is interesting to look at simulated results for the ice load acting on one frame (Su et al 2011). The ship used in simulation is MT Uikku as there exists measured results from her (Kotisalo and Kujala 1999). The histogram of calculated ice force peaks on one frame during a 10 min period with constant ice properties is shown in Fig. 6.19 compared with measurements with the same ship. The time histories of measured and simulated force on one frame in terms of the line load $q = F/s$ are both qualitatively and quantitatively very similar (F is the total force acting on the frame and s frame spacing), Fig. 6.19. This similarity is present in the distribution of the force peaks, which is shown in Fig. 6.20. These results suggest that the origin of statistics in ice force is both internal and external, where external

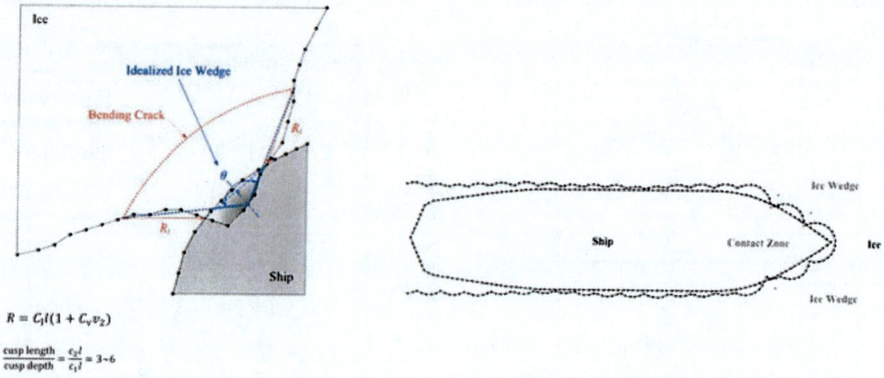

$R = C_l l(1 + C_v v_2)$

$\frac{\text{cusp length}}{\text{cusp depth}} = \frac{c_2 l}{c_1 l} = 3\text{–}6$

Fig. 6.18 The details for the simulation of ship progress in level ice. The force was assumed to be linear with speed and the wedge length depth ratio between 3…6 (Su et al. 2011)

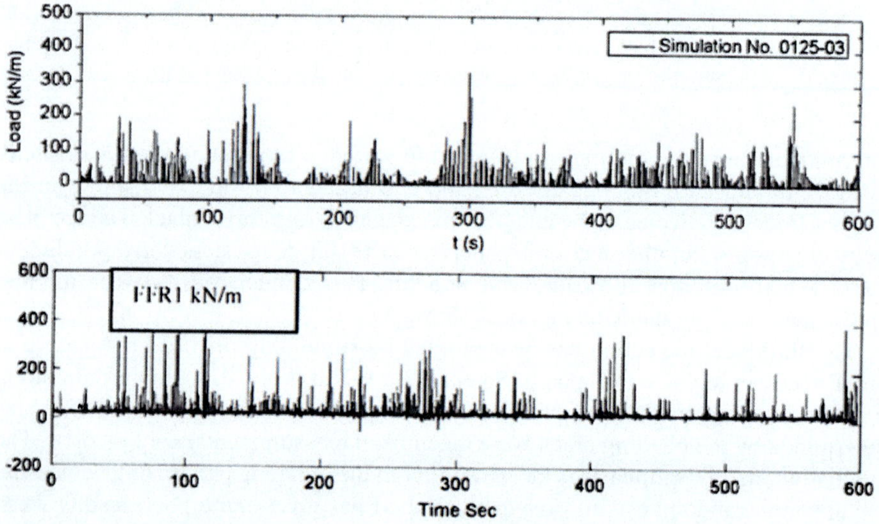

Fig. 6.19 Simulated (top) and measured (bottom) time histories of a frame ice force onboard MT Uikku (Su et al 2011)

refers to variation in ice properties (thickness, strength) and internal to the process itself. It is interesting that so large variation exists even in constant ice properties.

The description of theoretical calculations of ice action on ship hull shows clearly how the force value and size of the load patch are connected through ice pressure. Assumptions on ice pressure influence the resulting calculated load patch size at each ice force value. Actually the force on the load patch is the only reliable quantity that can be obtained from theoretical models—and even this depends in many models—like Popov's—on assumptions on ice bending failure. For offshore structures with

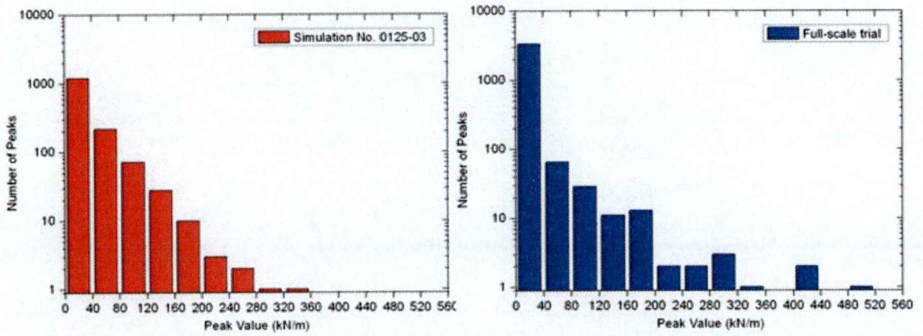

Fig. 6.20 The frequency distribution of frame ice load on MT Uikku, left simulated results and right measured results (Su et al. 2011)

vertical piles the load patch is evident a priori (ice thickness time pile diameter) but for ships the patch is very variable depending on the ice edge geometry. The theoretical models can give insight to parametric dependencies, but a large amount of uncertainty will remain in predicting local ice action on ship hull; the prediction for design has to be done using statistical means.

6.4 Measurement of Ice Action

The measurement of ice force—and accompanying quantities like ice pressure or load patch size—is not straightforward. When ship hull hits and indents an ice edge, the resulting load patch shape/size depends on the ice edge geometry which is very variable. Thus no ice action quantity—F_n, p_{av} nor A—is known a priori. This makes it impossible to deduct the average pressure on the whole load patch even if the total force could be measured. The best that can be done for ship ice action is to measure the force on some known area—this area is commonly the gauge area for the load panels used. The force is measured by measuring the response of the gauges or the ship structure and calibrating the response-force relationship with a known force either numerically or by a physical test. The questions in calibration are illustrated for one early ice pressure gauge, see Vuorio et al. (1979).

At present most ship ice action measurements apply the shear gauges on frames. These were first applied onboard IB Sisu (see Riska et al. 1983). The rationale of using shear gauges is that each frame can be assumed to be an ideal beam and shear strain is measured at two locations on the neutral axis. If these strains are converted to shear stresses τ_1 and τ_2 , then the force acting between these gauges (locations x_1 and x_2, x coordinate along the beam, see Fig. 6.21) is

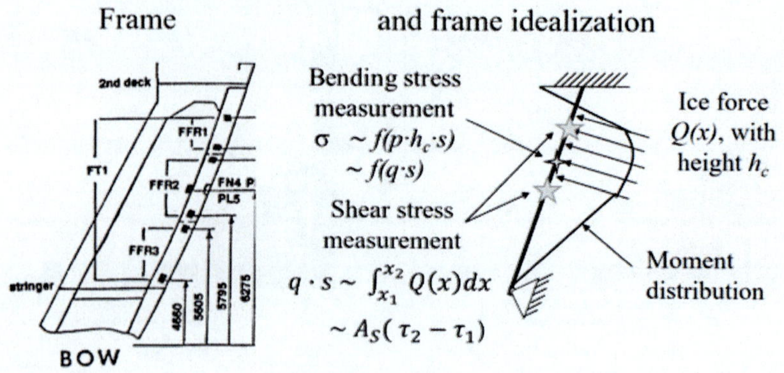

Fig. 6.21 Principle of the ice load measurement based on shear strains

$$F = \int_{x_1}^{x_2} Q(x)dx = A_s(\tau_2 - \tau_1) \tag{6.10}$$

where $Q(x)$ is the load on the beam (per one frame spacing s) and A_S the shear area of the beam. This method gives an estimate on the force on the frame between points x_1 and x_2 i.e. on the load height of $h = x_2 - x_1$. This is an estimate as the stress distribution on ship frames is not that of an ideal beam with an I cross section and also because the adjacent frames carry some of the load. Especially flat bar frames have very variable shear stress distribution along the frame height. An estimate is, however, obtained from the force on a load patch $s \, . \, h$.

The ice pressure on ship hull and also the shear stresses were measured first time onboard the icebreaker Sisu in late 1970s and early 80s (Riska et al. 1983; Kujala and Vuorio 1986). The ice pressure measurement was accomplished by stiffening much the shell structure and then measuring the shell displacement in the center of a circular supported area, see Fig. 6.22. The results of the measurement included statistical data on both local ice pressure and the force (line load q). Some results are shown in Fig. 6.23.

The Sisu results showed a high ice pressure, up to about 10 MPa, on a pressure gauge with a sensing area of about 0.01 m^2 (diameter 10 cm). This is about three times the value ice pressure was assumed to reach in the Baltic in the ship design routines of that day. Clearly some rethinking was required. At the same time the maximum frame line load measured was about 2 MN/m, similar to values obtained back-calculating from observed ship ice damages. The measured plate stress was lower than that given by uniform pressure, as calculated using FEM. The explanation for this was obtained by calculating what ice pressure distribution and load patch height give the measured stresses on plating and frames. The result of calculation is shown in Fig. 6.24. If the maximum pressure p_{max} is assumed to be about 9 MPa, the best fit to data is obtained

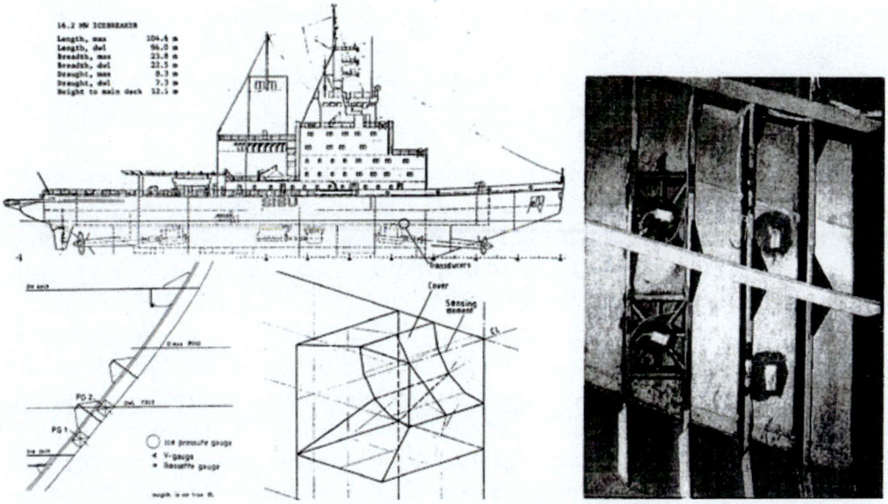

Fig. 6.22 Ice pressure gauges installed on IB Sisu (Kujala and Vuorio 1986)

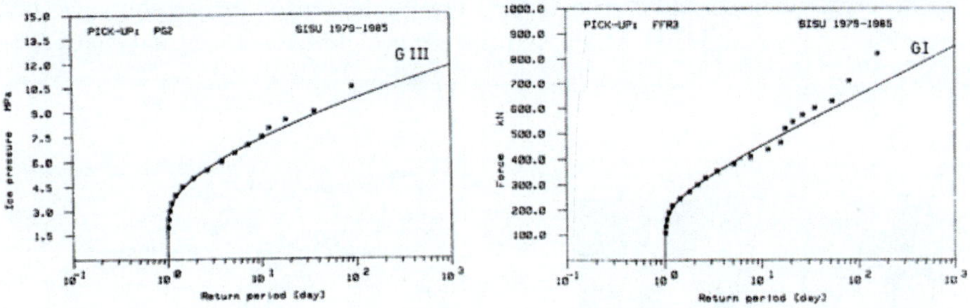

Fig. 6.23 Statistics of ice pressure and frame force measurements onboard IB Sisu—frame spacing in the shell structure of IB Sisu is 40 cm (Kujala and Vuorio 1986)

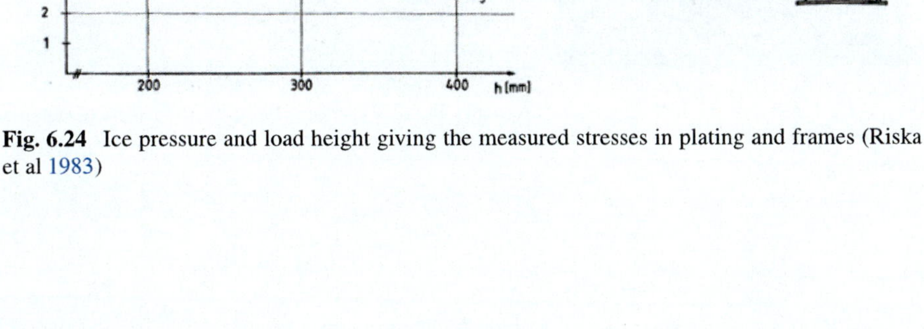

Fig. 6.24 Ice pressure and load height giving the measured stresses in plating and frames (Riska et al 1983)

by a triangular pressure distribution and quite small load height, less than 200 mm, see Fig. 6.24.

The high pressures measured onboard IB Sisu inspired some studies where small pressure gauges were used in laboratory pendulum tests (see Glen and Comfort 1983). Sea ice was used in these tests with high impact speeds, up to 5.6 m/s. The contact pressure was measured with pressure gauges with a sensing area of 1 inch in diameter. Maximum pressures measured were up to 40 MPa—so high that even the Korzhavin equation (Korzhavin 1962) cannot explain these. As mentioned, these were laboratory tests and it was considered sometimes that the result of high pressure is an artefact of laboratory conditions where ice is not resting on water.

The matter was resolved when local ice pressures were measured on IB Sampo in the northern Baltic (Riska et al. 1990). The measurement was carried out with pressure sensitive PVDF plate (Joensuu 1988) and at the same time the contact was filmed through a window at the waterline of the icebreaker. The measurement set-up is shown in Fig. 6.25 and pressure results and photographs of the contact in Fig. 6.26. The contact height was observed to be very small (line-like) and the local pressures on gauges with a small sensor area very high, up to 55 MPa. The maximum pressures measured showed also a dependency on the contact area, a phenomenon that has been also observed on larger areas. The dependency on the sensor area on ships was first time measured onboard CANMAR Kigoriak, see CANMAR (1982) and Riska et al. (1982).

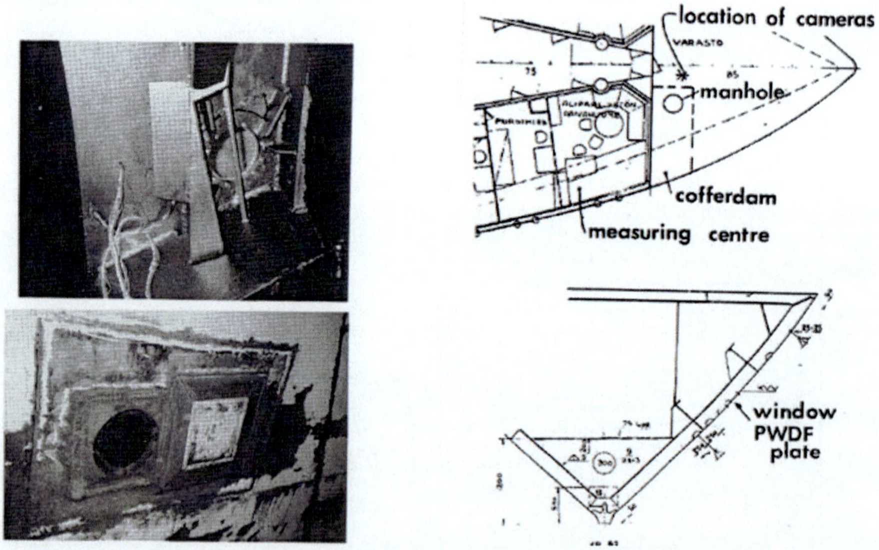

Fig. 6.25 Layout of the IB Sampo measurements and observations. Photo from inside and outside the ship of the window and pressure sensitive plate (left) and drawings of these (right) (Riska et al. 1990)

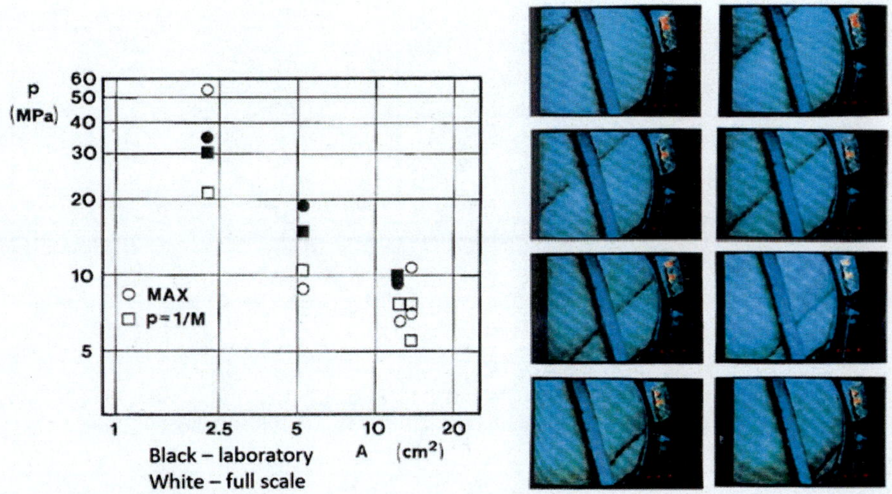

Fig. 6.26 Maximum ice pressures versus gauge area (left) and a collection of photographs of one contact event with about 4 ms between photos (right) (Riska et al. 1990). $p = 1/M$ refers to a predicted maximum from a fitted pressure peak distribution of M peaks

The measurement techniques have not brought any breakthrough for measuring the load patch area nor the ice pressure for ships—for laboratory tests this kind of breakthrough was obtained with tactile sensors, see for example Takeuchi et al. (1997). The emphasis in shipborne measurements has recently been—instead of improving the gauges—to obtain simultaneous data on ice action, ship shell response to ice action and knowledge about the ice causing the action. Video cameras are crucial in observing the ice; these have been calibrated to give the level ice thickness. Unfortunately, video cannot see the ice edge geometry very well so much remains still to be clarified. This pertains especially to the measured maximum values. In many measurements it has been noticed that the maximum measured peaks deviate from the other peaks when the peak frequency distribution is plotted, see Fig. 6.27. The very important question for design remains; what kind of ship-ice interaction caused these maxima?

Here it may be illuminating to perform some estimates. The maximum force measured onboard IB Sisu is on one frame (load length 40 cm) 820 kN and on five frames simultaneously (load length 2 m) 1700 kN. The maximum force measured at 30° frame inclination in a dedicated test (Varsta 1983) in 35 cm thick ice is 600 kN (load width 2 m). The maximum level ice thickness at sea in the northern Baltic is about 80 cm—thus if the result in 35 cm thick ice is upscaled to ice thickness of 80 cm using a typical thickness exponent of 1.7, the result is an extrapolation from 600 to 2400 kN. This reasoning suggests—if it is at all correct—that the maximum measured force onboard Sisu is low. This question remains to be clarified, and in the meanwhile the ship shell structure design is done using a line load value 2 MN/m.

The description of the ship ice action measurements has shown that:

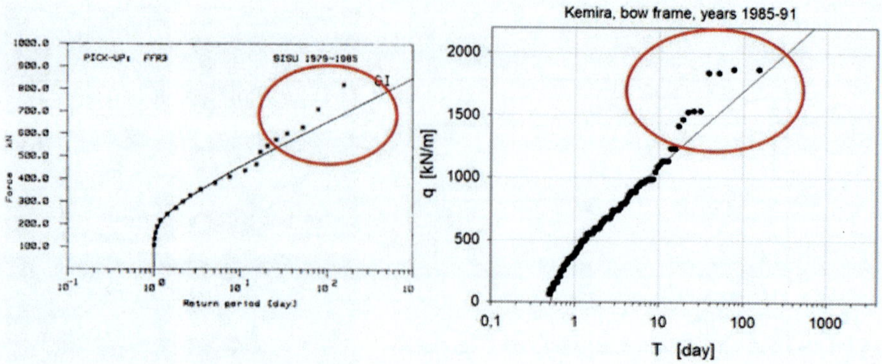

Fig. 6.27 The maxima from 24 h periods onboard IB Sisu (left) and 12 h periods onboard MT Kemira (right) plotted versus the return period (Kujala and Vuorio 1986, Muhonen 1991)

- Local ice pressure on ship can be high, about an order of magnitude higher than ice compressive strength;
- A small load patch height can be deduced from the measurements, clearly smaller than 200 mm;
- A reliable line load value can be obtained from measurements, to support the one deduced from ice damage surveys;
- Knowledge on the pressure versus area and line load versus load length is obtained;
- The effect of ice thickness or ship hull angles on the force magnitude is not clear;
- There still remains much uncertainty in ice action on ship hull, especially how to apply the pressure dependency on area in design (what design load patch to use?)

6.5 Summary

The description of different methods to clarify ship hull ice action has led to some conclusions. The items that are clearer at the moment are:

- The design load level can be considered reliable based on feedback from ship ice damage and (especially long-term) ice action measurements;
- An understanding for the description of ice action parametres, load height and ice pressure, exists. Less well known is the load patch width and also to some extent the total ice force;
- The rationale to theoretically model the ship-ice interaction to obtain the total force is available. This will be also reliable design tool once the dynamic bending strength of ice cover is clarified and the variation in the ice edge geometry can be taken into account;
- The design ice pressure remains to be clarified together with knowledge on how the ice edge fails. Here also the application of pressure-area concept for design requires improved foundation.

This description of methods for and results from ice action observations, calculations and measurements is necessarily subjective in sense what material is included. The author has tried to include material that has brought forward some answers or further questions. Several very interesting measurement campaigns have not been described (for example with the icebreakers Polar Seaand Louis St. Laurent). The author apologises for these omissions.

References

CANMAR, Final report on full scale measurements of ice impact loads and response of the Kigoriak. A report by Canadian Marine Drilling Ltd. to Coast Guard Northern, TP5871E (1982)

C. Daley, Energy based ice collision forces, in *Proceedings of the 15th POAC 1999* (Helsinki, Finland, 1999)

FSICR 2017, Ice class regulations and the application thereof, Act on the ice classes of ships and icebreaker assistance (1121/2005), section 4.1., TRAFI/494131/03.04.01.00/2016 (2017), 65 p.

I. Glen, G. Comfort, Ice impact pressure and load: investigation by laboratory experiments and ship trials, in *Proceedings of the 7th International Conference on Port and Ocean Engineering under Arctic Conditions (POAC), Helsinki*, vol. I (1983), pp. 516–533

W. Goldsmith, *Impact – The Theory and Physical Behaviour of Coliding Solids* (Edward Arnold Ltd., London, 1960), p. 379

S. Hänninen, Incidents and accidents in winter navigation in the Baltic Sea, winter 2002–2003. Winter Navigation Research Board, Res. Rpt. 54 (Helsinki, 2003), 39 p.

M. Hopkins, Numerical simulation of systems of multitudinous polygonal blocks. Tech. Rpt. 92–22, Cold Regions Research and Engineering Laboratory, CRREL (1992), 69 p.

A. Joensuu, Ice pressure measurements using PVDF film. *Proceeding OMAE* '88. Houston, Texas **4**, 153–158 (1988)

A. Joensuu, K. Riska, Structure/ice contact, measurement results from the joint tests with Wärtsilä Arctic Research Centre in Spring 1988. Helsinki University of Technology, Lab. of Naval Architecture and Marine Eng., Report M-88, Espoo (1989), 57 p. + 154 app. [in Finnish]

B. Johansson, On the ice-strengthening of ship hull. Int. Shipbuild. Prog. **14**(154), 231–245 (1967)

M. Jumeau, Ice interaction with ships - Calculation of the ice force in collision with a finite ice floe. Training report, Total E&P (Paris, France, 2017), 72 p.

C. Keijdener, H. Hendrikse, A. Metrikine, The effect of hydrodynamics on the bending failure of level ice. Cold Reg. Sci. Technol. **153**, 106–119 (2018)

C. Keijdener, The effect of hydrodynamics on the interaction between floating structures and flexible ice floes. Doctoral dissertation, Delft University of Technology, (2019), 149 p.

A.D. Kerr, The bearing capacity of floating ice plates subjected to static or quasi-static loads. J. Glaciol **17**(1976), 229–268 (1976)

K. Korzhavin, *Action of Ice on Engineering Structures*. Publishing House of Siberian Branch of USSR Academy of Sciences (1962). CRREL translation 1971, 321 p.

M. Kotilainen, J. Vanhatalo, M. Suominen, P. Kujala, Predicting ice-induced load amplitudes on ship bow conditional on ice thickness and ship speed in the Baltic Sea. Cold Regions Res. Technol. **135**(2017), 116–126 (2017)

K. Kotisalo, P. Kujala, *Analysis of ice load measurements onboard MT Uikku – Results from the ARCDEV-voyage to Ob Bay, April-May 1998*. Report no. D-50, Ship Laboratory (Helsinki University of Technology 1999)

P. Kujala, J. Vuorio, *Results and statistical analysis of ice load measurements on board icebreaker Sisu in winters 1979 to 1985*. Winter Navigation Research Board, Research report No. 43 (Helsinki, Finland, 1986), 131 p.

P. Kujala, *Damage statistics of ice-strengthened ships in the Baltic Sea 1984 – 1987*. Winter Navigation Research Board, Res. Rpt. 50, 67 p.

V.A. Kurdjumov, D.E. Kheisin, Hydrodynamic model of the impact of a solid on ice. Prikladnaja Mekhanika **12**, 103–109 (1976). ([in Russian])

A. Muhonen, *Ice Load Measurements onboard the MS. Kemira, Winter 1990*. Laboratory of Naval Architecture and Marine Engineering, Helsinki University of Technology, Report M-109 (1991), 24 p + app.

J. Popov, O. Faddeev, D. Kheisin, A. Yakovlev, *Strength of Ships Sailing in Ice*. Sudostroyeniye Publishing House Leningrad (1967), 223 p (Translated by U.S. Army Foreign Science and Technology Center 1968).

M. Rabatel, Modélisation dynamique d'un assemblage de floes rigides. Doctoral dissertation, Université de Grenoble (2015), 156 p.

K. Riska, S. Uto, J. Tuhkuri, Pressure distribution and response of multiplate panels under ice loading. Cold Reg. Sci. Technol. **34**, 209–225 (2002)

K. Riska, P. Frederking, Ice load penetration modelling, in *Proceedings of the Port and Ocean Engineering under Arctic Conditions*, vol. 1, Fairbanks, Alaska, August 17–21 (1987), pp. 317–328

K. Riska, J. Tuhkuri, Analysis of contact between level ice and a structure, in *Proceedings of the International Workshop on Rational Evaluation of Ice Forces on Structures, 2–4 February 1999* (Mombetsu, Japan, 1999), pp. 103–120

K. Riska, J. Matusiak, S. Rintala, J. Vuorio, *Measurements of Ice Pressures and Forces on CANMAR Kigoriak during the Repeated Trials in 1981*. Client report to Dome Petroleum, VTT/LAI-332/82 (1982)

K. Riska, P. Kujala, J. Vuorio, Ice load and pressure measurements on board IB Sisu, in *Proceedings of the POAC 1983* (Helsinki, Finland, 1983), pp. 1055–1069

K. Riska, H. Rantala, A. Joensuu, Full scale observations of ship-ice contact. Helsinki University of Technology, Lab. of Naval Architecture and Marine Eng., Report M-97, Espoo (1990), 54 p. + 293 pp.

K. Riska, Observations of the line-like nature of ship-ice contact, in *11th International Conference on Port and Ocean Engineering under Arctic Conditions, Proceedings*, vol. 2 (St. John's, Canada, 1991a), pp. 785 - 811.

K. Riska, Theoretical modelling of ice-structure interaction, in *Ice-Structure Interaction. IUTAM-IAHR Symposium, St. John's, Newfoundland, Canada*, ed, by S. Jones & R. McKenna & J. Tillotson & I. Jordaan (Springer, Berlin, 1991b), pp. 595–618

K. Riska, Lectures notes for the course 'Ship Design for Ice'. Norwegian University for Science and Technology (NTNU), Marine Technology Department (2018)

K. Riska, On the mechanics of the ramming interaction between a ship and a massive ice floe. Technical Research Centre of Finland, Publications no. 43 (Espoo, Finland, 1987), 93 p.

K. Riska, Models of ice-structure contact for engineering applications, in *Mechanics of Geomaterial Interfaces*, ed. by A. Selvadurai, M. Boulon (Elsevier Science B.V., 1995), pp. 77–103

D. Sodhi, T. Takeuchi, N. Nakazawa, S. Akagawa, H. Saeki, Medium-scale indentation tests on sea ice at various speeds. Cold Reg. Sci. Technol. **28**, 161–182 (1998)

D. Sodhi, T. Takeuchi, M. Kawamura, N. Nakazawa, S. Akagawa, Measurement of ice forces and interfacial pressure during medium scale indentation tests in Japan, in *Proceedings of the Port and Ocean Engineering under Arctic Conditions*, vol. 2 (Ottawa, Canada, 2001)

St. J. John, C. Daley, H. Blount, Ice loads and ship response to ice. Ship Structures Committee Report SR-1291, SSC-329 (1984), 332 p.

B. Su, K. Riska, T. Moan, Numerical study of ice-induced loads on ship hulls. Mar. Struct. **24**(2011), 132–152 (2011)

T. Takeuchi, T. Masaki, S. Akagawa, M. Kawamura, N. Nakazawa, T. Terashima, H. Honda, H. Saeki, K. Hirayama, *Medium-scale field indentation tests MSFIT: ice failure characteristics in ice-structure interactions*. Proceedings of the 7th International Offshore and Polar Engineering Conference. Honolulu, USA **II**, 376–382 (1997)

X. Tan, K. Riska, T. Moan, Effect of dynamic bending of level ice on ship's continuous-mode ice breaking. Cold Reg. Sci. Technol. **106–107**, 82–95 (2014)

TRAFI 2016, Ice Class Regulations and the Application There of. Finnish Transport Safety Agency, TRAFI/494131/03.04.01.00/2016 (2016), 65 p.

J. Tuhkuri, A. Polojärvi, A review of discrete element simulation of ice-structure interaction. Phil. Trans. R. Soc. A, **376**(2129)

A. Tunik, Dynamic ice loads on a ship. IAHR Ice Symposium, Hamburg, Germany **III**, 297–313 (1984)

P. Varsta, On the mechanics of ice load on ships in level ice in the Baltic Sea. D.Sc. thesis, Helsinki University of Technology, Espoo, Finland (1983) 93 p.

J. Vuorio, K. Riska, P. Varsta, Long term measurements of ice pressure and ice-induced stresses on the icebreaker SISU in Winter 1978. Winter Navigation Research Board, Res. Rpt. No. 28 (Helsinki and Stockholm, 1979), 56 p.

Chapter 7
Ice Interaction with Floating Structures

Sveinung Løset, Wenjun Lu, Marnix van den Berg, and Raed Lubbad

Abstract Ice-floating structure interaction involves several limiting mechanisms and multiple failure modes of the sea ice. Field observations indicate that ice failure modes coexist and compete with each other. In addition, the occurrence of different limiting mechanisms is clearly influenced by the physical states of the interactions (e.g., ice features, confinement, floe size, contact properties, and physical environmental driving forces). These processes involve quite different physical mechanisms, such as ice fracture mechanics, multibody dynamics and hydrodynamics. This paper introduces a novel simulation method that automatically handles the possible limiting mechanisms and simulates dominant ice failure modes. The algorithms behind this simulator are developed through extensive field studies to capture the major physical processes and relevant theoretical formulations in their respective subjects (i.e., ice fracture, multiple ice floes' interaction and hydrodynamics). These developed algorithms are implemented in the *Simulator for Arctic Marine Structures* (SAMS) and validation have been carried out in several engineering applications. In this paper, we demonstrate such a simulator's capability of simulating ice—floating structure interactions through calculating the drift ice action on the grounded trawler *Northguider*, in support of its salvage operation in the high North.

7.1 Introduction

The Arctic catches interest for a number of reasons: a vast area strategically located with huge resources related to fisheries, hydrocarbons and minerals in addition to maritime transport between the continents in the East and the West. Further, exotic cruise tourism is developing in these waters and the area is becoming a measure on the rate of global warming leading to increased research vessel traffic in the area.

S. Løset (✉) · W. Lu · M. van den Berg · R. Lubbad
Norwegian University of Science and Technology (NTNU), Trondheim, Norway
e-mail: sveinung.loset@ntnu.no

W. Lu
The Norwegian Academy of Science and Letters, Oslo, Norway

J. Tuhkuri and A. Polojärvi (eds.), *IUTAM Symposium on Physics and Mechanics of Sea Ice*, IUTAM Bookseries 39, https://doi.org/10.1007/978-3-030-80439-8_7

Ice—structure interaction studies play a key role in these engineering activities in the high North. Joining the current wave of 'digitalisation', great efforts have been made to numerically study the interaction between ice and structures. Given that background, this paper presents basic observations in the field and describes important physical processes that take place during a typical floe ice—sloping structure interaction: fracture of sea ice, multiple ice floes' interactions and hydrodynamics effect. Here, sloping structure encompasses all types of structures with a sloping contact surface against incoming ice floes. Typical sloping structures include for example the hull of ice-going vessels, conical shaped floating or fixed offshore structures, wide sloping coastal structures and bridge piers. In this paper, we focus on the floating type of sloping structures. Theoretical backgrounds of these processes are described with detailed presentation on their analytical and numerical implementation in the *Simulator for Arctic Marine Structure* (SAMS). However, no matter how 'fancy looking' a numerical simulation result can be, it is worthless without backing from field data inputs and field validation. Therefore, using our effort in studying ice fracture as an example, we present in this paper a 3-year field ice fracture campaign and some preliminary observations/results; in addition, this paper presents an engineering application using SAMS to support an Arctic engineering operation (i.e., ice actions on a grounded vessel) highlighting the excellent knowledge sharing loop of 'knowledge arising from field observations → theoretical formulations → serving back to field applications'.

7.2 Observation of Ice-Hull Interactions

7.2.1 Ice Action Considerations

There are many parameters that influence the ice action on a structure (see Fig. 7.1) (Løset et al. 2006). The first consideration is different ice features, among which,

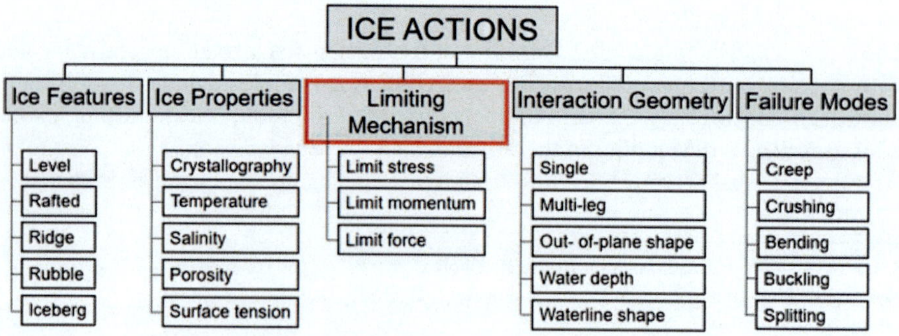

Fig. 7.1 Ice action considerations and three different limiting mechanisms

the level ice feature is often assumed and has been extensively studied back in time. However, the sea ice in the Arctic is not always an infinitely large flat intact ice sheet as would often be assumed as the 'level ice conditions (ISO19906 2019)'. Instead, a large part of the ice cover consists of discrete ice features such as ice floes and ice ridges. An ice floe's size ranges from about 10 m to around 10 km (Leppäranta 2011). Therefore, the concept 'floe ice' is a more general term that covers both the scenario of 'level ice' and the more often observed 'broken ice' conditions (Lu 2014). The ice floes can be as large (in the order of kilometres) as being considered as level ice and as small (in the order of metres) as being considered as rubble ice (Lu 2014).

Over the years, extensive researches have been carried out regarding 'level ice and structure/ship interactions', e.g., (Kotras et al. 1983; Naegle 1980; Lindqvist 1989; Kämäräinen 2007; Enkvist et al. 1979; Su et al. 2010b; 2010a; 2010c; Zhou et al. 2011; Tan et al. 2013; Lubbad and Løset 2011; Varsta and Riska 1977). However, floe ice interacting with structures are not as well studied as for the case of level ice. One reason is that apart from material properties, there are more parameters involved in describing a broken ice field, e.g., floe size, geometry, ice concentration, ice thickness and ice pressure. For level ice, apart from material properties, knowing the ice thickness is generally sufficient for ice loads calculations. The fracture patterns of various ice floes are more complicated than their level ice counterparts. Basically, all failure modes listed in Fig. 7.1 are present during floe ice—sloping structure interactions.

In addition to different ice features and different failure modes, the ice action is also under great influence of different possible limiting mechanisms (highlighted in Fig. 7.1). Three distinct limiting mechanisms can be identified and for conventional engineering practices, a presumption on the limiting mechanism is needed to proceed with ice action calculations (e.g., (Palmer and Croasdale 2013) and (ISO19906 2019)). This paper presents a numerical scheme that comprehensively accounts for most of the parameters in Fig. 7.1 without any presumption on limiting mechanisms and failure modes to calculate the ice actions in an efficient manner.

7.2.2 Characteristics of Floe Ice Interacting with Floating Structures

Three important processes should be considered regarding the interaction between a floating structure and floe ice. These are:

- The fracture of ice floes (i.e., failure modes)
- Multi-body interaction between all simulated ice floes (i.e., ice environmental conditions)
- Hydrodynamic effects on ice floes and the structure (i.e., wind, wave, and current environmental conditions).

7.2.2.1 The Hull—Ice Interaction Process with Floe Ice Fractures

For a sloping hull (e.g., the bow of an ice going vessel), the ice floe in contact normally fails either in a global splitting type and local bending type failure (see Fig. 7.2). A competition exists between the local bending failure and the global splitting failure. An example on documenting this competition is shown by a video camera system depicted in Fig. 7.3. The icebreaker primarily travelled within a large ice floe with continuous local bending failure (refer to Fig. 7.3a). When conditions permit (e.g., relatively small ice floes or an ice floe with minimal confinement), the global splitting failure tends to take place (see Fig. 7.3b and c) and alleviate the ice load. In following years, similar observations were further made using a 360-degree camera-based ice surveillance system (see Fig. 7.4).

Defining the ice floe as the reference plane, we can differentiate the out-of-plane failure (e.g., bending failure) and in-plane failure (e.g., splitting failure) of an ice floe. Extensive studies regarding the out-of-plane flexural type failure has been studied previous (e.g., see the general review work by Kerr (1976) and specific application by Lu et al. (2015c)). To calculate the out-of-plane flexural type failure, it is mainly the initiation of circumferential cracks that is of interests. This can be achieved via analytical solutions (Nevel 1961, 1972) or 'computational fracture mechanics'—based numerical simulations (e.g., (Lu et al. 2012)).

The in-plane splitting type failure, however, is not as thoroughly studied as the out-of-plane type flexural failure. Among many obstacles to achieve a general analytical solution is the arbitrary geometries of an encountered ice floe. Bhat (1988) and Bhat et al. (1991) offered Finite Element Method (FEM) results for the idealised square and disk-shaped ice floe's splitting. Later, Dempsey et al. (1993) offered the weight function based analytical solution to characterise the splitting of a square ice floe. The analytical solution was further extended to rectangular ice floes of arbitrary width to length ratio (Dempsey and Mu 2014). After summarizing previous research efforts

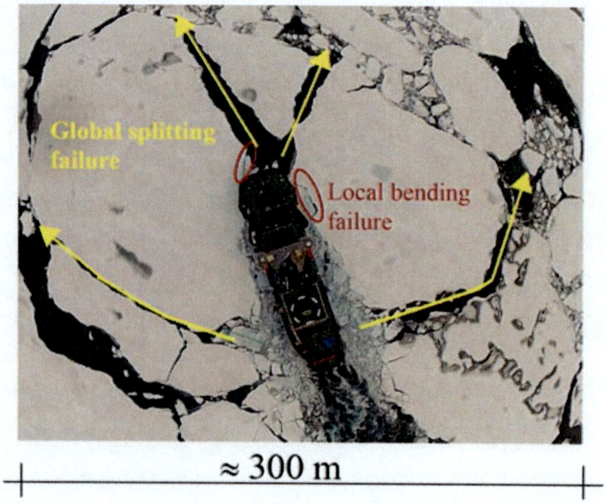

Fig. 7.2 Global splitting and local bending failure during the interaction between a ship hull (IB *Oden*) and a relatively small ice floe with mild confinement (from (Lu et al. 2015b))

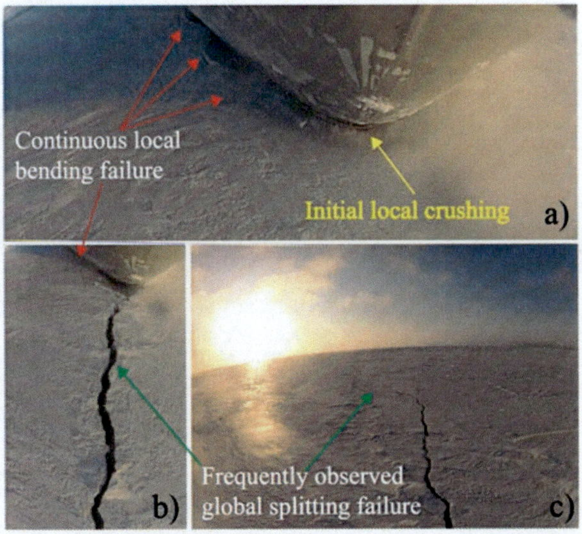

Fig. 7.3 **a** A close-by view showing the local bending failure, and **b**, **c** global splitting failure modes (the video demonstrates the competition between these two failure modes), originally from (Lu et al. 2016a)

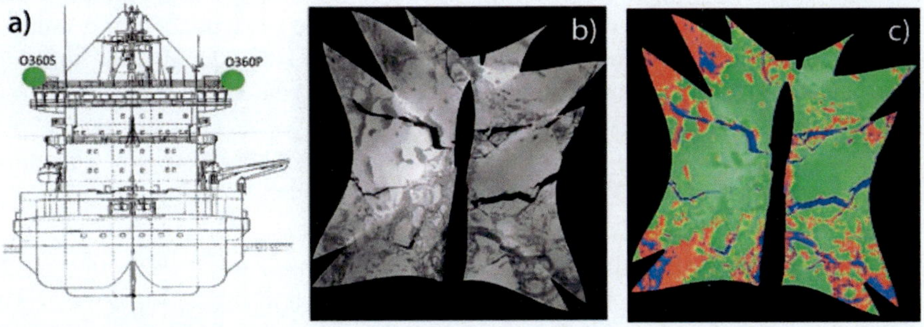

Fig. 7.4 **a** The 360—degree camera-based ice surveillance system on board IB *Oden*, showing **b** frequent splitting failures (in dark) of ice floes; **c** based on image processing, the splitting cracks are highlighted in blue colour, originally from (Heyn 2019)

(Lu et al. 2015b) and developing analytical formulas for crack kinking in off-centre splitting scenarios (Lu et al. 2018b, c), the analytical framework to characterise the splitting failure of an ice floe can be constructed. However, the utilised analytical solutions to describe the splitting of an ice floe of arbitrary geometry is not an exact solution; it is more an approximation. To achieve more accurate results on the splitting failure of an arbitrary ice floe, a computational approach can be adopted, e.g., by the extended Finite Element Method (XFEM) (Lu et al. 2018a). In this paper, we will only focus on the analytical fracture approach.

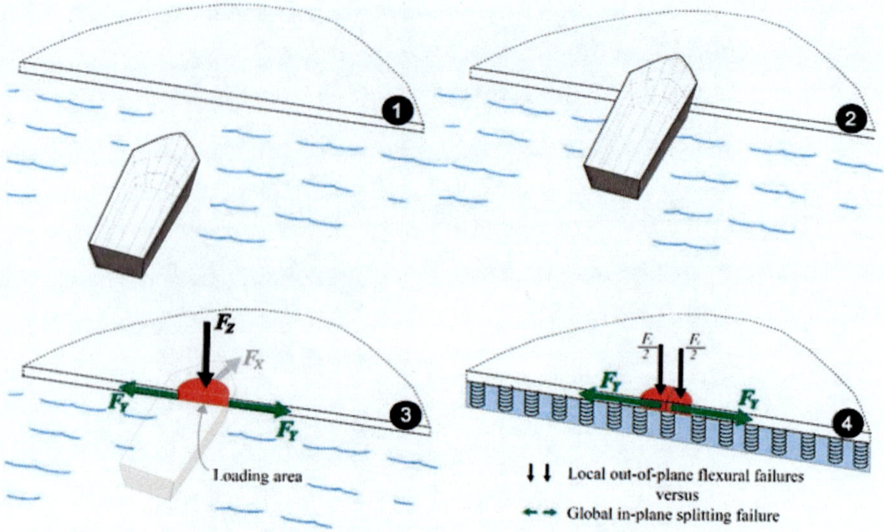

Fig. 7.5 Simplified interaction mechanism with contact force components

To achieve an analytical solution-based ice fracture calculation framework, the interaction mechanism between an ice floe and a floating structure can be simplified as shown in Fig. 7.5 together with related contact forces. The two types of failure modes, i.e., out-of-plane flexural failure by the vertical force F_Z and the in-plane splitting failure force by the horizontal 'wedge-out' force F_Y are considered separately in the following:

For the vertical direction contact force F_Z, which leads to the out-of-plane bending failure, the simplified formula by Nevel (1972) can be adapted to in Eq. (7.1).

$$F_Z = \begin{cases} \frac{\sigma_f h^2}{3} \tan(\frac{\alpha}{2})[1.05 + 2\frac{\delta}{\ell} + 0.5(\frac{\delta}{\ell})^2] & (\alpha \leq 90^\circ) \\ \frac{m\sigma_f h^2}{3} \tan(\frac{\alpha}{2m})[1.05 + 2\frac{\delta}{\ell} + 0.5(\frac{\delta}{\ell})^2] & (90^\circ < \alpha \leq 180^\circ) \end{cases} \tag{7.1}$$

in which,

σ_f is the flexural strength of sea ice, [kPa];

h is the thickness of sea ice, [m];

α is the angle of the wedge where the local bending failure occurs, in [°]. This angle is calculated by a contact detection algorithm (with edge tracking of two contacting bodies). After detecting the contact borders, the detailed calculation of α is similar to that of Su et al. (2010a);

δ is the maximum penetration depth and is considered equivalent to the loading area size at the wedge tip, [m];

m is the number of wedges that are bent off from the original ice sheet and is set as a random number between 2 and 3 (based on lab and field observations);

ℓ is the characteristic length [m] of sea ice and is calculated by Eq. (7.2).

$$\ell = \left(\frac{Eh^3}{12\rho_w g}\right)^{1/4} \tag{7.2}$$

in which, E, ρ_w and g are the Young's modulus [kPa], water density [kg/m^3] and gravitational acceleration [m/s^2], respectively.

For the splitting failure modes, two types of splitting scenarios can be considered: the analytical formulas to deal with the central splitting and off-centre splitting scenarios are presented in Eqs. (7.3) and (7.4) to calculate the 'splitting force term' F_Y (see Fig. 7.5).

$$F_Y(a) = \frac{hK_{IC}\sqrt{L}}{H(a,0)} \tag{7.3}$$

in which,

K_{IC}	is the fracture toughness of sea ice, [kPa$\sqrt{\text{m}}$];
$H(a, 0)$	is the weight function of a rectangular ice floe with a centred edge crack. Its formulation can be found in Dempsey and Mu (2014);
a	is the non-dimensional crack length given by $a = A/L$, whereas A is the crack length and L is the length of the idealised rectangular ice floe.

$$F_X = \frac{hK_{IC}\sqrt{W_1}}{\sqrt{[f_1^2(\frac{A_0}{W_1}, \beta_{YX}) + f_2^2(\frac{A_0}{W_1}, \beta_{YX})]}}$$

$$\beta_{YX} = \frac{F_Y}{F_X}$$

$$f_1(\frac{A_0}{W_1}, \beta_{YX}) = \sqrt{\frac{4\pi}{\pi^2 - 4}}(\frac{A_0}{W_1})^{-1/2}(\beta_{YX} - \frac{1}{\pi}) + \frac{2.0284(A_0/W_1) + 2.9890}{(A_0/W_1)^{-1.3569} + 1.0499}\beta_{YX} + \frac{1}{2}\frac{0.1856 - 0.2174(A_0/W_1)^{0.1635}}{(A_0/W_1)^{-6.2861} + 0.4092(A_0/W_1)^{0.1635}}$$

$$f_2(\frac{A_0}{W_1}, \beta_{YX}) = \frac{1.8134(A_0/W_1) + 0.5498}{(A_0/W_1)^{-1.4702} + 1.2041}\beta_{YX}$$

$$+\frac{1}{2}\frac{0.4051 - 1.3238(A_0/W_1)^{0.4339}}{(A_0/W_1)^{-0.9953} + 1.0099(A_0/W_1)^{0.4339}} \tag{7.4}$$

In Eq. (7.4),

A_0	is the length of the initial radial crack due to local bending failure, in [m]; it can be calculated as $A_0 = 2\ell$, with the characteristic length ℓ formulated in Eq. (7.2);
W_1	is the distance from the contact point to the closest free edge. The width of the bounding box of an arbitrary floe can be written as: $W = W_1 + W_2$ and $W_1 < W_2$. In general, it is required that $W_2 \geq 1.5 \cdot W_1$ for Eq. (7.4) to be valid. Otherwise, Eq. (7.3) is a better approximation;
$\beta_{YX} = F_Y/F_X$	represents the contact force ratio (see Fig. 7.5). The contact force F_X opposite to the ship's transit direction has a tendency to close the splitting crack.

More details on Eq. (7.4) can be found from Lu et al. (2018c). In order to calculate the off-centre splitting force F_Y, one must first identify the value of W_1, which gives a quantification of how 'off-centre' a crack is. Then one must formulate the contact properties based on the geometry of the contact point (i.e., to calculate the ratio β_{YX}, which reflects the ratio between the splitting force F_Y and impact force F_X). With known W_1, A_0, and the ratio β_{YX} (which is a geometrical quantification), one can calculate the values of f_1 and f_2. After all these preparation, the first formula in Eq. (7.4) can be used to calculate the exact value of F_X, which further leads to known F_Y.

The above analytical fracture based algorithms are implemented in SAMS, which is a three-dimensional (3D) multi-body dynamic based simulation tool (Lubbad et al. 2018).

7.2.2.2 Multi-Body Dynamics and Hydrodynamics

For ice—structure interactions, Fig. 7.1 shows that three interaction scenarios should be accounted for (ISO19906 2019; Løset et al. 2006). These are:

- Limiting stress
- Limiting force
- Limiting momentum.

The floe ice fracture process and calculations described in Sect. 2.2.1 are dealing only with the limiting stress scenario. To account for the other two limiting mechanisms, the interaction between ice floes and ambient environmental conditions (i.e., wind, wave and current forces) should be included in a complete simulation. These different force components can be accounted for by a multi-body dynamic based simulation platform. Figure 7.6 illustrates different force components and important physical processes that take place within this multi-body dynamic system. SAMS

Fig. 7.6 Different force components and physical processes involved during floe ice—structure interactions

will be utilised as an example here to describe the considered physicals and their implementation in a multi-body dynamics-based simulation platform.

SAMS is built upon a multi-body dynamic simulator, which detects the contacts, calculates the contact forces, and resolve potential overlaps among different bodies (Coutinho 2013). It is considered as the non-smooth Discrete Element Method (DEM) and has been used in ice—structure interaction studies by several researchers worldwide (Konno and Mizuki 2006; Konno 2009; Konno et al. 2011; Metrikin and Løset 2013; Lubbad and Løset 2011; Yulmetov et al. 2016; Septseault et al. 2014; Dudal et al. 2015; Daley et al. 2012). Comparing to the conventional DEM used in ice engineering (see the review work by Tuhkuri and Polojärvi (2018)), the non-smooth DEM is very efficient as it solves equations on the 'velocity level' not on the 'acceleration level', causing less equations to solve. This leads to the abrupt (non-smooth) velocity changes for impact bodies and thus obtained its name 'non-smooth DEM'. Because the solutions were performed at the 'velocity level', the corresponding contact 'force' term is not a force (i.e., $F = ma$) but an impulse (i.e., $P = mv$). However, as our simulation target is a 'force time history', it is necessary to convert impulses into forces. A rudimentary approach is to obtain the contact force via Eq. (7.5), i.e., dividing the contact impulse by the simulation time step (Δt).

$$F = P/\Delta t \quad (7.5)$$

However, this brings an issue that the calculated force is time step dependent. In addition, as the non-smooth DEM treats all the bodies as a rigid body, it does not reflect the real ice—structure interaction process (i.e., ice crushes and structures deform at the contact zone). Over the years, a new type of 'crushing contact' algorithm (Van den Berg et al. 2017, 2018) has been developed and implemented within SAMS to get rid of this limitation. The customarily developed ice—structure contact process (in form of ice crushing) within our model marks a significant difference compared to previously non-smooth DEM based simulators.

The crushing contact model in our model assumes constant energy absorption per unit crushed volume of ice, represented by a Crushing Specific Energy value (*CSE*). This is in accordance with previous research (Kim and Høyland 2014, Kinnunen et al. 2016). It is equivalent to assuming a constant contact pressure. The crushing contact model ensures that the energy absorbed in an ice-ice or ice-structure contact matches the change in overlap volume of the interacting bodies within each time step, where the overlap volume of interacting bodies represents crushed ice. The energy match is ensured by considering the projected area of a contact occurring, and the expected change of projected area. This is used to define a force-penetration gradient, as in Eq. (7.6).

$$\frac{\Delta F}{\Delta \delta} = \left(\frac{\left(A_{\text{proj}}^{\text{prop}} - A_{\text{proj}} \right) CSE}{\Delta \delta} \right) \tag{7.6}$$

where $A_{\text{proj}}^{\text{prop}}$ stands for the projected area when the interacting bodies are propagated with their current velocity, A_{proj} is the contact projected area at the beginning of the time step (these two terms can be formulated based on the discretised contact areas in the left of Fig. 7.7a), $\Delta\delta$ is the change of contact penetration (see the right plot of Fig. 7.7b), and $\Delta F/\Delta\delta$ is the force-penetration gradient. For clarity, 2-D sketches are used in this section, but the algorithm is implemented fully in 3-D. Contact crushing is represented by body overlap in the numerical simulation, under the assumption that the overlap volume that occurs in the numerical simulation represents crushed ice. The contact point is defined as the centre of the overlap volume. The contact normal direction is defined as the weighted average of the normal directions of the

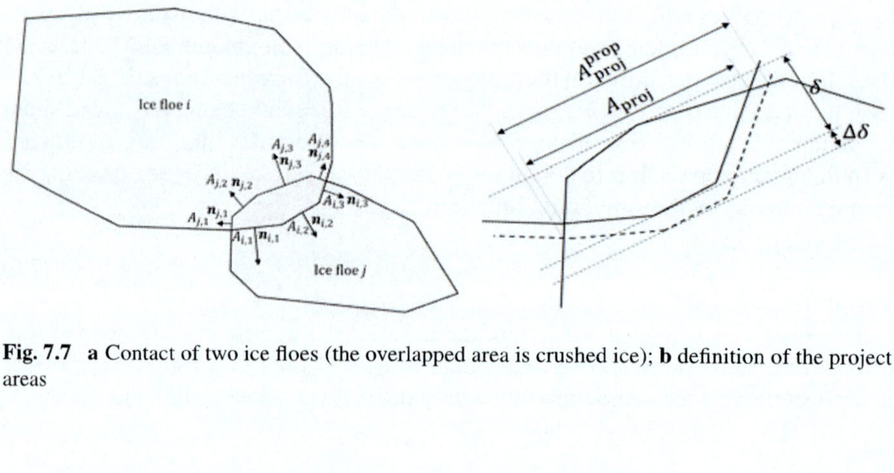

Fig. 7.7 **a** Contact of two ice floes (the overlapped area is crushed ice); **b** definition of the project areas

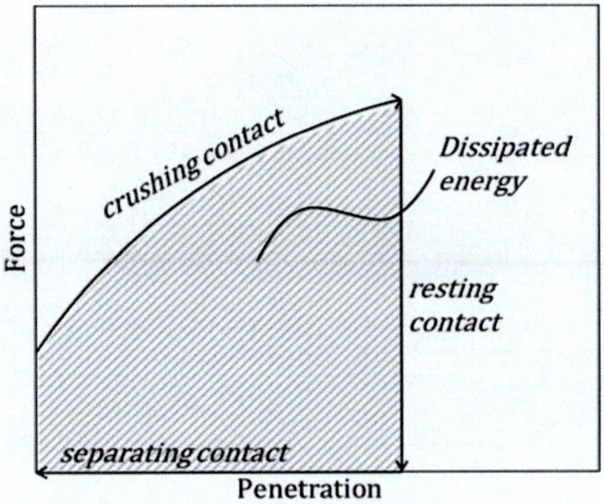

Fig. 7.8 Loading–unloading curve of ice-ice or ice-structure contact

sub contact areas related to one of the bodies (see Fig. 7.7a). Detailed formulations can be found in Berg et al. (2018). This force-penetration gradient is used together with the projected area at the beginning of the time step to define the compliance parameters that represent local ice crushing.

With the implementation in Eq. (7.6), the contact force in our model is no longer an impulse; it accounts for the ice crushing process in a smooth manner (see Fig. 7.8).

The hydrodynamic forces are separately calculated on each ice floe and the structure. For an arbitrary ice floe, the total hydrodynamic force is obtained by integrating the two drag components over the body's surface. A triangular mesh, as shown in Fig. 7.9, is used to discretize the body's surface, and the sum of the drag forces on the wholly immersed triangles is obtained as follows:

$$\mathbf{F}_{drag} = \sum_{k=1}^{M} \left[\rho_w C_i S^k |\mathbf{U}_{||}^k| \mathbf{U}_{||}^k - \left(\rho_w C_f S^k \left[(\mathbf{U}^k \cdot \mathbf{n}^k) \right]^2 \mathbf{n}^k \right) \Big|_{(\mathbf{U}^k \cdot \mathbf{n}^k) < 0} \right] \quad (7.7)$$

where $\mathbf{F}_{drag}$ is the total hydrodynamic force on the rigid body; M is the number of the wholly immersed triangles on the body's surface; S^k is the surface area of the k-th triangle; $\mathbf{U}^k$ is the relative fluid velocity at the geometrical centre of the triangle; $\mathbf{n}^k$ is a unit vector normal to the triangle and pointing outwards from the body's interior; $\mathbf{U}_{||}^k$ is the tangential velocity calculated as $\mathbf{U}_{||}^k = \mathbf{U}^k - (\mathbf{U}^k \cdot \mathbf{n}^k)\mathbf{n}^k$; and finally, C_i and C_f is the drag coefficients due to skin friction and pressure, respectively. The force is calculated individually for each triangle depending on the body's motion (Tsarau et al. 2017). The moment due to this force is obtained in a similar integration scheme with additional cross product between the force component and its corresponding arm length from the floe centroid. Detailed formulations can be found in Tsarau et al. (2017).

Fig. 7.9 Ice floes represented by triangular meshes

The total hydrodynamic drag force on a structure is calculated as the sum of the quadratic drag due to pressure, arbitrary linear drag (a term one can choose to conservatively predict drag force at low interaction speed range), and skin friction in the boundary layer. The following equations are used for the calculation:

$$\mathbf{F}_{dragStructure} = \mathbf{F}_{quadratic} + \mathbf{F}_{linear} + \mathbf{F}_{skin} \tag{7.8}$$

$$F^i_{quadratic} = \rho C^i_{pressure} S^i |U^i| U^i \tag{7.9}$$

$$F^i_{linear} = \rho C^i_{linear} S^i U^i \tag{7.10}$$

$$F^i_{skin} = \rho C^i_{skin} S^i_{wetted} |U^i| U^i \tag{7.11}$$

where U^i is the relative velocity, $i = 1,2,3$—corresponds to the x-, y- and z-directions, S^i is the projected area (approximated automatically if not specified by User) and S^i_{wetted} is the wetted-surface area.

7.3 Field Fracture Tests

The theoretical backgrounds of important physical processes are described in Sect. 7.2.2. Their formulations are mainly based on field observations and field experiments. As an example, this section describes the experimental effort in one of the three physical processes considered, i.e., ice fractures.

Section 7.2.2.1 describes the analytical and numerical efforts concerning the fracture of ice floes during the interaction. In addition to the analytical and numerical work, extensive experimental work has been carried out in the field for all related physical processes mentioned above. In this section, we present some preliminary results concerning ice fracture studies. From 2015 to 2018, a series of large-scale sea ice's fracture tests were performed at Spitsbergen.

The chosen test site is located at the Vallunden Lagoon in the Eastern end of the Van Mijen Fjord (Fig. 7.10a, b and c). The fjord is about 50 km long and penetrates into Spitsbergen from the West to the East at a latitude of about 77.8°N (Fig. 7.10b). The water depth at the chosen test site is in the range of 5 to 10 m.

Fig. 7.10 Overview of the chosen test field, Spitsbergen

7.3.1 Fracture Properties of Sea Ice

In Eqs. (7.3) and (7.4), there is one important material property, i.e., the fracture toughness of sea ice K_{IC} that needs to be determined. To measure this number, or more general, the fracture properties of sea ice, field ice fracture experiments were conducted. Most of the recent ice fracture experiments referred here are extensions of previous work carried in recognizing of the importance of size effects (Dempsey et al. 1999a, b; Dempsey 1991). The primary target has been to measure the fracture properties of sea ice (Lu et al. 2015a). The studies were made at different scales. In the larger scale, a number of ice floes in the range of 5 m by 10 m (see Fig. 7.11) were split by a screw-driven loading jack (see Fig. 7.12) providing a constant displacement rate.

After most, if not all, of the fracture experiments, the splitting crack passes through the initially stationed displacement sensor arrays (see Fig. 7.11b). The displacements ahead of the initial crack were measured in the course of the experiment together with the applied load (measured directly from the Jack shown in Fig. 7.12).

The field experiment successfully captured stable crack growth. This enabled us to stop the experiment at will and drill an ice core encompassing the crack tip (see Fig. 7.13) for further lab-scale investigations. This is still on-going research work by the time this paper is written. However, comparing with the rather 'deterministically' propagated large scale crack in Fig. 7.11b, the small-scale crack in Fig. 7.11c meanders and branches along its propagating path. This may potentially explain the mismatch of the fracture properties that are measured at lab scale and in full scale field studies (i.e., the measured sea ice fracture energy in the field is more than 10 times larger than in the laboratory (Lu et al. 2015a)).

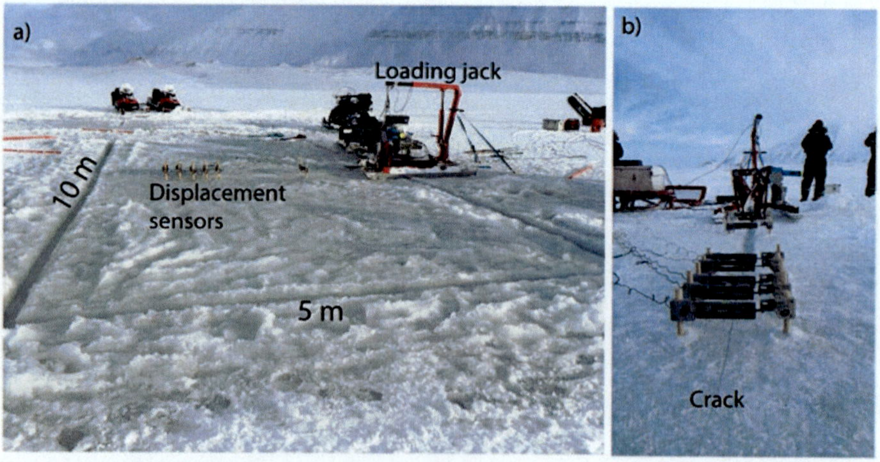

Fig. 7.11 **a** A photo and illustration of a large-scale ice fracture experiment, and **b** a crack passing through the displacement sensors after the experiments

Fig. 7.12 The screw-driven displacement-controlled loading device (i.e., the Jack)

7.3.2 Crack Kinking

Aside from the fracture properties tests, in 2018, a series of crack kinking tests were performed. This is in response to the long cracks that are often developed between parallel channels during an ice management operation (Lu et al. 2015d, 2016b). In addition, the 'crack kink' theory is also related to the off-centre ice splitting scenario described by Eq. (7.4). After thorough theoretical studies and operational experiments (see Fig. 7.14a), the idealised kink test was performed with existing fracture experimental instrumentations. Similar crack kinking which agrees well with theoretical prediction is observed and documented (see the red coloured highlighted crack in Fig. 7.14b). This signifies the validity of using classic fracture mechanics to deal with large scale ice fractures. This finding is important as not all ice fracture processes can be characterised by classic fracture mechanics with traditional concepts such as fracture toughness, especially when the fracture/crack takes place at smaller scales (Dempsey et al. 2018).

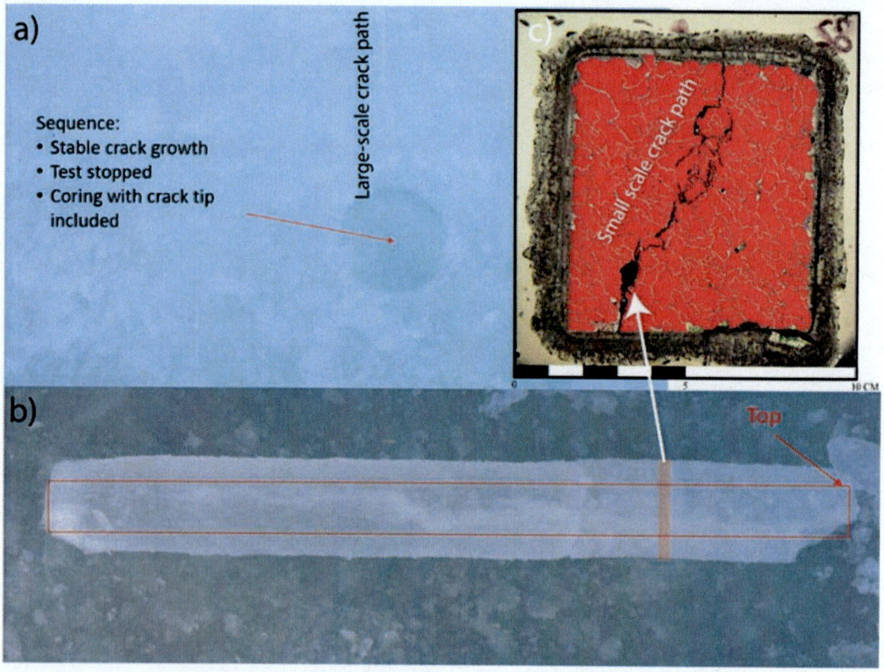

Fig. 7.13 **a** and **b** coring of crack tip; and **c** photo of a thick ice section (grain sizes 3–4 mm) including the fracture at lab scale (the planar size of the thick section is about 80 × 80 mm; plot by Shestov A.)

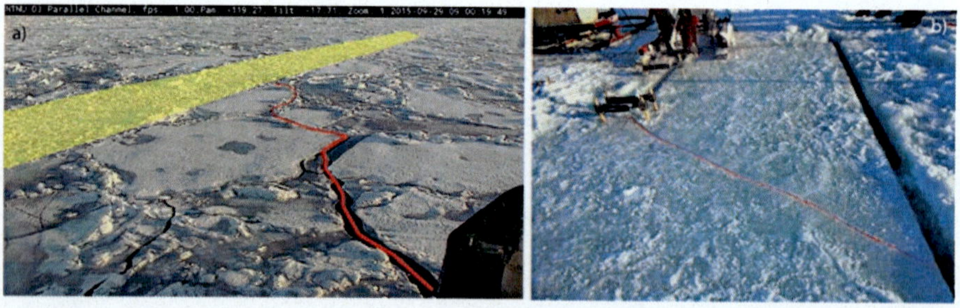

Fig. 7.14 **a** Field observed long crack kinking into the nearby parallel channel during an ice management operation; **b** idealised crack kinking test in the field

7.3.3 Ice Ridge Splitting Test

In 2018, in order to study the mechanical properties of a consolidated ice ridge, which was artificially created (Salganik et al. 2019; Lu et al. 2019), an ice ridge splitting experiment was performed with existing infrastructure. Based on a simple calculation to divide the work done by the Jack (i.e., WD) and an estimated crack surface area A^*,

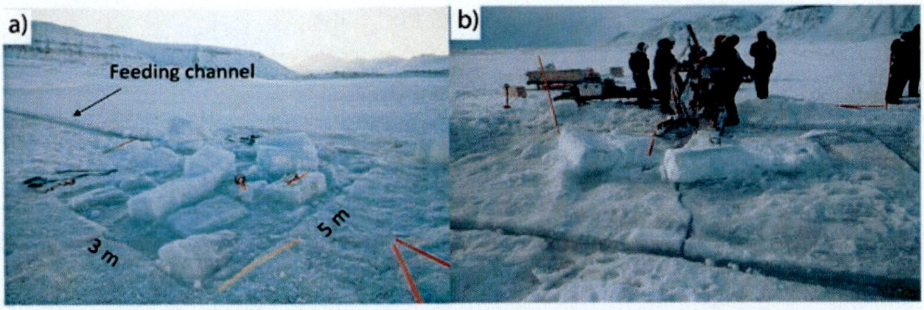

Fig. 7.15 **a** Creation of the artificial ice ridge with a feeding channel and manually-cut ice blocks; **b** splitting of the consolidated artificial ice ridge

$$G_F = WD/A^*(12), \quad (12)$$

the fracture energy of an ice ridge G_F was calculated to be about 12 J/m^2. At the same time, it is found that during the loading process, a significant amount of creep deformation is encountered by the Jack (Lu et al. 2019) (Fig. 7.15).

A summary to the preliminary results obtained from the ice fracture field experiments campaign can be made in the following:

- The global crack path is rather stable and can be predicted by classic fracture mechanics; however, the local crack path interacts with the grains and brine structures of sea ice, exhibiting a more tortuous nature. This might explain the mismatch between the lab scale and field scale ice properties' measurements. However, this is still on-going work and more research efforts here are needed.
- The crack kink theory (i.e., Eq. (7.4)) is largely based on classic fracture mechanics. Its prediction concerning the crack path matches well with field experiment (in Fig. 7.14). This signifies the usefulness of classic fracture mechanics in dealing with large scale ice fracture processes.
- The ice ridge splitting experiment preliminarily yield an ice fracture energy of 12 J/m^2. However, it also shows that there is a considerable amount of creep deformation taking place within sea ice under loading.

7.4 Simulation Examples—A Case Study

The previous sections introduced important field observations, their theoretical formulations, implementations and a description of field experiments on ice fracture. In this section, all these accumulated knowledge and implementation shall be applied in one of the recent engineering practices using SAMS to estimate the forces from the drifting ice acting on a grounded vessel. Before presenting the case study, it is worth mentioning that before this application, SAMS has been validated in several idealised situations, which include:

- A full-scale ship transit in a broken ice field (Lubbad et al. 2018).
- Back-calculating the icebreaker Oden's performance (i.e., h-v curve) in level ice (Raza et al. 2019).
- A multi-leg jacket type structure in model ice tests (van den Berg et al. 2019).
- A moored conical structure in model ice tests (Tsarau et al. 2018).

Because our model direct simulates all the three 'limiting mechanisms' considered, it becomes a natural choice to directly calculate ice actions on a structure with a given structure form, ice and environmental conditions (i.e., no intermediate assumptions such as limiting mechanisms or ice failure modes are needed). All the interaction processes, ice fracture included, are simulated in the time domain as sequential events. This is in line with the proposition of Daley et al. (1998) that treating ice failure as a process of discrete failure events is a best method to understanding ice load. This feature was utilized in real engineering practice in early 2019 with regard to calculating the drift ice action on the stranded trawler Northguider.

7.4.1 The Grounded Trawler Northguider

On December 28th, 2018, the trawler Northguider was grounded at Sparreneset in the Hinlopen Strait (see Fig. 7.16). The vessel has been abandoned and remains aground at Sparreneset on Nordaustlandet in the Hinlopen Strait, Spitsbergen at proximity to the shoreline in position 79° 53.9' N 018° 04.7' E (see Fig. 7.16b) on a Northeasterly heading.

The vessel is grounded at a location with a water depth of about 6.5 m, which is larger than the vessel's maximum draught (5.7 m). Only about 100–200 m away from the ground location, there is a sudden increase of water depth to about 400–500 m (NorwegianHydrographicService, 2018). Drifting sea ice is expected at the location

a) b)

Fig. 7.16 **a** Photo of the grounded trawler Northguider in early 2019 (photo: the Norwegian Coastal Administration); **b** the location of grounding (northern Svalbard)

of the grounded vessel. Sea ice can potentially exert significant forces on the vessel. Considering the sensitive location of the vessel, it is of great importance to know whether or not the vessel can be pushed by drifting sea ice into deeper waters before the planned salvage operation in the summer of 2019.

7.4.1.1 Ice Conditions

In general, there is no fast ice in the Hinlopen Strait, but mainly new-ice and first-year drifting sea ice with varying thickness depending on temperature and wind directions. At the grounded-vessel location, it is often drifting sea ice that is encountered. The origin of this ice is mainly the Wahlenbergfjorden and possibly the southern part of the strait, e.g., around Sørporten, under the influence of sea currents and southerly wind. The northerly wind will drive the drifting sea ice towards the Western part of the straight. The ice concentration information based on past four years' data is available through the Norwegian Polar Institute (NPI). The heaviest ice conditions are expected to be present during February and March.

As an example, Fig. 7.17 shows that the frequency of occurrence of 7/10 and 9/10 ice concentrations can reach up to 20% in the months of February and March. The ice thickness was estimated to be around 1.0 m based on our operational experiences.

However, drifting large ice floes can also be found in the region of the grounding. Openly available satellite images from the Sentinel 2 satellite were analysed in order to make an estimate of the floe size at the location of the stranded vessel. Sentinel 2 imagery is available for the spring seasons of 2016 and 2017. Sentinel 2 captures

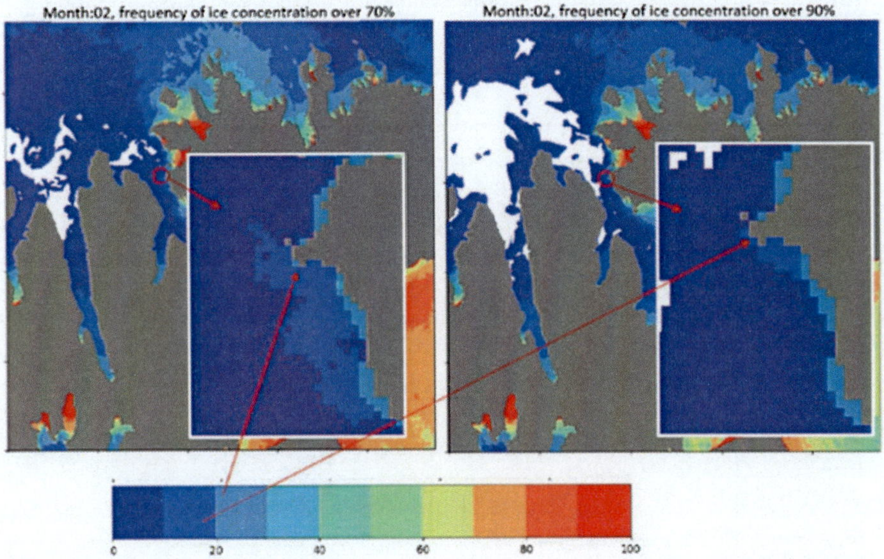

Fig. 7.17 Frequency of occurrence of 7/10 and 9/10 ice concentrations in the month of February

Fig. 7.18 Photo of the stranded trawler Northguider at Sparreneset in the Hinlopen Strait (March 18th, 2019)

optical imagery with a resolution of 10 m. The availability of images at the stranding location is limited by cloud cover and darkness during the winter. As an example, Fig. 7.18 shows the ice conditions with drifting large ice floes (up to 1 km across) near the location of the grounding.

Given the above information on ice conditions, we decided to model two types of drifting ice conditions, i.e., a broken ice field with various ice concentrations; and a single large ice floe interacting with the grounded vessel. All these ice features' drifting motion are driven by the ambient current and wind conditions (Fig. 7.19).

7.4.1.2 Current Conditions

No data are available on sea currents at Sparreneset. Based on own experience in the region and by consultation with the Norwegian Coastal Administration, who is actively surveying the site, we found a speed of 1.5 m/s as a reasonable estimation of sea currents at the site. The current changes direction under the influence of tide, resulting in currents coming from the North and from the South. Consequently, sea ice is expected to drift from both the North and the South directions. The ice drift from the North is not considered in this study because it can only push the vessel towards shore, which is a favourable scenario.

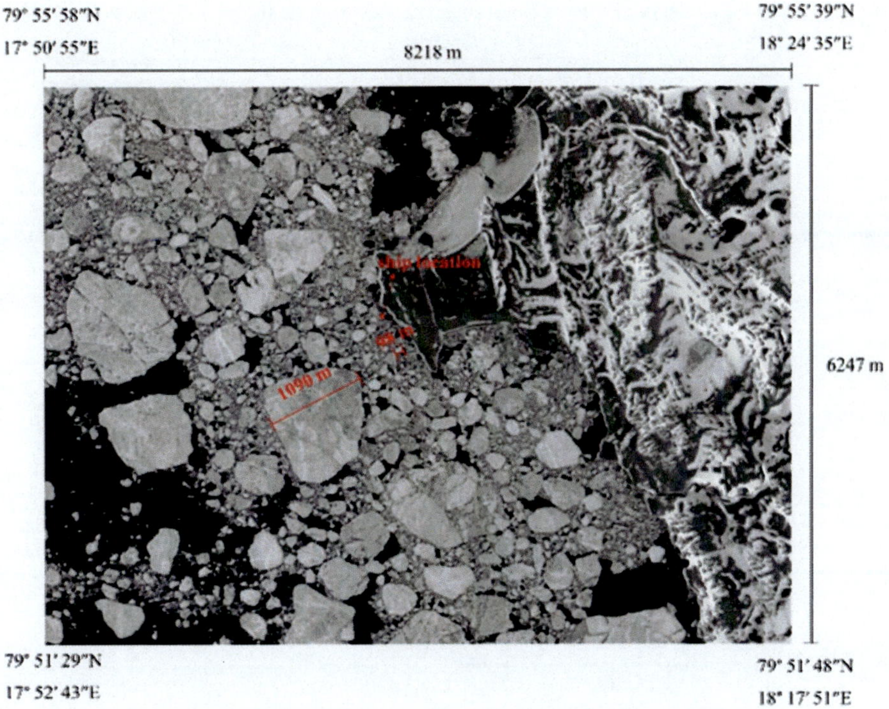

Fig. 7.19 Satellite photo of ice conditions at the stranding location, 25–6-2017 (Copernicus.eu)

7.4.1.3 Wind Conditions

Figure 7.20 shows the wind rose at Sparreneset, which is based on data from the Climate Forecast System Reanalysis (CFSR), NCEP NOAA, for the period from 01.01.1979 to 31.12.2017. It is evident that the wind speed may exceed 25.6 m/s and it is mostly southerly and south south-easterly wind.

7.4.1.4 SAMS Numerical Set-Up

The numerical set-up used for this case study follows from the environmental conditions as described in the previous section. For critical parameters with a high uncertainty (ice thickness and concentration, current conditions), two or more different values were used, and the used parameters varied from the expected value to a more conservative upper bound value. Simulation parameters related to the hydrodynamic drag, ice-ice and ice-vessel friction, and the ice material properties are the same in all the simulation runs. The constant parameters are listed in Table 7.1.

The grounded vessel Northguider was digitised from the structural drawings (see Fig. 7.21). There is very little information available on the contact properties between

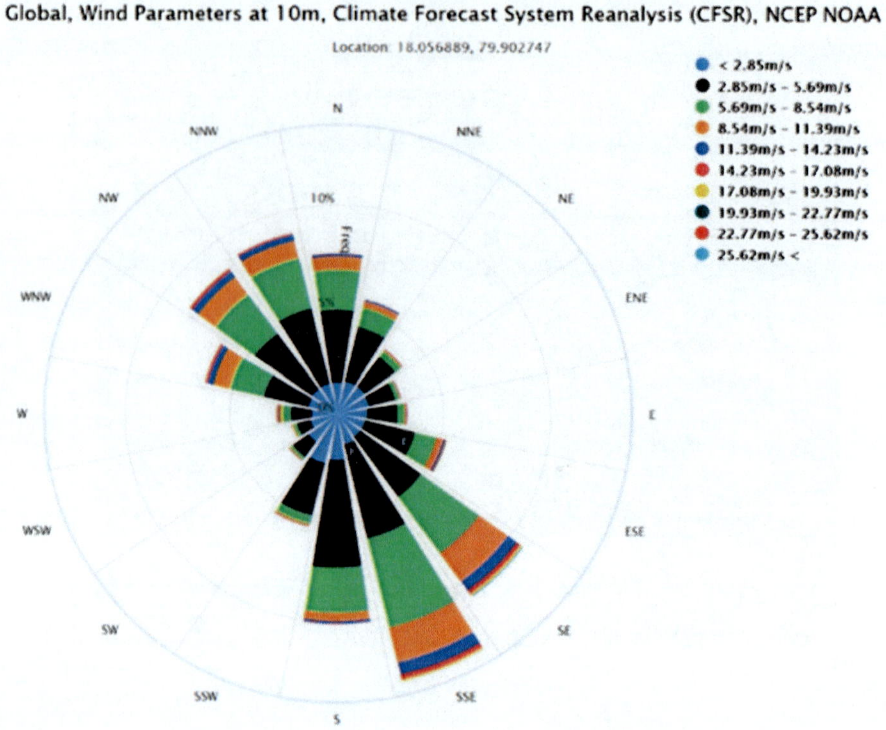

Fig. 7.20 A wind rose at Sparreneset based on data from CFSR, NCEP NOAA for the period from 01.01.1979 to 31.12.2017

Table 7.1 Simulation parameters

Name	Value
Ice form drag coefficient (horizontal, vertical)	0.1, 0.5 (-)
Ice-water skin friction coefficient	0.005 (-)
Ice-air skin friction coefficient	0.001 (-)
Ice-ice friction coefficient	0.15 (-)
Ice-vessel friction coefficient	0.15 (-)
Ice-environment friction coefficient	0.15 (-)
Ice density	910 kg/m^3
Crushing specific energy	2.0 MJ/m^3
Young's modulus	5 GPa
Poisson's ratio	0.3 (-)
Fracture toughness	150 kPa $\sqrt{\text{m}}$
Flexural strength	0.5 MPa
Tensile strength	0.5 MPa
Water density	1025 kg/m^3
Air density	1.04 kg/m^3

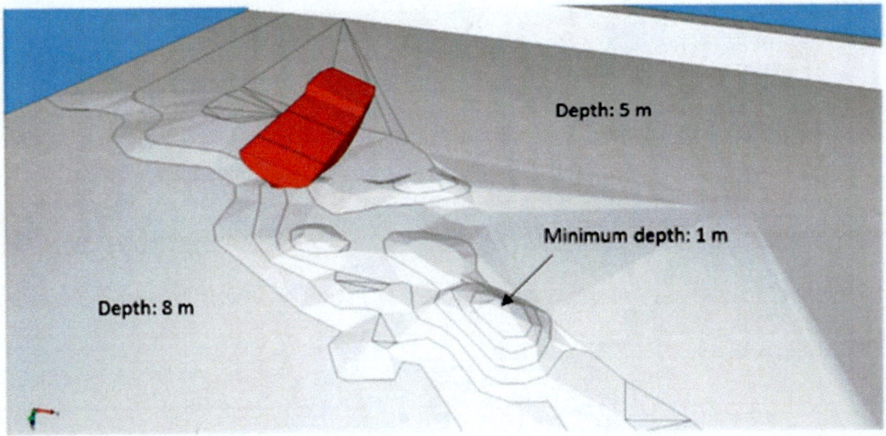

Fig. 7.21 Bathymetry and vessel position in the scene. The vessel angle of list is 17°. The detailed bathymetry is modelled in the vicinity of the vessel; the far away bathymetry is modelled as a flat bottom

the vessel and the sea bottom. Therefore, the vessel is assumed to be rigidly fixed to the seabed in all the performed numerical simulations. To the best of our knowledge, the vessel has at present a list angle of 17°. This reduces the ice forces on the vessel (as will be shown later) because it introduces a bending force component on the ice, resulting in bending failure.

The bathymetry in the vicinity of the vessel is included in the simulations. The bathymetry is relevant because there are shallow areas nearby (depth < 1.0 m) that may protect the vessel from more severe ice features. The bathymetry may also influence the ice clearance and grounding behaviour. The bathymetry as implemented in the simulation is shown in Fig. 7.21.

The coastline is modelled as a vertical wall as in Fig. 7.21. The distance between the vessel and the coastline, based on eyewitness reports and GPS coordinates, is 102 m (measured from the bow of the vessel) or 125 m (from the centre).

For the current, only a current speed of 1.5 m/s from the critical south-easterly direction is used in the modelling. Current and ice drift from the South-East would push the ship towards deeper water, reducing the ship-bottom resistance. Current from the North-West, on the other hand, would push the ship towards more shallow areas, resulting in a firmer grounding of the vessel. The current is assumed perpendicular to the vessel heading and parallel to the shoreline. In addition, a wind speed of 30 m/s is used in all the simulations. Conservatively, the wind direction is assumed to be aligned with the current direction.

As described in the previous section, two types of ice conditions are simulated. First, most of the simulations are performed with broken ice fields (i.e., the 1st ice condition), in which the largest floes have a Mean Calliper Diameter (MCD) of 50, 100 and 150 m. In the base case, the ice has a thickness of 1.0 m. An ice thickness of 1.0 m is identified as a possibly critical condition, since the ice draft is such that

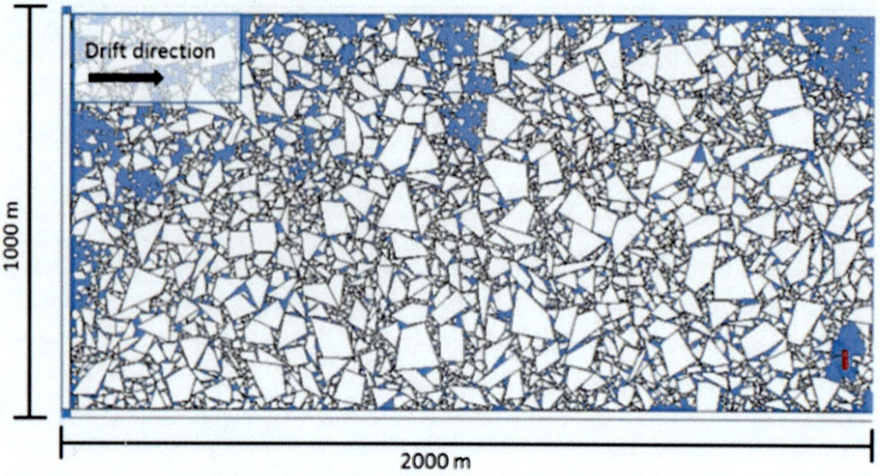

Fig. 7.22 Simulation domain and the broken ice field (the grounded vessel in "red" colour)

the ice will not (initially) ground on the shallow areas towards the South-East of the vessel. The ice concentration, the ice thickness and the floe size were varied in order to study the influence of these parameters on the ice force. The ice concentration was varied between 50 and 80%. The ice thickness was varied from 0.5 m to 1.5 m. The maximum floe size was varied from 50 to 150 m MCD. Lastly, a scenario of broken ice fields with imbedded thick ice features are simulated. The randomly seeded thick ice feature is assigned a thickness of 2.0 m with limited area coverage percentage (= 20%). The majority of the broken ice floes are with a thickness of 1.0 m. This case is to mimic scenarios, in which, a thick ice feature (e.g., second year ice or ice ridge) may potentially encounter the vessel.

The broken ice floe fields were initialized in a 1000 m × 2000 m domain. The domain is shown in Fig. 7.22. The domain 'width' (1000 m) is chosen such that the boundary conditions (except from the boundary representing the shore) do not influence the simulation outcome. The domain 'length' (2000 m) is chosen such that there is sufficient simulation time to capture the important processes that determine the ice force on the vessel.

For the 2nd ice condition with a large ice floe impacting the vessel, we conducted two sets of simulations. One set of simulations is performed with a 1500 m × 1500 m square floe and another set of simulations is performed with a circular floe with a radius of 846 m resulting in the same floe area as the square floe. The initial conditions of the simulation with a single large floe are shown in Fig. 7.23.

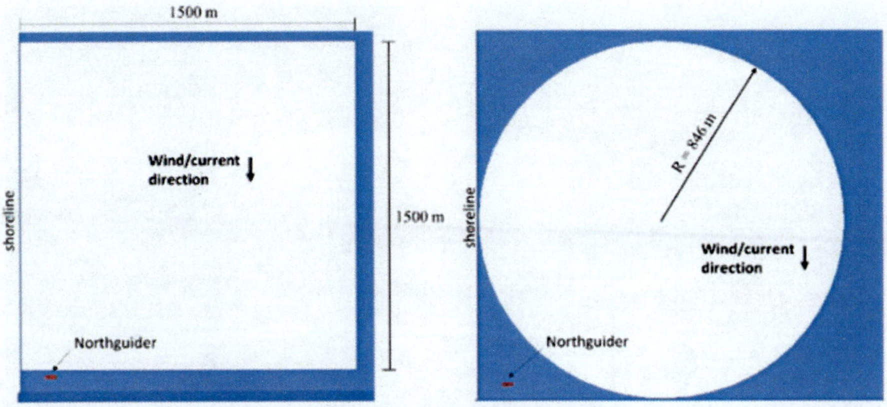

Fig. 7.23 Initial conditions of the simulations with large floes

7.4.2 *Results of the Drift Ice Action*

Extensive simulations are performed by SAMS. For the broken ice field interacting with the grounded vessel, a simulation snapshot is presented in Fig. 7.24. Visually, it shows that the ice floes impact the grounded vessel, fracture and accumulate in front of the vessel. After studying the motion of the drifting ice floes, it clearly shows that the major driving sforce arises from the wind's and current's fetch on a large area of ice floes that were 'blocked' by the grounded vessel.

A typical ice load history for the case with 80% ice concentration and an average floe size of 100 m is shown in Fig. 7.25. For the entire simulation period (1200 s), a major part of the ice load is lower than around 300 ton (an estimated value for ground friction at this location). After investigating the duration of exceedance to this 300 ton criteria, it shows that only 0.23% of the time the ice force level exceeds that criteria. Moreover, each duration of the exceeding event is only about 0.94 s. This means, even if the force has exceeded the vessel-to-ground friction level, there is generally not enough time to move the vessel off location.

The same trend as above has been observed for other broken ice conditions. The general conclusion is that the ice force can be larger than 300 ton, however, with limited time duration to be of no concern. In the following, we shall present different limiting mechanisms, which are natural outputs from the simulations.

7.4.2.1 Limit Force Scenario

Most periods with high ice forces in the broken ice conditions are caused by floes build-up upstream of the vessel. This belongs to the limiting force scenario, meaning that the driving forces from wind and current drag as well as frictional interaction between ice floes are the main factor determining the ice force on the vessel (as

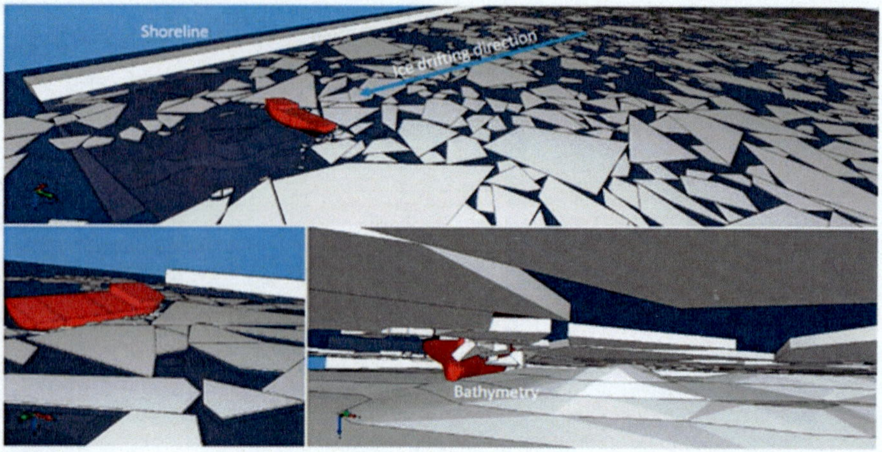

Fig. 7.24 A simulation snapshot in SAMS during the grounded vessel interacting with the drifting broken ice

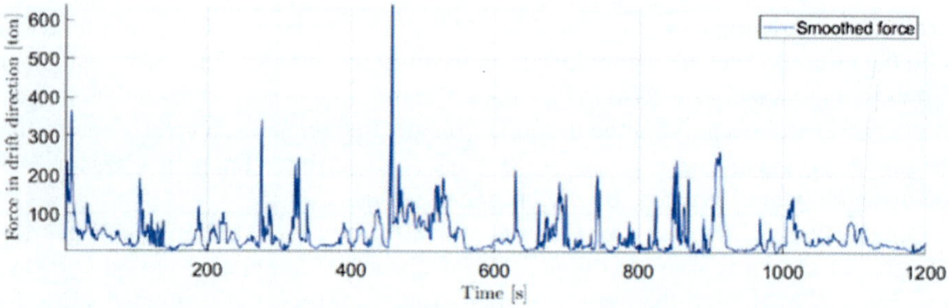

Fig. 7.25 Ice force history (in the drifting direction) for a typical broken ice field

opposed to stress and momentum limited loading mechanisms, where it is the ice failure strength or the momentum of ice features, respectively, that represent the main factors influencing the ice force). A typical force limited loading mechanism as occurs in the simulations is shown in Fig. 7.26.

7.4.2.2 Limit Momentum Scenario

High loads with a short duration are often caused by the limit momentum events, where the kinetic energy of an ice floe is dissipated by local crushing against the vessel. An example of a momentum limited loading event is shown in Fig. 7.27. The floes upstream from the vessel dissipate their kinetic energy in 6 s by a combination of crushing and bending failure against the vessel.

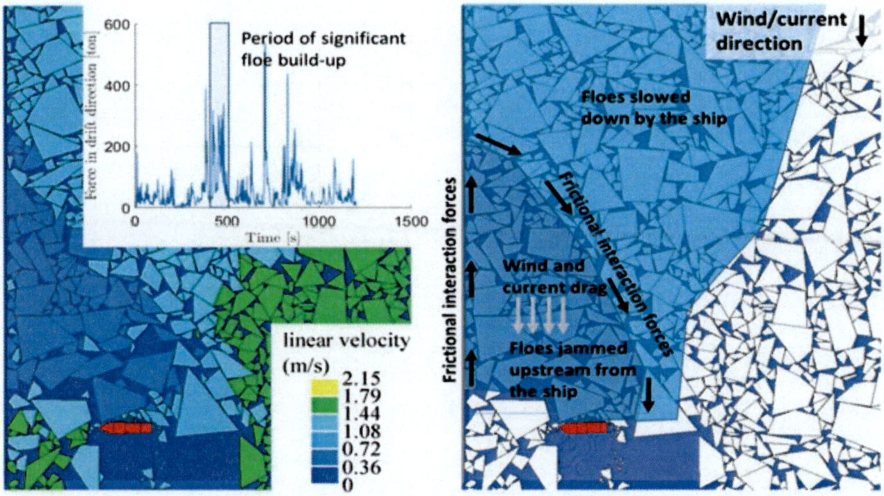

Fig. 7.26 Force limited loading by ice floe build-up upstream from the grounded vessel (in "red" colour)

7.4.2.3 Limit Stress Scenario

For the 2nd ice condition, i.e., large ice floes interacting with the grounded vessel, the interaction mechanism gives a quite different picture. Figure 7.28 shows a simulation snapshot during the interaction of the rectangular ice floe (1500 m in length) with the vessel. It is mainly ice fractures (in forms of splitting cracks) that limits the force level. The major driving force that transmits to the vessel is through the shearing and compression forces along the splitting crack as depicted in Fig. 7.28.

Fig. 7.27 Momentum limited loading event. Left: Start of interaction, floes are moving with a velocity between 1.0 and 1.4 m/s. Right: End of interaction, floes have stopped, and the kinetic energy is dissipated by a combination of local crushing and splitting failure

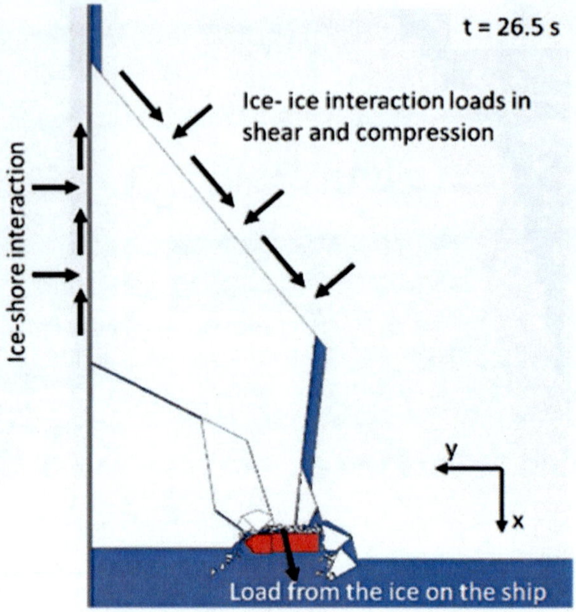

Fig. 7.28 Snapshot of interacting mechanisms between a large drifting ice floe and the grounded vessel

The force level can be rather high in this interaction scenario. Figure 7.29 shows the interaction force in the initial interaction stage. The force reduces after the initial splitting failures. However, in the very beginning, force levels in the drifting direction (-*x* direction) and the -*y* direction can be well beyond the 300-ton vessel-to-ground friction level with a rather long duration.

Given the simulation results, we recommended to the Norwegian Coastal Administration that:

- For the drifting broken ice condition, the simulated peak forces are often of too short duration and low exceedance time to be of a practical concern.
- If an interaction between a large ice floe and the vessel would occur, it is expected that the vessel will not remain in its current position. However, the chance for a large ice floe approaching the grounded vessel (see Fig. 7.18) and again travels at least in parallel to the shoreline, is considered rather low.

Given our above quantitative recommendations and other practical matters, the Norwegian Coastal Administration decided to wait until the summer of 2019 to perform the salvage operation.

Several months later, when the salvage operation started, it was found that the Northguider stayed in its originally grounded location, only with a larger list angle (see Fig. 7.30).

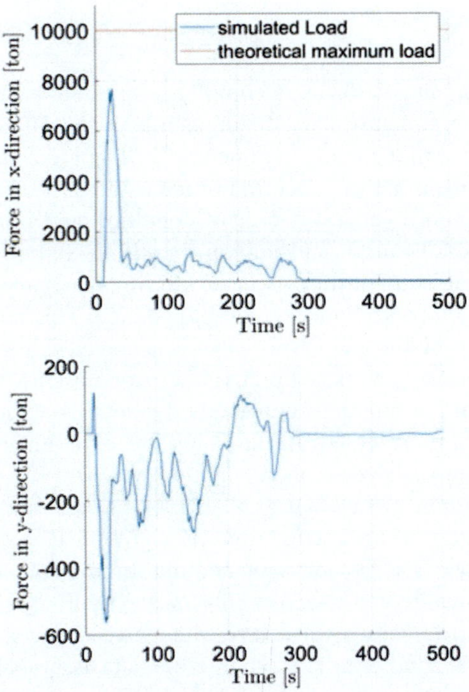

Fig. 7.29 Force in x- and y- direction from a large square ice floe, directions as indicated in Fig. 7.28

Fig. 7.30 The Northguider in the summer of 2019 (photo: Norwegian Coastal Administration)

7.5 Conclusions

Starting from the pursuit of offering a complete numerical package to calculate the ice action on floating structures, we distinguish between different parameters that influence ice actions. Among many parameters, different ice features, failure modes and limiting mechanisms are of concern. For ice features, we limit our presentation in this paper to the more general 'floe ice' condition and propose the frequently observed 'out-of-plane flexural' and 'in-plane splitting' failure mode. This sets the limiting stress mechanism. For the other two limiting mechanisms, the ambient ice, wind, wave and current conditions have to be modelled. This is a typical multi-body dynamic simulation problem and in this paper is achieved via implementing the ice–ice and ice—structure contact based on the Non-smooth DEM platform. The theoretical formulation of ice fracture, multi-body dynamics (i.e., ice crushing contact) and hydrodynamics are presented. Additionally, the experimental effort behind the ice fracture modelling is also described.

With the observation, theory and experiment—backed simulator, we present a case study concerning the drift ice action on a grounded vessel (i.e., the trawler Northguider). Because SAMS' implemented algorithms start from basic physical processes (i.e., ice fracture, multi-body contacts, and hydrodynamics), it enables ice actions calculations from ice on a structure with a given structure shape (geometry), ice and environmental conditions (i.e., no intermediate assumptions such as limiting mechanisms or ice failure modes are needed). Different limiting mechanisms and failure modes are natural output from the simulations. Given our simulated ice action, a sound engineering judgement was made on the salvage operation of the trawler Northguider. It was later shown that our recommendation is valid and endured the test of real life. To this end, a knowledge loop that starts from the field and back to the field is achieved. In this process, aside from the numerical simulation tool (SAMS) that was developed, many research results are also found. These include:

- Contrary to the conventional design approach with the preselection of one idealised limiting mechanism, in real life, all three limiting mechanisms are present during an ice—structure interaction process. It is important to capture all possible scenarios without presumption to give a less conservative ice load design.
- Different failure modes can co-exist, and they compete with each other as a load releasing mechanism.
- Different ice features can give quite different limiting mechanisms and consequentially different ice action levels. For example, it is mainly the limiting force and limiting momentum mechanism that are dominant for the broken ice condition; whereas it was the limiting force mechanism that dominates for a single large ice floe condition.

When it comes to the described ice fracture experiments, some preliminary results/observations can be made in the following:

- The global crack path is rather stable whereas the local crack path is more prone to be influenced by the grains and brine structures of sea ice, exhibiting a more tortuous nature.
- The crack kink theory (i.e., Eq. (7.4)) is largely based on classic fracture mechanics. Its prediction concerning the crack path matches well with field experiments (in Fig. 7.14). This signifies the usefulness of classic fracture mechanics in dealing with large scale ice fracture processes.
- The ice ridge splitting experiment preliminarily yield an ice fracture energy of 12 J/m^2. However, it also shows that there is a considerable amount of creep deformation taking place within sea ice under loading.

Acknowledgements The authors would like to thank the Research Council of Norway through the research centre of SAMCoT CRI for financial support in carrying out the experiment. The second author also thanks VISTA—a basic research programme in collaboration between The Norwegian Academy of Science and Letters, and Equinor for financial support issn writing this paper.

References

A. Dudal, C. Septseault, P.-A. Beal, S. L. Yaouanq, B. Roberts, A new arctic platform design tool for simulating ice - structure interaction, in *The 23rd International Conference on Port and Ocean Engineering under Arctic Conditions* (Trondheim, Norway, 2015)

A. Kinnunen, M. Tikanmäki, J. Heinonen, An energy model for ice crushing in ice-structure impact, in *23 rd IAHR International Symposium on Ice* (Ann Arbor, Michigan, 2016)

A. Konno, T. Mizuki, numerical simulation of pre-sawn ice test of model icebreaker using physically based modelling, in *Proceedings oh the 18th IAHR International Symposium on Ice* (Sapporo, Japan, 2006)

A. Konno, O. Saitoh, Y. Watanabe, Numerical investigation of effect of channel condition against ship resistance in brash ice channels. *Proceedings of the 21st International Conference on Port and Ocean Engineering under Arctic Conditions* (Montréal, Canada, 2011)

A. Konno, Resistance evaluation of ship navigation in brash ice channels, in *Proceedings of the 20th International Conference on Port and Ocean Engineering under Arctic Conditions* (Luleå, Sweden, 2009)

A. Tsarau, M. Van den Berg, W. Lu, R. Lubbad, S. Løset, Modelling results with a new simulator for arctic marine structures-SAMS, in *ASME 2018 37th International Conference on Ocean, Offshore and Arctic Engineering, 2018* (American Society of Mechanical Engineers, 2018), V008T07A035-V008T07A035

B. Su, K. Riska, T. Moan, Numerical simulation of local ice loads in uniform and randomly varying ice conditions. *Cold Regions Science and Technology* (2010)

B. Su, K. Riska, T. Moan, Numerical simulation of ship turning in level ice, in *Proceedings of the ASME 29th International Conference on Ocean, Offshore and Arctic Engineering, 2010*, vol. 4 (American Society Mechanical Engineers, 2010c), pp. 751–758

BHAT, S. U. , Analysis for splitting of ice floes during summer impact. Cold Reg. Sci. Technol. **15**, 53–63 (1988)

M. van den Berg, R. Lubbad, S. Løset, The effect of ice floe shape on the load experienced by vertical-sided structures interacting with a broken ice field. Mar. Struct. **65**, 229–248 (2019)

S.U. Bhat, S.K. Choi, T. Wierzbicki, D.G. Karr, Failure analysis of impacting ice floes. J. Offshore Mech. Arct. Eng. **113**, 171 (1991)

C. Septseault, P.-A. Béal, S. Le Yaouanq, A. Dudal, B. Roberts, A new ice simulation tool using a multi-model program, in *Offshore Technology Conference* (2014)
C. Daley, S. Alawneh, D. Peters, B. Quinton, B. Colbourne, GPU modeling of ship operations in pack ice, in *International Conference and Exhibition on Performance of Ships and Structures in Ice, Banff Alberta, Canada September* (2012), pp. 20–23
D.E. Nevel, The narrow free infinite wedge on an elastic foundation. U. S. Army Snow Ice and Permafrost Research Establishment, Corps of Engineering (1961)
D.E. Nevel, The ultimate failure of a floating ice sheet, in *International Association for Hydraulic Research, Ice Symposium* (1972), pp. 17–22
C. Daley, J. Tuhkuri, K. Riska, The role of discrete failures in local ice loads. Cold Reg. Sci. Technol. **27**, 197–211 (1998)
J. Dempsey, D. Cole, S. Wang, E. Sciences, Tensile fracture of a single crack in first-year sea ice. J Philos Trans R Soc: Math, Phys **376**, 20170346 (2018)
J.P. Dempsey, S.J. Defranco, R.M. Adamson, S.V. Mulmule, Scale effects on the in-situ tensile strength and fracture of ice part i: large grained freshwater ice at spray lakes reservoir, Alberta. Int. J. Fract. **95**, 325–345 (1999)
J.P. Dempsey, Z. Mu, Weight function for an edge-cracked rectangular plate. Eng. Fract. Mech. **132**, 93–103 (2014)
E. Enkvist, P. Varsta, K. Riska, The ship-ice interaction, in *Proceedings of the 5th International Conference on Port and Ocean Engineering under Arctic Conditions., 1979 Norwegian Institute og Technology* (Trondheim, Norway, 1979), pp. 977–1002
E. Kim, K. Høyland, Experimental investigations of the energy absorption capacity of ice during crushing: is the specific energy scale independent? *Proceedings of the 22nd IAHR International Symposium on Ice* (Singapore, 2014)
E. Salganik, K. Høyland, A. Shestov, S. Løset, A.-N. Heijkoop, Medium-scale consolidation of artificial ice ridge – Part I: surface temperature, thickness and mechanical properties, in *Proceedings of the 25th International Conference on Port and Ocean Engineering under Arctic Conditions* (Delft, The Netherlands, 2019)
G. Lindqvist, A straightforward method for calculation of ice resistance of ships, in *Proceedings of POAC 1989* (1989), pp. 722–735
H.-M. Heyn, Motion sensing on vessels operating in sea ice: a local monitoring system for transit and stationkeeping operations under the influence of sea ice (2019)
I. Metrikin, S. Løset, Non-smooth 3D discrete element simulation of a drillship in discontinuous ice, in *Proceedings of the 22nd International Conference on Port and Ocean Engineering under Arctic Conditions* (Espoo, Finland, 2013)
ISO19906, Petroleum and natural gas industries - Arctic offshore structures,International Standard, International Standardization Organization (Geneva, Switzerland, 2019)
J.N. Naegle, *Ice-Resistance Prediction and Motion Simulation For Ships Operating in The Continuous Mode of Icebreaking* (The University of Michigan, 1980)
J.P. Dempsey, S.J. Defranco, D. Blanchet, A. Prodanovic, Splitting of ice floes, in *The 12th International Conference on Port and Ocean Engineering under Arctic Conditions (POAC 93), 17–20 August 1993* (Hamburg, 1993), pp. 17–22
J.P. Dempsey, R.M. Adamson, S.V. Mulmule, Scale effects on the in-situ tensile strength and fracture of ice. Part II: first-year sea ice at Resolute, NWT. Int. J. Fract. **95,** 347-366 (1999a)
J.P. Dempsey, The fracture toughness of ice. Ice-Struct. Interact. 109–145 (1991)
Kerr, A. D. , The bearing capacity of floating ice plates subjected to static or quasi-static loads. J. Glaciol. **17**, 229–268 (1976)
Kämäräinen, J. , *Theoretical Investigation on the Effect of Fluid Flow Between the Hull of a Ship and Ice Floes on Ice Resistance in Level Ice* (Helsinki University of Technology, Department of Mechanical Engineering, Laboratory for Mechanics of Materials, Helsinki, Finland, 2007)
L. Zhou, B. Su, K. Riska, T. Moan, Numerical simulation of moored structure station keeping in level ice, in *Cold Regions Science and Technology* (2011).

LU, W. , *Floe Ice - Sloping Structure Interactions* (Norwegian university of science and technology, Doctor of Philosophy Doctoral, 2014)

Leppäranta, M. , *The Drift of Sea Ice* (Springer, Helsinki, 2011)

W. Lu, H.-M. Heyn, R. Lubbad, S. Løset, A large scale simulation of floe-ice fractures and validation against full-scale scenario. Int. J. Naval Architect Ocean Eng. **10**, 393–402 (2018)

W. Lu, R. Lubbad, S. Løset, In-plane fracture of an ice floe: A theoretical study on the splitting failure mode. Cold Regions Sci. Technol. (CRST) **110**, 77–101 (2015)

W. Lu, R. Lubbad, S. Løset, Out-of-plane failure of an ice floe: Radial-crack-initiation-controlled fracture. CRST **119**, 183–203 (2015)

W. Lu, R. Lubbad, S. Løset, M. Kashafutdinov, Fracture of an ice floe: Local out-of-plane flexural failures versus global in-plane splitting failure. CRST **123**, 1–13 (2016)

R. Lubbad, S. Løset, A numerical model for real-time simulation of ship-ice interaction. Cold Reg. Sci. Technol. **65**, 111–127 (2011)

M. Van Den Berg, R. Lubbad, S. Iøset, Accuracy of a non-smooth time stepping scheme with non-rigid contacts for ice-structure interaction, in *Proceedings of the 24th International Conference on Port and Ocean Engineering under Arctic Conditions, June 11–16, 2017* (Busan, Korea, 2017)

M. Berg, R. Lubbad, S. Løset, An implicit scheme and an improved contact model for ice-structure interaction simulations, in *CRST* (2018), pp. 193–213

M.G. Coutinho, *Guide to Dynamic Simulations of Rigid Bodies and Particle Systems* (Springer, Berlin, 2013)

N. Raza, V.D.M. Berg, W. Lu, R. Lubbad, Analysis of oden icebreaker performance in level ice using simulator for arctic marine structures (SAMS), in *Proceedings - International Conference on Port and Ocean Engineering under Arctic Conditions* (Delft, The Netherlands, 2019)

Norwegianhydrographicservice, The Norwegian pilot sailing directions Svalbard and Jan Mayen (Stavanger, 2018)

P. Varsta, K. Riska, Failure process of ice edge caused by impact with ship's side, in *From the Symposium on Ice, Ships and Winter Navigation* (1977), pp. 16–17

A.C. Palmer, K.R. Croasdale, *Arctic Offshore Engineering* (World Scientific, 2013)

R. Lubbad, S. Løset, W. Lu, A. Tsarau, M. Van den Berg, An overview of the Oden Arctic Technology Research Cruise 2015 (OATRC2015) and numerical simulations performed with SAMS driven by data collected during the cruise, in *Cold Regions Science and Technology* (2018)

S. Løset, K. Shkhinek, O. Gudmestad, K. Høyland, *Actions from Ice on Arctic Offshore and Coastal Structures* (Krasnodar, St Petersburg, Russia, 2006)

B. Su, K. Riska, T. Moan, A numerical method for the prediction of ship performance in level ice. Cold Reg. Sci. Technol. **60**, 177–188 (2010)

T.V. Kotras, A.V. Baird, J.N. Naegle, *Predicting Ship Performance in Level Ice* (New York, SNAME, 1983), pp. 329–349

X. Tan, B. Su, K. Riska, T. Moan, A six-degrees-of-freedom numerical model for level ice–ship interaction. Cold Reg. Sci. Technol. **92**, 1–16 (2013)

A. Tsarau, R. Lubbad, S. Løset, A numerical model for simulating the effect of propeller flow in ice management. Cold Reg. Sci. Technol. **142**, 139–152 (2017)

J. Tuhkuri, A. Polojärvi, A review of discrete element simulation of ice–structurcxse interaction. Philos. Trans. R. Soc. a: Math., Phys. Eng. Sci. **376**, 20170335 (2018)

W. Lu, R. Lubbad, S. Løset, K.V. Høyland, Cohesive zone method based simulations of ice wedge bending: a comparative study of element erosion, CEM, DEM and XFEM, in *The 21st IAHR International Symposium on Ice, June 11–15, 2012*, ed. By Li & Lu (Dalian, China, 2012), pp. 920–938

W. Lu, S. Løset, A. Shestove, R. Lubbad, Design of a field test for measuring the fracture toughness of sea ice, in *The 23rd International Conference on Port and Ocean Engineering under Arctic Conditions, June 14–18, 2015*, ed. by E. Kim, W. Lu, K. Høyland (Trondheim, Norway, 2015a)

W. Lu, R. Lubbad, S. Løset, Tentative fracture mechanisms of the parallel channel effect during ice management, in *The 23rd International Conference on Port and Ocean Engineering under Arctic Conditions, 14–18 June 2015*, ed. by E. Kim, W. Lu, K. Høyland (Trondheim, Norway, 2015d)

W. Lu, R. Lubbad, S. Løset, R. Skjetne, Parallel channel tests during ice management operations in the arctic ocean, in *Arctic Technology Conference 2016* (St. John's, Newfoundland and Labrador, 2016b)

W. Lu, R. Lubbad, S. Løset, Parallel channels' fracturing mechanism during ice management operations. Part II: experiment, in *Cold Regions Science and Technology* (2018b)

W. Lu, R. Lubbad, A. Shestov, S. Løset, Parallel channels' fracturing mechanism during ice management operations. Part I: theory. *Cold Regions Science and Technology* (2018c)

W. Lu, A. Shestov, S. Løset, E. Salganik, K.V. Høyland, Medium-scale consolidation of artificial ice ridge – part ii: fracture properties investigation by a splitting test, in *Proceedings - International Conference on Port and Ocean Engineering under Arctic Conditions* (Delft, The Netherlands, 2019)

R. Yulmetov, R. Lubbad, S. Løset, Planar multi-body model of iceberg free drift and towing in broken ice. Cold Reg. Sci. Technol. **121**, 154–166 (2016)

Part III
Waves and Ice

Chapter 8
Modeling and Observations of Wave Energy Attenuation in Fields of Colliding Ice Floes

Agnieszka Herman, Sukun Cheng, and Hayley H. Shen

Abstract Interactions between sea ice and waves in the marginal ice zone (MIZ) have attracted much attention in the last years. However, many important aspects of those interactions remain poorly understood. This paper summarizes recent results by the authors related to the role of floe collisions in shaping sea ice dynamics and wave energy attenuation in the MIZ. The analysis is based on simulations with a discrete-element sea ice model coupled to a wave energy transport model, as well as on laboratory observations of wave attenuation in broken sea ice. We analyze collision patterns for different floe size/wave forcing combinations and their phase-averaged effects. Combined processes of floe collisions and ice–water drag are shown to lead to floe-size-dependent wave energy attenuation rates. Importantly, we demonstrate that several different combinations of model parameters satisfactorily reproduce the observed variability of attenuation rates, underlining the need for more observations covering multiple processes that would allow to constrain those parameters.

8.1 Introduction

One of the defining features of the marginal ice zone (MIZ) are mutual interactions between the sea ice cover and waves. Waves entering the ice from the open ocean create a highly dynamic environment in which ice floes undergo collisions, breaking, rafting, overwash, etc. In turn, the presence of ice influences wave propagation and leads to the attenuation of wave energy through non-dissipative scattering and several dissipative mechanisms, including viscous deformation of the ice, generation of tur-

A. Herman (✉)
Institute of Oceanography, University of Gdansk, Gdansk, Poland
e-mail: oceagah@ug.edu.pl

S. Cheng
Nansen Environmental and Remote Sensing Center, Bergen, Norway
e-mail: Sukun.Cheng@nersc.no

H. H. Shen
Department of Civil and Environmental Engineering, Clarkson University, Potsdam, NY, USA
e-mail: hhshen@clarkson.edu

J. Tuhkuri and A. Polojärvi (eds.), *IUTAM Symposium on Physics and Mechanics of Sea Ice*, IUTAM Bookseries 39, https://doi.org/10.1007/978-3-030-80439-8_8

bulence, friction between ice floes and at the ice–water interface, ice fragmentation, inelastic floe collisions, and many more.

Numerical studies with large-scale sea ice models show that the simulated sea ice volume and extent are sensitive to the wave attenuation rates in the MIZ (Bateson et al. 2019). Thus, reliable parametrizations of processes contributing to wave attenuation are important for the performance of sea ice and, more generally, climate models. However, the development of those parametrizations is hindered by, first, limited availability of observational data for model calibration and validation, and second, our incomplete understanding of the details of the relevant processes and their relative contribution to wave attenuation in different sea ice types and different wave conditions. The observed attenuation rates are highly variable (e.g., Stopa et al. 2018b) and difficult to interpret due to lack of accompanying information on ice properties and their spatial variability. Therefore, not only the magnitude of wave attenuation, but also its functional form is usually hard to determine. As discussed in detail by Squire (2018), the change of wave amplitude a with distance along the propagation direction x can be written as:

$$\mathrm{d}a/\mathrm{d}x = -\alpha a^n, \tag{8.1}$$

where the attenuation coefficient α depends on wave frequency: $\alpha = \alpha(\omega)$. In the majority of studies (see, e.g., Stopa et al. 2018a, b; Ardhuin et al. 2018, and references therein), $n = 1$ is adopted in Eq. (8.1), i.e., exponential attenuation is assumed, $a = a_0 \exp(-\alpha x)$, and no alternative forms of $a(x)$ are considered. The above-mentioned lack of sufficient observations might be used as an argument in favor of the simplest, linear model. However, it is known that several physical processes result in nonlinear attenuation, with $n \neq 1$, for which the solution of (8.1) is $a^{1-n} = a_0^{1-n} - (1-n)\alpha x$, with $a_0 = a(x = 0)$ (Shen and Squire 1998; Kohout et al. 2011). It is very likely that different values of n are suitable for different situations and/or different parts of the MIZ, depending on dominating dissipation mechanisms. Similarly, the form of $\alpha(\omega)$ is intrinsically related to individual physical processes responsible for attenuation. Although, again, the span of observed $\alpha(\omega)$ is large, most observations suggest a power law relationship within the high-frequency tail, $\alpha \sim \omega^m$, with m between 2 and 3, which is lower than predicted by many models (Meylan et al. 2018).

This paper reviews recent results by Herman (2018) and Herman et al. (2019a, b) related to numerical modeling and laboratory observations of the kinematics and dynamics of ice floes on waves and wave–ice interactions leading to the attenuation of wave energy in MIZ. Contrary to most recent studies on the subject, related to situations with very small ice floes and high effective viscosity of the ice (i.e., grease/pancake ice in freezing conditions; see, e.g., Doble et al. 2015; Rogers et al. 2016; Cheng et al. 2017; Ardhuin et al. 2018; De Santi et al. 2018; Voermans et al. 2019), the works described below concentrate on relatively large ice floes, with sizes comparable with wavelength. In the following, we describe the main features of a coupled sea ice–wave model, based on a discrete-element model (DEM) of sea ice and a wave energy transport model (Sect. 8.2). In Sect. 8.3 we present a summary of the results related to the floe collision patterns in different model configurations (a stand-alone DEM with prescribed wave forcing; coupled model runs for an idealized MIZ), followed by a theoretical analysis of wave attenuation due to ice–water drag in confined sea ice (Sect. 8.4). A case study of wave attenuation in colliding ice floes, based on laboratory observations and numerical modeling, is discussed in Sect. 8.5. Finally, Sect. 8.6 contains a brief summary and a discussion of effects associated with interactions between scattering and dissipation, not included in the hitherto analysis. More details related to the contents of Sects. 8.2–8.5 can be found in Herman (2018), Herman et al. (2019a, b).

8.2 Coupled Sea Ice–Wave Model

8.2.1 Basic Assumptions

In the analysis in this paper, linear, unidirectional, propagating waves are assumed, with surface elevation $\eta = a(x)\cos\theta$, where a denotes (space-varying) amplitude, $\theta = kx - \omega t$ denotes phase, x the propagation direction and t time. The angular frequency ω and the wavenumber k are related by the following dispersion relation (see, e.g., Fox and Squire 1990):

$$\omega^2 \left(1 + \beta_1 k \tanh[kh]\right) = (g + \beta_2 k^4) k \tanh[kh], \tag{8.2}$$

with $\beta_1 = h_i \rho_i / \rho_w$ and $\beta_2 = E h_i^3 / (12 \rho_w (1 - \nu^2))$, where ρ_w and ρ_i are water and ice density, respectively, g denotes acceleration due to gravity, h_i ice thickness, E elastic modulus, ν the Poisson's ratio, and h the mean water depth. Depending on the values of the inertial coefficient β_1 and the flexural rigidity β_2, the dispersion relation (8.2) corresponds to the elastic plate model ($\beta_1 \neq 0$ and $\beta_2 \neq 0$) or reduces to the mass loading model ($\beta_1 \neq 0$, $\beta_2 = 0$) or open water waves ($\beta_1 = \beta_2 = 0$). As discussed in more detail in Herman et al. (2019a), Eq. (8.2) is suitable for the type of sea ice considered in this analysis (as opposed to several other studies concentrating on grease/pancake ice, in which the viscous parameter of the ice influences dispersion), and, importantly, it is consistent with the assumptions underlying the DEM.

Apart from the surface elevation η, the horizontal component of water velocity u_w, and the dynamic pressure p are important for the further analysis. They are given by:

$$u_w(x, z, t) = a(x)\omega \frac{\cosh[k(z + h)]}{\sinh[kh]} \cos\theta, \tag{8.3}$$

$$p_w(x, z, t) = a(x)\rho_w g \frac{\cosh[k(z + h)]}{\cosh[kh]} \cos\theta. \tag{8.4}$$

It is assumed that the waves are propagating through a series of $i = 1, \ldots, N_f$ identical, rectangular ice floes, with length $L_x = 2r$ (along the wave propagation direction x), width L_y (along the perpendicular direction y) and mass m_i. Archimedean balance is assumed, i.e., the submerged part of the ice floes has thickness $h_i \rho_i / \rho_w$. The state of each floe is described by its instantaneous position along the x axis, x_i, and its horizontal (surge) velocity u_i.

8.2.2 The DEM Sea Ice Model

The DEM model is that of Herman (2016), with extensions described in Herman (2018). In the simulations discussed here, a very simple model configuration is used, in accordance with the assumptions described above and the experimental setup presented in Sect. 8.5. Thus, the DEM solves a one-dimensional linear-momentum equation for each ice floe, with four types of forces:

$$m_i \frac{\mathrm{d}u_i}{\mathrm{d}t} = F_{\mathrm{w},i} + F_{\mathrm{v},i} + F_{\mathrm{d},i} + F_{\mathrm{c},i}, \qquad i = 1, \ldots, N_f, \tag{8.5}$$

where $F_{\mathrm{w},i}$ denotes the wave-induced force (Froude–Krylov force), $F_{\mathrm{v},i}$—the virtual (or added) mass force, $F_{\mathrm{d},i}$—the drag force, and $F_{\mathrm{c},i}$—the sum of Herzian contact forces from all collision/contact partners of floe i. A detailed discussion on the formulation of these forces can be found in Herman (2018). Importantly, the forces are frequency-dependent. Both $F_{\mathrm{w},i}$ and $F_{\mathrm{d},i}$ are computed by integrating the local force over the submerged surface area of each floe, S_i, as opposed to earlier similar models, in which forcing was specified at the floe's center (e.g., Shen and Ackley 1991; Hopkins and Shen 2001; Meylan et al. 2015; Yiew et al. 2017). Thus:

$$F_{\mathrm{w},i} = -\mathbf{n}_x \int_{S_i} p_w \mathbf{n} ds \tag{8.6}$$

and

$$F_{\mathrm{d},i} = \int_{S_i} \tau_w ds, \tag{8.7}$$

where $\mathbf{n}_x$ is a unit vector in the x direction, $\mathbf{n}$—a unit vector perpendicular to the local floe's surface and directed outwards, p_w is given by (8.4), and τ_w denotes the module of the local ice–water stress, computed from the quadratic drag law, i.e., $\tau_w = \rho_w C_{\mathrm{sd}} u_{\mathrm{rel}}^2$, with a constant drag coefficient C_{sd} and the module of the local, instantaneous ice-water velocity difference $u_{\mathrm{rel}} = |u_i(t) - u_w(x, 0, t)|$. Thus, both $F_{\mathrm{w},i}$ and $F_{\mathrm{d},i}$ are functions of the relative floe size kr. Moreover, the contact model used to compute $F_{\mathrm{c},i}$ is suitable for both short, energetic collisions and a prolonged, semi-permanent contact between ice floes (as opposed to, for example, Herman (2011); Yiew et al. (2017), where infinitely short collision duration was assumed). As discussed in detail in Herman (2018), those two features make the model suitable for studying the surge motion of floes with sizes comparable with wavelength, at high ice concentration.

The DEM can be run as a stand-alone model, with prescribed wave forcing (as in Herman 2018), or coupled to the wave energy transport model (as in Herman et al. 2019a, b). In a coupled mode, $F_{w,i}$ and $F_{d,i}$ are computed at each DEM time step Δt from the instantaneous wave phase and from wave amplitudes that are averaged over a coupling time step $\Delta t_c \gg \Delta t$. For regular waves considered here, $\Delta t_c = n_T T$ is used, with n_T an integer and the wave period $T = 2\pi/\omega$. Crucially, all scattering effects are disregarded in the coupled mode, i.e., the presence of the ice does affect the wave amplitude and dispersion relation, but only the travelling component is considered in computation of the forces in (8.5). Possible consequences of this simplification are discussed in Sect. 8.6.2.

8.2.3 *Wave Attenuation Model*

The stationary, one-dimensional wave energy transport equation solved by the coupled wave–ice model has the form:

$$\frac{\mathrm{d}}{\mathrm{d}x}(c_g E_w) = S_{\mathrm{sd}} + S_{\mathrm{ow}}, \tag{8.8}$$

where the wave energy $E_w = \frac{1}{2}\rho_w g a^2$, $c_g = \mathrm{d}\omega/\mathrm{d}k$ denotes the group velocity, computed from the dispersion relation (8.2), and the two source terms on the right-hand-side of (8.8) describe energy dissipation due to skin drag at the ice–water interface (S_{sd}) and due to overwash (S_{ow}).

As described, e.g., in Shen and Squire (1998), S_{sd} can be obtained from:

$$S_{\mathrm{sd}} = -\frac{1}{\Delta t_c}\frac{1}{A_{\mathrm{bot}}}\int_{t_0}^{t_0+\Delta t_c}\int_{A_{\mathrm{bot}}} \tau_w u_{\mathrm{rel}} \mathrm{d}s\mathrm{d}t, \tag{8.9}$$

where $A_{\mathrm{bot}} = L_x L_y$, and τ_w is computed in the same way as in (8.7). For the S_{ow} source term, a simple parametrization based on Skene et al. (2015, 2018) is developed, in which overwash is treated as a shallow water wave propagating over the upper ice surface (see Herman et al. 2019a for details).

8.3 Wave-Induced Collisions of Finite-Size Floes

Based on a scaling analysis of the model equations and results of simulations with a stand-alone DEM, Herman (2018) identified the following set of parameters influencing the patterns of wave-induced motion and collisions of ice floes: ice concentration c, relative floe size kr, wave steepness ka, ice restitution coefficient ε, nad drag

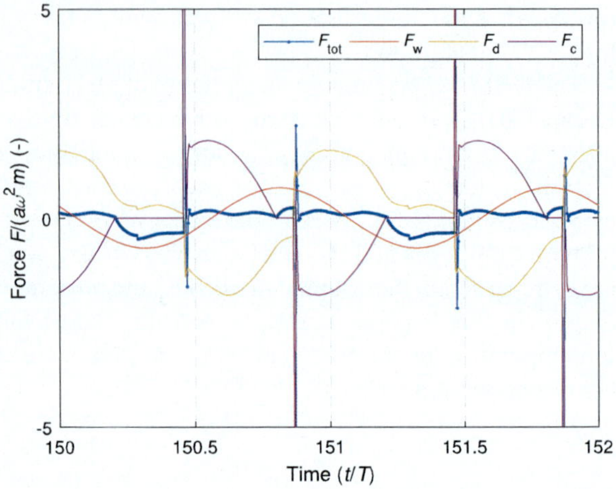

Fig. 8.1 Time series of scaled forces (F_{tot}—total force, F_w—wave-induced force, F_{sd}—drag force, F_c—contact force) acting on a selected floe from a simulation with $c = 0.99$, $kr = 0.6\pi$, $ka = 0.10$, $\varepsilon = 0.1$, $C_{sd} = 0.5$. All forces are scaled with $a\omega^2 m_i$, time is given in wave periods

coefficient C_{sd}. For a single ice floe, its size relative to wavelength, kr, determines the amplitude of its horizontal motion, which, for a square floe, is proportional to $\sin(kr)/(kr)$, i.e., it decreases with increasing floe size. An important consequence is that large floes cannot follow the motion of the surrounding water and experience significant ice–water drag, even if the value of C_{sd} is not very high. For multiple floes at an ice concentration sufficiently high for collisions to occur, large floes tend to undergo collisions with prolonged contact, occupying a substantial fraction of the wave period—which additionally increases the relative ice–water velocity and contributes to increased drag force. Typical time series of forces acting on a large ice floe colliding with its neighbors are shown in Fig. 8.1 (see Herman 2018 for more examples and for analogous time series of floe velocities). Characteristically, after a high peak of the contact force corresponding to the initial contact, F_c drops, but stays relatively high as the floes move together, balanced by the opposing drag force F_{sd}. In the example in Fig. 8.1, collisions are very regular, with each floe colliding with one of its two neighbors during every wave period. In general, for a given set of c, ka and kr, collision patterns are governed by the combination of ε and C_{sd}, as illustrated schematically in Fig. 8.2. Regular collision patterns develop only at sufficiently low ε and high C_{sd}; at the other extreme, high ε and low C_{sd}, floes move irregularly, as they undergo high accelerations after each collision and the coupling with the underlying water is too weak to quickly reduce the excess velocity back to the pre-collision level. Similarly, prolonged floe–floe contact requires sufficiently low restitution coefficients and does not occur at very high values of ε (unlikely to occur in real sea ice).

As shown in Herman (2018), the results summarized above not only provide interesting details about kinematics and dynamics of floes moving on waves, but suggest an important mechanism of wave energy attenuation in fragmented ice, related to combined effects of collisions and ice–water drag. Indeed, simulations with a coupled model (as well as observations presented further in Sect. 8.5) seem to confirm that this

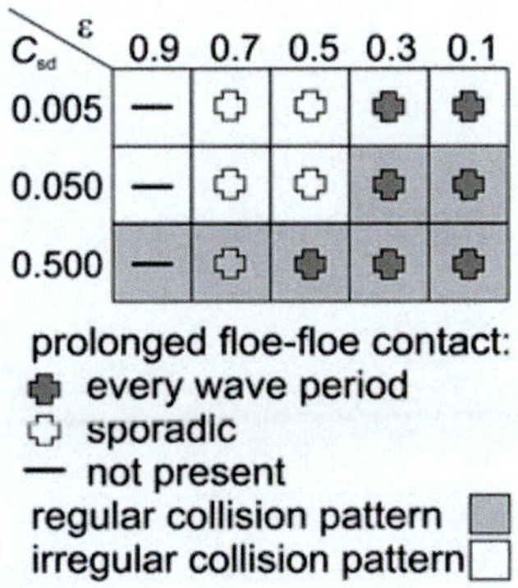

Fig. 8.2 Diagram showing motion/collision patterns in simulations with ice concentration $c = 0.99$, relative floe size $kr = 0.6\pi$, wave steepness $ka = 0.10$, for different combinations of the restitution coefficient ε (columns) and drag coefficient C_{sd} (rows). Cases with regular, repeatable collision patterns are marked with gray color; cross symbols mark presence of collisions with prolonged contact between floes. (Reproduced with permission from Herman 2018)

mechanism might be very effective. Although, obviously, the details of any particular situation depend on ice and wave properties, the general model behavior exhibits a consistent, repeatable pattern (Fig. 8.3). The $a(x)$ curve reflects the existence of two zones with very different "regimes" of ice–wave interactions. In the (usually very narrow) zone at the ice edge the wave amplitude is high and, accordingly, the floes undergo energetic collisions, leading to high dissipation rates by ice–water drag, which in turn results in high $\mathrm{d}a/\mathrm{d}x$ and strong wave drag acting on the ice, additionally contributing to overwash and, possibly, rafting of the ice floes. In this outer zone, due to high granular pressure, the ice concentration is slightly lowered, contrary to the inner zone, in which floes are densely packed, stay in (semi)permanent contact and undergo only limited fluctuations in the horizontal direction. In this zone the wave attenuation is significantly lower and can be estimated theoretically, as described further in Sect. 8.4. Hence, floe collisions are likely to occur—and thus to contribute to wave attenuation—only within a limited region at the ice edge. On the other hand, however, the presence of that region is very important for the dynamics of the MIZ as a whole, because it significantly reduces the incoming wave energy, i.e., acts as a "buffer" zone for the remaining ice cover. It is remarkable that, qualitatively, the zonal division of the MIZ described above corresponds very well to field observations of Squire and Moore (1980), who analyzed propagation and attenuation of wind waves in a marginal ice zone of the Bering Sea. In spite of several important differences in the sea ice and forcing conditions (variable floe size, irregular waves in the field *versus* identical floes and monochromatic waves in the model), the clear division of the ice cover into a lower-concentration, very dynamic outer zone and a calmer inner zone is present in both cases. Full-scale DEM simulations with irregular wave forcing would be necessary for a more direct, quantitative comparison with the data from Squire and Moore (1980) and later, similar field campaigns.

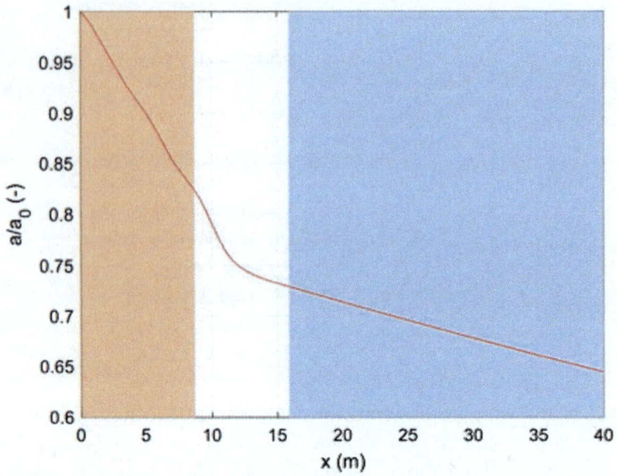

Fig. 8.3 A typical wave amplitude profile $a(x)$ in simulations with a coupled sea ice–wave model, with a zone of strong wave attenuation and energetic floe collisions at the ice edge (brown) and an inner zone of weaker attenuation and only sporadic collisions (blue). See text and Herman et al. (2019a) for more details

A very important aspect of the results is the sensitivity of the simulated wave attenuation rates to floe size. As shown in Fig. 8.4, the dependence of $a(x)$ on floe size is related to two combined effects. First, collisions lead to higher relative ice–water velocities when the floes are small, and the number of collisions per distance is larger. And second, as follows from Eq. (8.8), $\mathrm{d}E/\mathrm{d}x \sim c_g^{-1}$, i.e., the wave energy attenuation is very sensitive to changes in wave group velocity c_g (independently of the dissipation mechanism(s), i.e., the number and formulation of the source terms in the energy transport equation). The observations described in Sect. 8.5 indicate that, in the ice cover type analyzed here, the wavenumber k increases with decreasing floe size (Cheng et al. 2018; Herman et al. 2019b), so that the mass loading model seems more appropriate for the smallest floes considered ($L_x = 0.5$ m), and the elastic plate model more appropriate for the largest floes ($L_x = 3.0$ m and 6.0 m). As can be seen in Fig. 8.4, the two effects together produce a drastic change in attenuation (compare the dashed yellow and the continuous blue line). In summary, the wave dissipation mechanism analyzed here is strongly floe-size dependent. Anticipating discussion in Sect. 8.6.2, scattering effects, not included in the results presented so far, might further enhance that dependence.

8.4 Theoretical Analysis of Attenuation in Compact Sea Ice

A limiting case of the MIZ setup discussed in the previous section is a situation of confined ice, $u_i \equiv 0$. As discussed above, this condition is fulfilled in the inner MIZ (blue region in Fig. 8.3) or in any other situation where the horizontal mobility of the ice is limited. In this special case, and disregarding overwash effects, the wave energy transport equation can be solved analytically, producing:

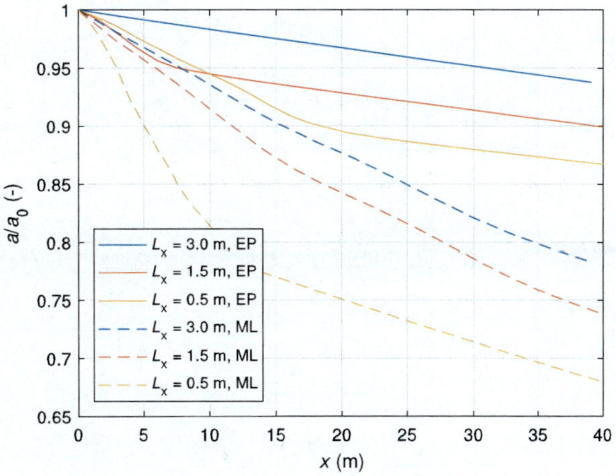

Fig. 8.4 Simulated wave amplitude profiles $a(x)$ for three different floe lengths L_x and two different dispersion relations: elastic plate (EP) and mass loading (ML) model. (Reproduced from Herman et al. 2019a)

$$\frac{\mathrm{d}a}{\mathrm{d}x} = -\alpha_c a^2 \quad \text{with} \quad \alpha_c = \frac{4C_{\mathrm{sd}}}{3\pi g} \frac{\omega^3}{c_g \tanh^3[kh]}. \tag{8.10}$$

Thus, attenuation has the form (8.1) discussed by Squire (2018), with $n = 2$. The solution of (8.10) is:

$$a(x)/a_0 = (a_0\alpha_c x + 1)^{-1}. \tag{8.11}$$

As mentioned earlier, α_c is proportional to c_g^{-1}, and the overall dependence of α_c on wave frequency is given by $\omega^3/(c_g \tanh^3[kh])$. For the open water dispersion relation, often adopted in studies of wave attenuation in MIZ (see, e.g., Meylan et al. 2018), this gives $\alpha_c \sim \omega^4$ (in deep water). With the mass loading dispersion relation the increase of α_c with frequency becomes significantly faster (red and violet curves in Fig. 8.5), with the elastic plate dispersion relation—it becomes slower (blue and yellow curves in Fig. 8.5). Notably, in the last case the power law relationship within the high-frequency tail is $\alpha_c \sim \omega^m$ with m dependent on ice properties, but in most cases in the 2.0–2.5 range, i.e., in agreement with many field observations (Meylan et al. 2018).

Importantly, the attenuation rates in the inner regions of the MIZ, simulated with the coupled wave–ice model (Figs. 8.3, 8.4), are very close to those derived theoretically, underlying the very small role of collisions there.

8.5 Laboratory Case Study of Wave Attenuation in Fragmented Ice

As part of the Hydralab+ Transnational Access project "Loads on Structure and Waves in Ice" (LS-WICE), performed in the Large Ice Model Basin (LIMB) of

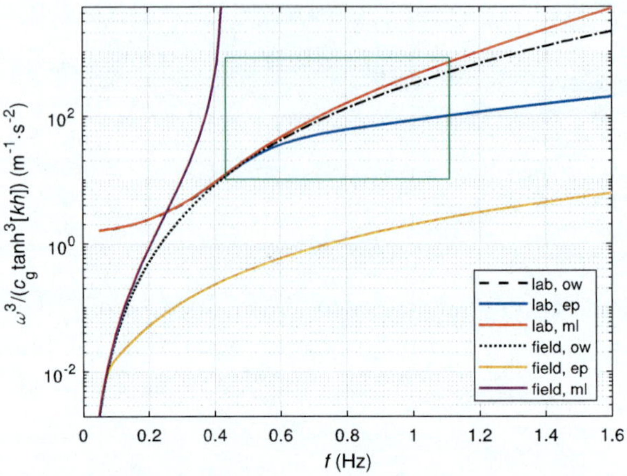

Fig. 8.5 The ratio $\omega^3/(c_g \tanh^3[kh])$ versus wave frequency $f = \omega/(2\pi)$, computed for thin laboratory ice (LS-WICE experiment; see Sect. 8.5) and for field conditions with $h = 1000$ m, $\rho_w = 1025$ kg/m^3, $\rho_i = 910$ kg/m^3, $E = 6 \cdot 10^9$ Pa, $h_i = 1.5$ m. The corresponding open water solutions are shown in black (note that the two black curves overlap for $f > 0.5$ Hz). Green rectangle marks region covered by the LS-WICE experiment. (Reproduced from Herman et al. 2019a)

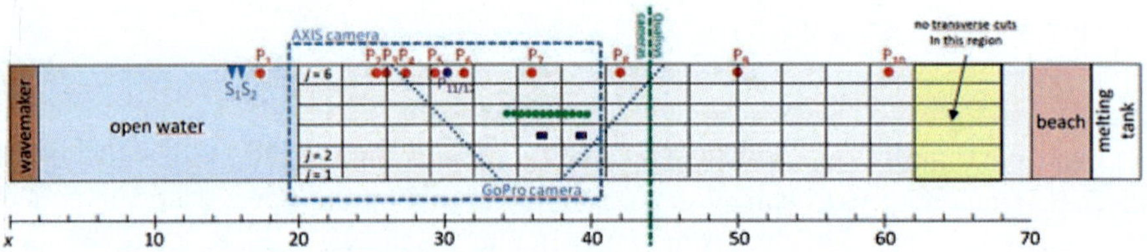

Fig. 8.6 Setup of the LS-WICE experiment. The tank is 72 m long, 10 m wide, and 2.5 m deep. The ice edge is located at $x_0 = 20$ m, the gray area represents the ice sheet (at the stage when it was cut into floes with $L_x = 3$ m). Dots numbered $P_1, \ldots, P_{12}$ show positions of pressure sensors, S_1 and S_2—positions of ultrasound sensors. The dashed and dotted blue lines mark the fields of view of video cameras. Green dots mark the Qualisys markers, violet rectangles—inertial measurement units. (Reproduced from Herman et al. 2019b)

the Hamburg Ship Model Basin (Hamburgische Schiffbau-Versuchsanstalt, HSVA), a set of experiments was conducted related to wave propagation through broken ice fields. One of the goals was to investigate how the wave energy attenuation depends on wavelength and floe size. The setup of the experiment is shown in Fig. 8.6. It has been described in detail in Cheng et al. (2018) and Herman et al. (2019b); an analysis of floe kinematics based on the same data can be found in Li and Lubbad (2018). Here, only the most essential information is provided.

The experiment consisted of several groups of tests, with floe length L_x equal to 0.5, 1.5, 3.0 and 6.0 m. The wavemaker generated regular, low-amplitude waves with periods ranging from 0.9 to 2.3 s (see Herman et al. 2019b for a full list of tests).

The vertical deflection of the ice along the tank was measured with pressure sensors installed along one of the tank side walls, and (in some tests) with the Qualisys motion capture system, with markers placed along the central axis of the ice sheet (Fig. 8.6). Additionally, video material was recorded to qualitatively document floe motion, collisions and the occurrence of overwash.

Figure 8.7 shows two examples of the observed wave attenuation, together with the results of a least-square fit of equation (8.1) with $n = 1$ and $n = 2$, and with the results of the scattering model of Kohout et al. (2007), Kohout (2008), based on the Matched Eigenfunction Expansion Method (MEEM). As expected, scattering processes do not explain the observed wave attenuation; however, the MEEM model is very helpful in understanding the influence of scattering on the spatial variability of the amplitudes of vertical deflection of the ice and the related scatter of the observational data, strongly dependent on the location of sensors relative to the edges of the ice floes. That scatter, together with the lack of sensors in the zone close to the ice edge ($x < 25$ m in Fig. 8.7), contribute to difficulties in interpretation of the observed variability of attenuation rates. Although, undoubtedly, attenuation is strongly frequency dependent, the least-square fit of equation (8.1) with $n = 1$ and $n = 2$ produces similar results, which in both cases strongly depend on assumptions regarding the incident wave amplitude (continuous *vs* dashed lines in Fig. 8.7). The fact that distinguishing between exponential and non-exponential attenuation is so difficult under controlled conditions in a laboratory illustrates how large is the uncertainty of analogous analyses in field conditions, when the number of data points usually is much smaller, and the amount of (spatially variable) factors influencing wave propagation is much larger.

Considering the above-mentioned limitations of observational data, it is not surprising that numerical modelling does not provide clear, unequivocal answers to the mechanisms responsible for wave energy attenuation in any particular situation. In the case study analyzed here, the coupled sea ice–wave model was run for all combinations of floe sizes L_x and wave periods T, with several combinations of adjustable parameters not known from measurements: the restitution coefficient ε, the drag coefficient C_{sd}, and two parameters used in the overwash parametrization. Based on an assumption that those parameters did not significantly change between different tests (i.e., that the ice properties remained approximately constant throughout the experiment), the goal was to find a single combination of parameters that best reproduces the observational data for all L_x and T. As discussed in detail in Herman et al. (2019b), more than one combination of coefficients produced satisfactory results, with very similar model performance measures, making the selection of "the best" model setup impossible.

8.6 Conclusions and Discussion

In this section, we summarize the most important aspects of the results described above, important for the analysis of wave attenuation data, and we sketch some

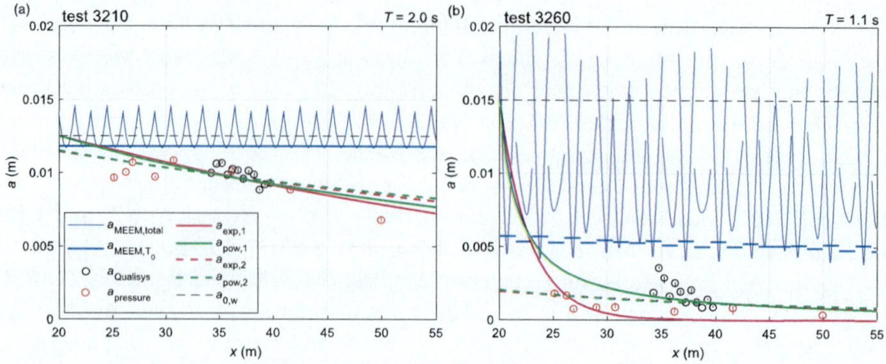

Fig. 8.7 Amplitude of the vertical ice deflection along the tank in two selected tests from the LS-WICE experiment (see Herman et al. 2019b for all tests). The measured amplitude is shown with red (pressure sensors) and black (Qualisys) circles (note that, in most cases, the error bars, marked with vertical lines, are shorter than the size of the symbols). Thin blue lines show the total amplitude from the MEEM model; the corresponding amplitudes of the transmitted propagating mode (T_0) are shown with thick light blue lines. The black dashed line shows the wavemaker amplitude $a_{0,w}$. Green and magenta lines are least-square fits of the data with Eq. (8.1) for $n = 1$ and $n = 2$, respectively (continuous lines: prescribed a_0, dashed lines: fitted a_0). See text for details. (Reproduced from Herman et al. 2019b)

perspectives for the follow-up research, related to combined effects of dissipation and scattering, treated separately in the present study.

8.6.1 Analysis and Interpretation of Wave Attenuation Data

The results of our studies (Herman 2018; Herman et al. 2019a, b) illustrate several aspects of wave–ice interactions in the MIZ that are very important in analyzing wave attenuation and in identifying particular processes that dissipate the wave energy. As already mentioned, observations are usually limited to a few locations at which information on wave height is available. For example, in the majority of cases analyzed in an extensive study by Stopa and Colleagues (2018b), the attenuation coefficient α was estimated from just two data points, one within the ice cover and one in the open ocean. Obviously, any attenuation curve can be adopted with that data. With an assumption of an exponential attenuation, made by the authors, the resulting values of α represent an "average" linear attenuation over the distance separating the two points. As our results suggest (Figs. 8.3, 8.4, 8.7), the actual attenuation might be spatially nonuniform, with very strong dissipation (and thus high $\mathrm{d}a/\mathrm{d}x$) close to the ice edge and weaker dissipation further inside the MIZ. Obviously, many data points are necessary to capture that variability and to determine the functional form of attenuation in different sections of the MIZ.

Moreover, as shown with our LS-WICE case study, measuring of the wave amplitude alone is not sufficient to attribute the observed attenuation to individual dissipative processes, as different combinations of those processes might produce similar attenuation patterns. Although analyzing attenuation over a wide range of wave frequencies (as opposed to considering the significant wave height, and thus the total wave energy alone) enables narrowing the list of candidate processes, because they are associated with a certain $\alpha(\omega)$ behavior, this might be extremely difficult in situations with several dissipation mechanisms acting simultaneously. Thus, it is very important to constrain the values of coefficients in parametrizations of those processes used in numerical models, including those describing the mechanical properties of the ice, ice–water drag, under-ice turbulence, etc.

8.6.2 Perspective: The Role of Scattered Modes in Modifying Ice–Water Drag and Wave-Induced Force

In the analysis in Herman (2018), Herman et al. (2019a, b), it was assumed that the wave motion under the ice was limited to the propagating component, as described in Sect. 8.2.1. The MEEM results of Herman et al. (2019b) showed that, although the contribution of scattering to the overall wave attenuation in LS-WICE was small, the damped propagating (both reflected and transmitted) components reached substantial amplitudes (blue curves in Fig. 8.7), thus modifying the instantaneous pressure and water velocity around the ice floes. Obviously, this affects the ice–water drag τ_w, and thus F_{d} and S_{sd}, as well as the wave-induced force F_{w}, and thus might influence the motion and collisions of ice floes and the resulting wave attenuation.

Combining the scattering and dissipation in one model without violating the formal assumptions underlying both parts is far from trivial. The existing scattering models, including the MEEM, are valid for non-dissipative, potential waves. The wave energy dissipation, on the other hand, is modelled for travelling waves, disregarding the existence of several additional modes with different wavelengths and propagation speeds. Accordingly, in continuum sea ice models, based on a full version of the wave energy transport equation (8.8), E and c_g represent the energy and group velocity of the propagating waves, and the two sea ice-related source terms, describing energy dissipation and its redistribution in spectral space due to scattering, respectively, are treated separately (see, e.g., Shen 2019 for a review).

With the type of a wave—ice model discussed here, a possible way to proceed that seems reasonable in situations with relatively weak dissipation, is to assume that the relative contribution of the different modes to the overall wave amplitude is unaffected by dissipation. That is, if the surface elevation under the ith floe disregarding dissipation is given as a sum of several transmitted and reflected modes, $\eta_i(x,t) = \sum_m (\eta_{i,m,\mathrm{T}}(x,t) + \eta_{i,m,\mathrm{R}}(x,t))$, then the analogous surface elevation with dissipation is $\eta_i(x,t) = \gamma_i \sum_m (\eta_{i,m,\mathrm{T}}(x,t) + \eta_{i,m,\mathrm{R}}(x,t))$, with the coefficient γ_i dependent on energy dissipation, but common for all modes. Importantly, this require-

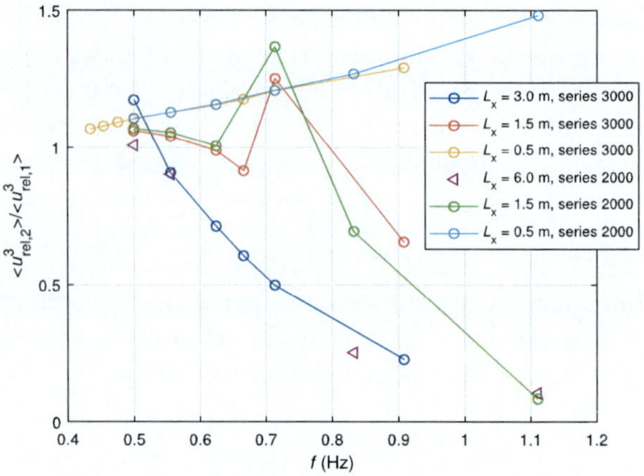

Fig. 8.8 The ratio $\langle u^3_{\mathrm{rel},2}\rangle / \langle u^3_{\mathrm{rel},1}\rangle$ in function of wave frequency f (Hz) for different floe sizes used in the LS-WICE experiment

ment ensures fulfillment of all necessary boundary conditions at floes' edges (continuity of velocity potential and its derivatives, etc.). Without formulating the details of the modified model—which is a subject for subsequent research—it is instructive to analyze whether the above assumption significantly modifies the terms in the model equations. In particular, as S_{sd} depends on the floe- and phase-averaged product $\langle \tau_w u_{\mathrm{rel}}\rangle \sim \langle u^3_{\mathrm{rel}}\rangle$, it is useful to compare that average computed for a traveling wave, $\langle u^3_{\mathrm{rel},1}\rangle$, and obtained from the full MEEM solution, $\langle u^3_{\mathrm{rel},2}\rangle$. The ratio of the two, $\langle u^3_{\mathrm{rel},2}\rangle / \langle u^3_{\mathrm{rel},1}\rangle$, is shown in Fig. 8.8 for the data from the LS-WICE experiment. Remarkably, the scattered modes increase dissipation due to the ice—water drag in small ice floes, but decrease dissipation in large ice floes, especially at high wave frequencies, suggesting one more mechanism leading to floe-size-dependent wave attenuation. Further research will show how other terms are affected, and how all those effects combine in coupled wave—ice simulations.

Acknowledgements The work of A.H. has been supported by the Polish National Science Centre research grant No. 2015/19/B/ST10/01568 ("Discrete-element sea ice modeling—development of theoretical and numerical methods"). Coauthors SC and HHS are supported in part by ONR grant No. N00014-17-1-2862.

References

F. Ardhuin, G. Boutin, J. Stopa, F. Girard-Ardhuin, C. Melsheimer, J. Thomson, A. Kohout, M. Doble, P. Wadhams, Wave attenuation through an arctic marginal ice zone on 12 October 2015: 2. Numerical modeling of waves and associated ice breakup. J. Geophys. Res. **123**, 5652–5668 (2018). https://doi.org/10.1002/2018JC013784

A. Bateson, D. Feltham, D. Schröder, L. Hosekova, J. Ridley, Y. Aksenov, Impact of floe size distribution on seasonal fragmentation and melt of Arctic sea ice. The Cryosphere Discuss (2019). https://doi.org/10.5194/tc-2019-44

S. Cheng, W. Rogers, J. Thomson, M. Smith, M. Doble, P. Wadhams, A. Kohout, B. Lund, O. Persson, C. Collins III., S. Ackley, F. Montiel, H. Shen, Calibrating a viscoelastic sea ice model for wave propagation in the Arctic fall marginal ice zone. J. Geophys. Res. **122**, 8740–8793 (2017). https://doi.org/10.1002/2017JC013275

S. Cheng, A. Tsarau, K.U. Evers, H. Shen, Floe size effect on gravity wave propagation through ice covers. J. Geophys. Res. **124**, 320–334 (2018). https://doi.org/10.1029/2018JC014094

F. De Santi, G. De Carolis, P. Olla, M. Doble, S. Cheng, H. Shen, P. Wadhams, J. Thomson, On the ocean wave attenuation rate in grease-pancake ice, a comparison of viscous layer propagation models with field data. J. Geophys. Res. **123**, 5933–5948 (2018). https://doi.org/10.1029/2018JC013865

M. Doble, G. De Carolis, M. Meylan, J.R. Bidlot, P. Wadhams, Relating wave attenuation to pancake ice thickness, using field measurements and model results. Geophys. Res. Lett. **42**, 4473–4481 (2015). https://doi.org/10.1002/2015GL063628

C. Fox, V. Squire, Reflection and transmission characteristics at the edge of shore fast sea ice. J. Geophys. Res. **95**, 11629–11639 (1990). https://doi.org/10.1029/JC095iC07p11629

A. Herman, Molecular-dynamics simulation of clustering processes in sea-ice floes. Phys. Rev. E **84**, 056104 (2011). https://doi.org/10.1103/PhysRevE.84.056104

A. Herman, Discrete-element bonded-particle Sea Ice model DESIgn, version 1.3a – model description and implementation. Geosci. Model Dev. **9**, 1219–1241 (2016). https://doi.org/10.5194/gmd-9-1219-2016

A. Herman, Wave-induced surge motion and collisions of sea ice floes: finite-floe-fize effects. J. Geophys. Res. **123**, 7472–7494 (2018). https://doi.org/10.1029/2018JC014500

A. Herman, S. Cheng, H. Shen, Wave energy attenuation in fields of colliding ice floes. Part A: discrete-element modelling of dissipation due to ice–water drag. Cryosphere **13**, 2887–2900 (2019a). https://doi.org/10.5194/tc-13-2887-2019

A. Herman, S. Cheng, H. Shen, Wave energy attenuation in fields of colliding ice floes. Part B: a laboratory case study. Cryosphere **13**, 2901–2914 (2019b). https://doi.org/10.5194/tc-13-2901-2019

M. Hopkins, H. Shen, Simulation of pancake-ice dynamics in a wave field. Ann. Glaciol. **33**, 355–360 (2001)

A. Kohout, Water wave scattering by floating elastic plates with application to sea-ice. Ph.D. thesis, University of Auckland, New Zealand (2008). 188 pp

A. Kohout, M. Meylan, D. Plew, Wave attenuation in a marginal ice zone due to the bottom roughness of ice floes. Ann. Glaciol. **52**, 118–122 (2011)

A. Kohout, M. Meylan, S. Sakai, K. Hanai, P. Leman, D. Brossard, Linear water wave propagation through multiple floating elastic plates of variable properties. J. Fluids Struct. **23**, 649–663 (2007). https://doi.org/10.1016/j.jfluidstructs.2006.10.012

H. Li, R. Lubbad, Laboratory study of ice floes collisions under wave action, in *Proceedings of the 28th International Ocean and Polar Engineering Conference. ISOPE-2018* (Sapporo, Japan, 2018)

M. Meylan, L. Bennetts, J. Mosig, W. Rogers, M. Doble, M. Peter, Dispersion relations, power laws, and energy loss for waves in the marginal ice zone. J. Geophys. Res. **123**, 3322–3335 (2018). https://doi.org/10.1002/2018JC013776

M. Meylan, L. Yiew, L. Bennetts, B. French, G. Thomas, Surge motion of an ice floe in waves: comparison of a theoretical and an experimental model. Ann. Glaciol. **56**, 155–159 (2015). https://doi.org/10.3189/2015AoG69A646

W. Rogers, J. Thomson, H. Shen, M. Doble, P. Wadhams, S. Cheng, Dissipation of wind waves by pancake and frazil ice in the autumn Beaufort Sea. J. Geophys. Res. **121**, 7991–8007 (2016). https://doi.org/10.1002/2016JC012251

H. Shen, Modelling ocean waves in ice-covered seas. Appl. Ocean Res. **83**, 30–36 (2019). https://doi.org/10.1016/j.apor.2018.12.009

H. Shen, S. Ackley, A one-dimensional model for wave-induced ice-floe collisions. Ann. Glaciol. **15**, 87–95 (1991)

H. Shen, V. Squire, Wave damping in compact pancake ice fields due to interactions between pancakes. Antarctic Sea Ice: Phys. Process., Interact. Vari. **74**, 325–341 (1998)

D. Skene, L. Bennetts, M. Meylan, A. Toffoli, Modelling water wave overwash of a thin floating plate.015 J. Fluid Mech. **777**, R3 (2015). https://doi.org/10.1017/jfm.2015.378

D. Skene, L. Bennetts, M. Wright, M. Meylan, Water wave overwash of a step. J. Fluid Mech. **839**, 293–312 (2018). https://doi.org/10.1017/jfm.2017.857

V. Squire, A fresh look at how ocean waves and sea ice interact. Phil. Trans. R. Soc. A **376**, 20170342 (2018). https://doi.org/10.1098/rsta.2017.0342

V. Squire, S. Moore, Direct measurement of the attenuation of ocean waves by pack ice. Nature **283**, 366–368 (1980). https://doi.org/10.1038/283365a0

J. Stopa, F. Ardhuin, J. Thomson, M. Smith, A. Kohout, M. Doble, P. Wadhams, Wave attenuation through an Arctic marginal ice zone on 12 October 2015. 1. Measurement of wave spectra and ice features from Sentinel 1A. J. Geophys. Res. **123** (2018a). https://doi.org/10.1029/2018JC013791

J. Stopa, P. Sutherland, F. Ardhuin, Strong and highly variable push of ocean waves on Southern Ocean sea ice. Proc. Nat. Acad. Sci. **115**, 5861–5865 (2018b). https://doi.org/10.1073/pnas.1802011115

J. Voermans, A. Babanin, J. Thomson, M. Smith, H. Shen, Wave attenuation by sea ice turbulence. Geophys. Res. Lett. **X**, xx–xx (2019). https://doi.org/10.1029/2019GL082945

L. Yiew, L. Bennetts, M. Meylan, G. Thomas, B. French, Wave-induced collisions of thin floating disks. Phys. Fluids **29**, 127102 (2017). https://doi.org/10.1063/1.5003310

Chapter 9
Wave-Ice Interaction Models and Experimental Observations

Hayley H. Shen

Abstract Arctic sea ice reduction has increased the fetch for wave generation as well as facilitated shipping opportunities. The region facing the most significant impact is the marginal ice zone, where the ice covers are more dynamic and the wave-ice interactions are stronger than in the central Arctic. The need for reliable forecasts of the wave and ice conditions in this partially ice-covered region has motivated a rapid increase of wave-in-ice studies. Mathematical models for wave-in-ice have been developed for over a century. Earlier models were based on simple physical concepts. The complexity of these models increased over time. Field data have been difficult to obtain because of the scale of the phenomenon and the harsh environment. Laboratory experiments have shown promise for obtaining meaningful data to check the theories. However, scaling up to field conditions is a challenge. With the improvements of instrumentation coupled with remote sensing capabilities, large amount of field data have become available. Laboratory studies have also increased in recent years. This paper provides a brief overview of the theories, laboratory and field experiments. Knowledge gaps and the outlook for future development are discussed.

9.1 Introduction

Arctic sea ice reduction provides increased fetch for wave generation. As a result, the wave intensity has become stronger even without notable change in the atmospheric forcing. At the same time, sea ice reduction has also facilitated shipping opportunities across the Arctic. The majority of the shipping routes are through the marginal ice zone (MIZ). Because of the lighter ice conditions than the central Arctic, the ice covers in this zone are more dynamic, and the wave-ice interactions are stronger than in the central Arctic. To meet the navigation needs, reliable forecasts of the wave and ice conditions in this partially ice-covered region are required. Thomson et al., (2018) summarized the climatology of the western Arctic from 1992–2017, focusing

H. H. Shen (✉)
Department of Civil and Environmental Engineering, Clarkson University, Potsdam, NY 13699-5710, USA
e-mail: hhshen@clarkson.edu

J. Tuhkuri and A. Polojärvi (eds.), *IUTAM Symposium on Physics and Mechanics of Sea Ice*, IUTAM Bookseries 39, https://doi.org/10.1007/978-3-030-80439-8_9

on the wind and wave characteristics. They found that wind, the source of waves, has not changed over the two and a half decades, but the significant wave height has increased by over 10% and the peak wave period by over 20%. The reduction of the ice cover and the associated increase of fetch have caused this change. To forecast wave conditions in ice-covered seas is challenging because of the lack of understanding of wave-ice interactions, which depend on the highly variable ice types.

The first wave-ice interaction workshop that the author participated was held in 1991, at the Scott Polar Research Institute, co-organized by Peter Wadhams and Vernon Squire (Wadhams et al., 1992). There were only a handful of participants, yet they were from five nations representing three continents. From the start, this research field has been international and consisted of people from diverse disciplines. At that time, the study of wave-ice interaction was mainly curiosity-driven. Things have changed due to the warming of the Arctic. In the closing report of the 1991 wave-ice interaction workshop, Wadhams et al., (1992) summarized as follows:

> *(In pack ice, the questions include) How waves propagate in an ice field composed of broken floes, how waves are scattered and attenuated in such an ice field, how waves cause the floe size distribution itself to become modified by flexural break up, how waves contribute to ice margin dynamics, and how attenuation is balanced with wave generation when ice field is diffuse...(In continuous sea ice) Outstanding problems include the physics of the generation mechanisms, the dependence of the dispersion upon ice properties, and the nature of attenuation mechanism.*

The questions raised then have remained central to the wave-in-ice research since. In this paper, the development of theoretical and experimental studies addressing the above questions is examined. In the decades following that workshop, this field has expanded a great deal in the number of researchers and the resulting literature. The author will focus on the MIZ and studies that emphasize wave dispersion and attenuation. Floe fracture and wave scattering will only be briefly mentioned, where extensive literature may be found in the references listed in Squire (2007, 2018, 2020) and therein.

9.2 The Marginal Ice Zone and the Wave Model

The MIZ is a dynamic zone where both mechanical and thermodynamic processes continuously change the ice concentration, thickness, and its type. The ability to forecast wave conditions in such a zone depends on accurate information on the ice cover. This information may be acquired through reliable ice models or direct observations, both are under active research. The spatial and temporal variability of ice covers in the MIZ can be seen in satellite images. Figure 9.1 shows an example from Landsat 8 with a 15 m resolution. Landsat 8 uses visible and infrared sensors with a 16-day revisit period. Its ability to monitor the rapidly changing MIZ is limited by the cloud cover and the low temporal resolution. These remotely sensed images provide important information on the ice cover over a large area. Details

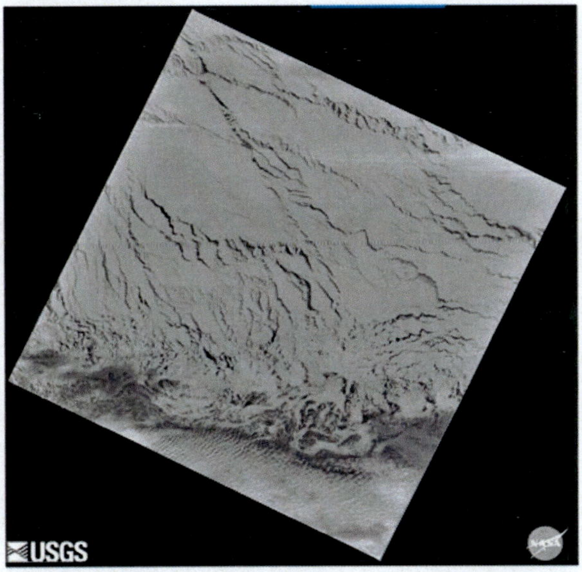

Fig. 9.1 Landsat 8 image taken on 22 Oct. 2015. Scene ID: LC80770092015295LGN01 with center location at 72.2 N, 150.3 W. The size of the image is 180 by 185 km (https://landsat.gsfc.nasa.gov/landsat-8/landsat-8-overview/)

of the ice cover morphology, i.e. the type of ice and floe size distribution down to meter-scale or less, require further algorithm development using compound sensors of different frequency bands or high-resolution satellite imagery. Kwok (2014) highlighted several applications of the Medea (1995) 1 m resolution imagery. In which, the floe size distribution, the lateral melt rate, and the interpretation of radar backscatter are of particular interest to the MIZ processes.

The current techniques of using remote sensing to determine ice types relevant to wave propagation is insufficient, particularly in areas of newly formed thin ice (e.g. Wiebe et al., 2009; Zhao et al., 2015; and Zhang et al., 2018). Figure 9.2 consists of two aerial photos obtained during a recent field experiment (Thomson et al., 2018), each corresponds to one of the two circles shown in Fig. 2a. It is important to point out that one of the images, Fig. 2b, corresponding to the blue circle was ~20 km outside the "ice edge" commonly defined by the 15% ice concentration contour. As shown in this photo, the surface in view was nearly entirely covered by pancake ice, which we now know is effective in damping high frequency waves. Missing such in-situ ice type information significantly reduces the wave forecast accuracy, as we will see later when discuss the wave attenuation. Figure 2c shows a very different ice type, taken eight days after Fig. 2b, approximately 100 km inside the ice edge. Over this period, the ice edge advanced roughly 100 km southward according to the AMSR2 data. Such rapid change of the ice cover morphology in the MIZ is challenging to model or monitor.

The opening of the Arctic has not only increased the dynamic nature of the ice cover in the MIZ, it also has changed the population of the ice types. The formerly more prevalent large floes that fringed the pack ice zone still dominate the MIZ in the summer melting season (Lee et al., 2016), but in the fall freezing season,

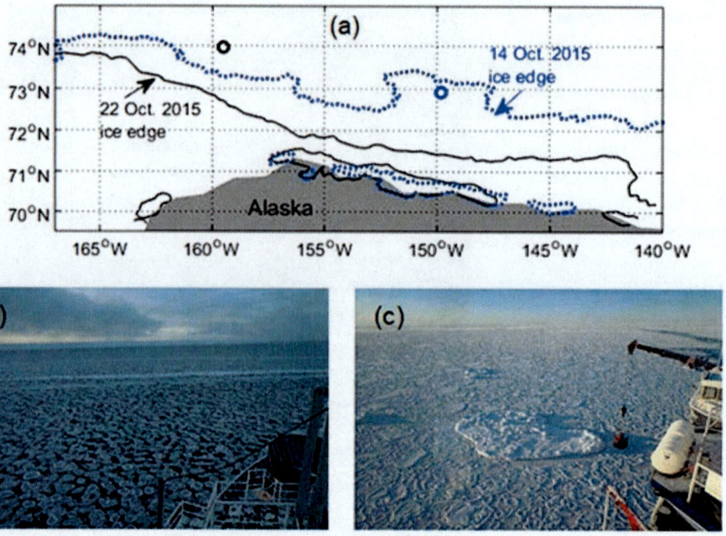

Fig. 9.2 Change of ice cover between 14 and 22 Oct. 2015. The ice edge contours correspond to the 15% ice concentration based on the AMSR2 data (the advanced microwave scanning radiometer 2. https://www.ifm.uni-hamburg.de/en/workareas/remote/remotesensing/seaice/ice-concentration.html) **b** location shown as the blue circle in the map, and **c** as the black circle

frazil/grease/pancake ice have become widespread (Thomson et al., 2017, 2018). The formation process of these types of ice, long recognized as important in the MIZ of the Southern Ocean, is now an integral part of wave-ice interaction in the Arctic. To use remotely sensed imagery to detect this type of ice (e.g. Mitnik et al., 2016) is important both for future improvements of the ice dynamics models such as CICE (https://github.com/CICE-Consortium/CICE/wiki) and for wave forecast models.

WAVEWATCHIII® (WW3) is the first to include ice effects for global wave forecasts (The WAVEWATCH III® Development Group, 2019). The governing equation for the wave propagation is the conservation of the directional wave energy density $F(\boldsymbol{x}, t, k, \theta)$ as shown below, where $\boldsymbol{x}$ is the spatial coordinate, t is time, k is the wavenumber, θ is the wave direction.

$$\frac{\partial}{\partial t}\left(\frac{F}{\sigma}\right) + \nabla_x \cdot \dot{\boldsymbol{x}}\left(\frac{F}{\sigma}\right) = \frac{S}{\sigma}; \dot{\boldsymbol{x}} = \boldsymbol{c}_g + \boldsymbol{U} \tag{9.1}$$

In the above, the bold-face quantities are vectors, and $\sigma = 2\pi f$ is the angular frequency, $\boldsymbol{c}_g$ is the group velocity, and $\boldsymbol{U}$ is the current. The wavenumber k and wave frequency f are related through a dispersion relation. The dispersion for open water wave is known, but for ice-covered region, it depends on the model used. Equation (9.1) is for regions away from the coastal zone so that additional effects from depth gradient and strong current variations may be ignored. The source term

$S = S_{in} + S_{ds} + S_{nl} + S_{ice}$ includes: the wind energy input S_{in}, dissipation (whitecapping and turbulence) S_{ds}, nonlinear transfer between frequencies (or equivalently wavenumbers) S_{nl}, and the effect due to the ice cover S_{ice}. The ice effect term is divided into a conservative directional redistribution due to scattering, $S_{ice,s}$, and a non-conservative dissipation, $S_{ice,d}$:

$$S_{ice,s} = -c_g\alpha_s(\boldsymbol{x}, t, k, \theta)F(\boldsymbol{x}, t, k, \theta) + c_g\int_0^{2\pi} S_\sigma(\boldsymbol{x}, t, k, \theta, \theta')F(\boldsymbol{x}, t, k, \theta)d\theta' \tag{9.2}$$

$$S_{ice,d} = -c_g\alpha_d(\boldsymbol{x}, t, k)F(\boldsymbol{x}, t, k) \tag{9.3}$$

The integral term in Eq. (9.2) represents the directional redistribution from θ, the direction of $\boldsymbol{k}$ (where $\boldsymbol{k} = k$), to other directions θ'. The scattering and dissipative attenuation coefficients are α_s and α_d, respectively. To determine α_s, α_d, and the k-σ relation is central to the wave-ice interaction models.

The most recent WW3 version 6.07 includes both dissipative and scattering redistribution components. For each component, there are different "switches" the users may choose to account for the respective physical process. Each switch is based on a different set of assumptions used to describe the ice effects. A thorough study of how these different switches perform against each other over a large domain and under realistic wind forcing has not yet been performed. A preliminary comparison of several switches addressing only the dissipative mechanism was reported in Rogers and Zieger (2014). Their hindcasts of the Beaufort and Chukchi Seas for August 2012 showed remarkable differences in the resulting wave height from different switches. Calibration and validation of these switches are challenging due to limited data.

No matter which switch is used, the most important input for wave forecasts is the ice condition. Apparently, accurate wave forecasts rely on accurate ice forecasts. The reverse is also true because waves can break existing ice covers to alter the floe size distribution, and determine the morphology of new ice formed in open water, i.e. nilas, frazil, grease, or pancake ice. Thus, ice and wave dynamics intricately depend on each other. In the following sections ice effects on waves will be discussed first, followed by wave effects on ice.

9.3 Ice Effects on Waves

Theoretically modeling wave propagation through ice covers started as early as the nineteenth century (Greenhill, 1886). The ice cover was idealized to be pure elastic over inviscid water. The thin plate assumption was applied to the coupled ice-water system. The resulting Euler–Bernoulli equation governed the motion of the ice cover. The dispersion relation was obtained from solving this equation for its eigenvalues. The resulting dispersion relation showed the combined effects of flexural rigidity and the inertia of the ice cover on wavelength and wave speed. There was no wave attenuation from such a model because the ice-water system conserved energy. Over half a century later, a mass loading theory based on a completely different idealization was developed. In which, the ice cover was assumed to consist of disjoint mass points (Weitz and Keller, 1950; Keller and Weitz, 1953). Nearly forty years after that, it was pointed out that under the zero flexural rigidity condition, the thin elastic plate theory reduced to the mass loading theory (Squire 1993).

Wave attenuation was not considered in the early theories. However, it was recognized early on that when waves entered an ice cover from open water, a portion of the energy would be reflected due to the discontinuity at the leading edge. Although the total wave energy remained conserved, the forward going wave energy would drop. Wadhams (1973) laid the foundation for applying the thin elastic plate theory to construct an exponentially decaying and frequency dependent attenuation theory based on this energy scatter mechanism. This idea has been fully developed through many researchers, from the early works of Fox and Squire (1990, 1994) to the most recent one by Meylan and Bennetts (2018). This body of literature based on wave scattering began with semi-infinite slabs (a two dimensional problem) and expanded to circular and arbitrary shapes (a three dimensional problem). This scattering effect has also been implemented in the operational wave model WW3 and a coupled ice-wave model (Williams et al., 2017). Reviews and recent perspectives of this large body of literature may be found in Squire (2007, 2018, 2020).

Scattering attenuates the forward-going wave energy by redistributing the energy to other directions. This diffusion process conserves the total energy. The first wave-in-ice theory that included energy dissipation was a direct extension of the thin elastic plate model. Squire and Allan (1980) introduced a viscoelastic thin plate theory in order to describe the hysteresis effect in ice. This theory considered dissipation processes in the ice cover itself. The idea of damping from the water body under ice came later. Two theories considering damping from the water body were developed almost simultaneously (Weber 1987; Liu and Mollo-Christensen 1988). Weber (1987) assumed that the ice cover was nearly rigid with an extremely high viscosity and Liu and Mollo-Christensen (1988) assumed a thin elastic plate. The dissipation from both theories came from the dissipation under the ice cover. Returning to dissipation within the ice cover itself, Keller (1998) took a different approach from the thin plate theories. In Keller's model the ice cover was assumed to be purely viscous and the water body remained inviscid. Modeling ice as a viscous material was motivated by a different field condition from those of the elastic thin plate or viscoelastic thin plate theories. In early years, field observations of wave propagation through ice covers were mostly over continuous ice sheets, or broken fields of large floes on the order of hundreds of meters. Brash and grease ice attracted attention later when more field experiments on wave-ice interactions became possible. These "soft" ice types are very effective in damping high frequency waves. To model these types of ice covers consisting of small ice particles, Keller (1998) removed the thin plate assumption so that the ice cover was allowed to deform in both horizontal and vertical directions as a pure viscous material. As pointed out in a recent study (Zhao et al., 2017), the relaxation of the thin plate assumption produced many complex wave modes, even under the simplest linear viscous constitutive law. The mode that was closest to the open water dispersion was recommended to be the dominant one.

All of these early theories addressed the one-sided effect of ice on waves except Fox and Squire (1991), who recognized the implications of waves on break up of ice covers and thus studied the strain field over a semi-infinite uniform ice cover. Studies of wave effects on ice covers are discussed in the next section.

The gaps between births of new wave-in-ice theories shortened after 2000. In an effort to bridge the pure viscous and pure elastic theories, Wang and Shen (2010a) extended the viscous layer theory of Keller (1998) to a viscoelastic layer. Assuming that the material property of an ice cover could be described by a Voigt model, with parallel linear spring and dashpot, Wang and Shen (2010a) obtained the dispersion relation between wave frequency and the complex wave number. The real part of this complex number determined the wave speed and the imaginary part determined the attenuation. This dispersion relation was shown to converge to the pure elastic thin plate or the pure viscous layer results under proper limits. It is important to note that there are many different viscoelastic models (see e.g. Flügge, 1975; and Malvern, 1969). Although it makes physical sense to view ice covers as a material that can simultaneously store and dissipate energy, so that they may be represented as a viscoelastic material, it is an open question if any of the infinite choices of viscoelastic constitutive laws is appropriate to describe all types of ice covers. Intuitively, the Voigt model in Wang and Shen (2010a) agrees with the physical picture of a composite ice cover, which consists of distinct elastic elements (floes or pancakes) mixed with a viscous slurry (grease or brash ice). Hence the volume average of the total internal stress against deformation is the sum of the elastic (floes) and viscous (slurry) components, as defined in a Voigt constitutive law. However, whether such models are indeed applicable to various ice covers has not been proven. A comparison of dispersion relations from three different viscoelastic theories is given in Mosig et al., (2015). Regardless, when formulating an ice cover as a viscoelastic continuum the chosen constitutive law must satisfy the frame-indifference criterion (e.g. Malvern, 1969). This is a strong constraint that will help build dispersion relations in the future.

More recently, De Santi and Olla (2017) proposed another model targeting grease/pancake ice covers. Their study differed from previous ones by considering a collection of small, thin, nearly rigid, elements floating over a highly viscous slurry above an inviscid water body. This picture more closely resembled the grease/pancake ice cover observed in the laboratory and the field than the previous picture of a depth-wise homogeneous material. In their limit of a close-packed agglomerate of small ice floes, their model conceptually approached the limiting case of a nearly rigid ice cover in Weber (1987). In both cases energy dissipation originated from the water body beneath an immobile ice cover. Wave damping initiated in the water body also received attention from other studies. De Carolis and Desiderio (2002) expanded the theory of Keller (1998) to include the viscosity of water under the ice cover. Zhao and Shen (2018) proposed a three-layer model with a viscoelastic ice layer adjacent to a turbulent boundary layer over inviscid water. In proper limits, this theory converged to thin elastic plate, viscoelastic layer, and eddy viscosity models. All of these theories contain parameters that need to be calibrated with data.

The common goal of all the above theories is to derive the dispersion relation under the assumption of a linear wave that follows the form of $e^{i(kx-\sigma t)}$, where $k = k_r + ik_i$ is the complex wave number and σ is the corresponding angular frequency. The attenuation is embedded in k_i, which introduces the exponential decay $e^{-k_i x}$ to the underlying wave profile $e^{i(k_r x-\sigma t)}$. To simplify the notations without losing the essential information, we continue by assuming the deep water condition ($k_r H > \pi$, H : water depth). In open water when damping is absent, $k_i = 0$ and k_r follows the simple dispersion relation shown below.

$$\sigma^2 = Qgk_r;\ Q = 1 \tag{9.4}$$

Depending on the assumptions for the physical properties of the ice cover, the dispersion relation changes according to the complexity of Q. Several examples of Q are given in Table 9.1. The spectral behaviors of these dispersion relations differ from each other. Applying a leading order analysis in the limit of $k_r h \ll 1$ (long waves relative to ice thickness h) to the dispersion relations listed in Table 9.1, one may solve k_i in terms of k_r. The viscoelastic thin plate theory predicts $k_i \propto k_r^{5.5} \propto \sigma^{11}$, the pure viscous and the viscoelastic layer theory both predict $k_i \propto k_r^{3.5} \propto \sigma^7$, and the eddy viscosity theory predicts $k_i \propto k_r^{1.75} \propto \sigma^{3.5}$. The same results were shown in

Table 9.1 Dispersion relations from various theories

Model	Q
Mass loading	$Q_{ml} = \frac{1}{1+\rho kh}$
Thin elastic plate	$Q_{te} = \frac{(\frac{Gh^3}{3\rho_{water} g} k^4 + 1)}{1+\rho kh}$
Viscous layer	$Q_{vl} = 1 + \rho \frac{g^2k^2 S_k C_\alpha - (N^4 + 16k^6\alpha^2\nu^4) S_k S_\alpha - 8k^3\alpha\nu^2 (C_k C_\alpha - 1)}{gk(4k^3\alpha\nu^2 S_k C_\alpha + N^2 S_\alpha C_k - gk S_k S_\alpha)}$ where $\alpha^2 = k^2 - i\sigma/\nu; N = \sigma + 2ik^2\nu$
Viscoelastic thin plate	$Q_{vt} = \frac{h^3(\frac{G}{\rho_{water} g} - \frac{\sigma\rho\nu}{g} i)}{6(1-\upsilon)} k^4 + 1 - \frac{\rho h \sigma^2}{g}$
Viscoelastic layer	$Q_{vel} = 1 + \rho \frac{g^2k^2 S_k C_\alpha - (N^4 + 16k^6\alpha^2\nu_e^4) S_k S_\alpha - 8k^3\alpha\nu_e^2 (C_k C_\alpha - 1)}{gk(4k^3\alpha\nu_e^2 S_k C_\alpha + N^2 S_\alpha C_k - gk S_k S_\alpha)}$ where $\nu_e = \nu + iG/\rho_1\sigma; \alpha^2 = k^2 - i\sigma/\nu_e; N = \sigma + 2ik^2\nu_e$
Eddy viscosity	$Q_{te} = \frac{(\frac{Gh^3}{3\rho_{water} g} k_r^4 + 1)}{1+\rho kh}$ $k_i = \frac{k_r\sqrt{\nu\sigma}}{2\sqrt{2}} \frac{d\sigma}{dk_r}$
Notes	In the above, $k = k_r + ik_i$; σ : angular frequency; $\rho = \frac{\rho_{ice}}{\rho_{water}}$; G : shear modulus;ν : kinematic viscosity; h : ice thickness. Other theoretical dispersion relations mentioned do not have closed-form solutions $S_k = \sinh kh$, $S_\alpha = \sinh \alpha h$, $C_k = \cosh kh$, $C_\alpha = \cosh \alpha h$

Meylan et al. (2018), who calculated k_i from the energy loss rate directly. Using field data to test these theories is not straightforward. While short waves (less than 6 s) typically have attenuation rates on the order of 10^{-4}–10^{-3} m^{-1}, long waves (more than 12 s) typically have attenuation rates below 10^{-6} m^{-1}. These low attenuation rates can only be determined by monitoring wave amplitude change over a considerable distance, especially for long waves. However, over a long distance other source/sink terms in addition to ice effects increasingly affect data interpretation. This situation is especially challenging when trying to establish a trend of wave attenuation over a large spectral domain. On one hand, high frequency components are strongly affected by wind input over a large distance, which offsets the dissipation so that the apparent attenuation can be much lower than the ice-induced attenuation (Li et al., 2017). On the other hand, the nonlinear wave energy transfer over a large distance may be significant for low frequency components. The ice type variation over a large distance is an additional factor to keep in mind for data interpretation for all frequencies.

The dispersion relation from viscoelastic thin plate, viscoelastic layer, and eddy viscosity theories have all been implemented in WW3, as three separate switches (in the most recent version 6.07, they are labelled as IC5, IC3, and IC2, corresponding to viscoelastic thin plate, viscoelastic layer, and eddy viscosity theories, respectively).

Experimental studies on wave propagation through ice covers are scarce both in the field and in laboratories. Early field experiments of wave propagation through sea ice successfully measured wave speed change and attenuation using series of pitch-roll buoys and accelerometers (Squire and Moore, 1980; Wadhams et al., 1986, 1988) or pairs of strain gauges (Squire et al., 1994). The data demonstrated frequency dependent wave attenuation and directional spreading as predicted by the scattering theory. These earlier field studies were conducted in pack ice or the MIZ consisting of floes of sizes on the order of 100 m. Wave propagation through grease/pancake ice was first reported for the Antarctic MIZ (Doble et al., 2003; Doble and Wadhams, 2006), in which six different formation processes of pancakes were described, from simple cyclic accretion of frazil to rafting and agglomeration of individual pancakes. This observation expanded the understanding of "pancake ice cycle" first reported in Lange et al., (1989), a formation process under waves distinctly different from the congelation ice formed under calm conditions. Waves measured from this same field experiment were analyzed in Doble et al., (2015). The level of attenuation was found to be above those from the earlier results with much larger floes. The smallness of pancake floes (no more than meters in size) significantly reduced the scattering effect, hence additional damping mechanisms were dominant in this type of ice covers. Kohout et al., (2014) conducted a seventeen-day long measurement of waves in the Antarctic MIZ with sensors deployed on ice floes. Continuous wave spectra were obtained along the floe drift tracks over hundreds of kilometers, where the pancake size ranged from 2–3 m near the ice edge to 10–20 m some 200 km into the ice cover. The datasets from Doble et al. (2003) and from Kohout et al., (2014) were further analyzed to address the partition of wind input, nonlinear transfer, and the net attenuation of wave energy in the spectral space (Li et al., 2017). More recently, several field experiments targeting the Arctic MIZ in the summer (Lee et al., 2016) and the fall (Thomson et al., 2018) provided a broad range of multi-sensor data. These two journal special issues addressed a host of wave-ice interaction questions, including scattering, attenuation, dispersion, model calibration, and floe size distribution. The observations on wave attenuation raised an important question about the general behavior of the measured spectral attenuation of waves. As discussed in Meylan et al., (2018), the drop-off of the attenuation at low frequencies from the data was much slower than predicted by the viscoelastic loss in the ice cover alone. Other types of energy loss laws which produced attenuation trends that agreed better with the data were thus examined. Further investigation of the turbulence dissipation under ice covers indicated that a properly parameterized eddy viscosity model might fit the attenuation of the most energetic portion of the wave spectra (Voermans et al., 2019). These studies cannot conclusively determine which model is physically more realistic. At present, we only know that simple models with simple parameterization cannot universality explain all available observations so far.

Laboratory experiments were conducted to study wave propagation through grease, pancake and fragmented ice covers using wave tanks from 4 to 72 m long (Newyear and Martin, 1999; Wang and Shen 2010b; Zhao and Shen, 2015; Cheng et al., 2019). Limited by the size of the wave tanks, the wavelengths studied were much shorter than in the field. Scaling up the results is difficult because different mechanisms may dominate at different scales. Nonetheless, laboratory studies provided controlled environment to observe phenomena and test theories. These experiments showed that wave attenuation depended on the ice type and ice thickness. Data from these experiments were used to calibrate the equivalent mechanical properties in the viscous layer (Newyear and Martin, 1999) and viscoelastic layer model (Zhao and Shen, 2015). Results suggested that the equivalent elasticity of the grease ice was negligible and fragmented ice had the highest equivalent elasticity, but still much lower than the intrinsic elasticity of sea ice reported in Weeks and Assur (1967). The equivalent viscosity of grease ice was roughly four orders of magnitude above that of water. Those for the pancake and fragmented ice covers were higher than for grease ice. The reduction of equivalent elasticity in fragmented ice covers was first studies by Sakai and Hanai (2002). They used polyethylene as surrogate ice and measured wave speed through the cover, started from a continuous cover and followed by systematically reduced segment sizes. A drop of wave speed as the size of the segments dropped was observed. The wave speed of all cases was bounded by the thin elastic plate theory as the upper limit and the mass loading theory as the lower limit. Using the thin elastic plate dispersion relation to inversely determine the equivalent elasticity of the segmented cover, Sakai and Hanai (2002) proposed a simple empirical relation to estimate the equivalent elasticity from the intrinsic elasticity of the material and the segment size. This experiment was repeated with real ice and a wider range of wave frequencies (Cheng et al., 2019). The gradual reduction of wave speed as the segment size decreased was again observed. However, the data set with expanded wave frequency range and lower elasticity showed that the empirical formula from Sakai and Hanai (2002) needed to be modified. Using the multiple scattering theory and the associated numerical procedure described in Kohout et al., (2007) and Kohout (2008), Cheng et al. (2019) proposed an expanded empirical equation for the equivalent elasticity of a segmented ice cover spanning wavelengths from laboratory to field scale. In fact, reduction of the equivalent elasticity of a fragmented ice cover is expected from physical considerations. Since the edges of each floe are free from bending, the elastic energy stored in an aggregate of floes is reduced from that in a continuous ice sheet. When the floe size drops, the free edge effect increases proportionally, eventually approaching the mass loading model where no elastic energy is stored in the ice cover. A rigorous derivation of the equivalent elasticity for an aggregate of elastic floes using the above energy considerations awaits to be developed.

Laboratory studies have also been used to isolate various mechanisms of ice effects on waves. Montiel et al., (2013a, b) examined the surface deformation of and the wave scattering by a single PVC disk constrained from drifting. Bennetts and Williams (2015) extended that study to an array of disks. Yiew et al. (2016) investigated the bodily motion of a free disk. This type of studies were useful in validating theories and identifying missing mechanisms. For instance, spilling of water over the floating disk and collisions between disks were shown to be significant damping mechanisms that should be included in models (Bennetts and Williams, 2015). Other laboratory studies have been conducted in recent years that suggested jet formation between colliding floes as yet another possible damping mechanism (Rabault et al., 2019). Sree et al., (2018) measured wave attenuation under a continuous PDMS (Polydimethylsiloxane) sheet. The material was viscoelastic with properties directly measured using a viscometer. The data showed significantly higher attenuation above the dissipation from the material damping alone. Careful analysis of wall friction, water viscosity, and all other possible energy sinks pointed out that the boundary layer under the cover was likely a major missing factor. Inclusion of such factor has been considered in theories as mentioned earlier. The difficulty is to quantify this factor under different wave and ice conditions. Old eddy viscosity data obtained under pack ice were cited in Liu and Mollo-Christensen (1988). Recent field data have been used to estimate the turbulence dissipation under a pancake ice field (Voermans et al., 2019).

The existing laboratory studies demonstrated a whole host of floe-scale processes that dissipate wave energy. Each of which may exist under field conditions. In a recent study, Herman et al., (2019a, b) used the Discrete Element Method to numerically reproduce the laboratory experiment reported in Cheng et al. (2019). Herman et al., (2019a, b) tried to include several floe scale processes such as floe collisions, wave scattering, drag between floes and water, and spilling of water over floes. They showed that many different combinations of the model parameters could reproduce the measured wave attenuation. Hence, to determine the relative importance of all potential damping effects, wave attenuation data is not enough. Simultaneous measurements of many processes are necessary. Building a theory that includes all possible mechanisms is a daunting task. Simplified theories that focus on dominating mechanisms need to be established for applications.

9.4 Wave Effects on Ice Covers

We now turn to the other side of the story. Waves affect the existing ice covers by fracturing them to reduce the floe sizes and herd the floes to compress the ice edge or diffuse it to form ice bands. They also influence the type of ice formation from open water and the rate of ice growth. Presently the most widely used ice model CICE (https://github.com/CICE-Consortium/CICE/wiki) determines the growth/decay and transport of ice by considering all energy and momentum fluxes from the atmosphere and the ocean (CICE documentation, 2019), but no wave effects yet. More research-oriented ice models have begun including some wave effects, most notably the fracturing process and the associated floe size distribution (Williams et al., 2013, 2017, Roach et al., 2018b).

In the sea ice community, ice fracture has been a central research topic, due to the many applications from submarine surfacing to offshore structure and icebreaker designs. Fracturing itself is a complicated mechanical process that concerns numerous engineering fields. For wave-ice interactions the problem is actually simpler, because instead of a combination of many failure mechanisms, the process is dominated by the flexural failure. Fox and Squire (1991) provided the theoretical basis to determine the maximum strain in a uniform ice sheet under wave actions. Herman (2017) used the Discrete Element Method to validate this theory and determined the floe size breaking from the ice sheet. To apply such theories to field conditions is not simple. The thickness and material property variations in a natural ice cover influence the fracture pattern. For applications a much simpler assumption is adopted, where the floe size breaking from wave flexing is assumed to be half of the dominant wavelength (Williams et al., 2013, 2017).

Both ice drift and formation of ice bands have been observed at the ice edge. Perrie and Hu (1997) showed that combined wind and wave forcing could double the ice drift from wind forcing alone. Ice band formation under wave action was explained in Wadhams (1983). These phenomena are locally confined to the ice edge. However, for navigation purposes such local phenomena are important to be included in the forecasts.

The presence or absence of waves also affect the type of ice formed from open water. Ice forms when the thermal flux at the water surface combining all sources and sinks becomes negative. However, there are no theoretical guidelines to determine which type of ice would form and what the initial thickness of the ice cover is. Ample observations have shown that under calm conditions, a thin, uniform layer of ice, called skim ice or nilas, forms first (Wadhams, 2000). This thin layer then becomes a platform under which congelation ice develops. On the other hand, under wavy conditions, frazil ice evolves into grease or pancake ice (Lange et al., 1989). Laboratory experiments showed that the size of pancake ice was related to the wave condition (Shen et al., 2004). Two simple equations were proposed to determine the maximum pancake diameter:

$$\frac{D_{max}}{\lambda} \approx \frac{c_1 \lambda}{A} \; or \; \frac{D_{max}}{\lambda} \approx \left(\frac{c_2}{A}\right)^{1/2} \tag{9.5}$$

where λ is the wavelength and A is the amplitude, $c_1 = C_1/(2\pi^2 E)$; $c_2 = 2C_2/(\pi^3 g\rho_{ice})$, E is the Young's modulus, C_1 and C_2 are both constants with the unit of N/m^2. They represent the freezing bond strength between pancakes, controlled by temperature and salinity. The first equation came from a bending failure criterion and the second from the tensile failure. Very limited laboratory (Shen et al., 2004) and field data (Roach et al., 2018a) are available to check these theories. The current thinking favors the tensile failure theory for initial pancakes grown in open water. Smith and Thomson (2019) adopted the tensile failure theory to show that C_2 dropped with rising skin temperature of the ocean, as suggested based on physical arguments. Thus it is expected that the more energetic the waves the smaller the pancake size. However, laboratory experiment also showed that when waves were exceedingly high, grease ice persisted without developing into pancakes. Presently, no field observations or theoretical studies are available to determine the conditions that dictate which of the three ice types would form from open water: nilas, pancakes, or grease ice.

Formation of new ice is also controlled by the oceanic heat flux. As shown in Smith et al. (2018) a pancake ice field can quickly melt in growing storms when the mixed layer temperature is still above freezing. The scenario described in their paper corresponded to a field near the ice edge where pancakes already formed under gentle waves when the skin temperature was below freezing. As the waves grew under an intensifying storm, the stored heat in the mixed layer was stirred up to melt the pancakes and returned the surface to open water. Their study showed that near the ice edge the wave-ice interaction must also include the oceanographic processes to fully capture the dynamical changes properly.

9.5 Conclusions

This paper reviewed the theoretical and experimental studies concerning wave-ice interactions. The pace of study in this field has increased dramatically in recent years, driven by the accelerated changes in the Arctic Ocean. As a research field, the more we look into a problem, the more complicated it becomes. Facing the rapidly changing Arctic, practical models are urgently needed for engineering, environmental, and geopolitical applications. Such a model depends on the temporal and spatial scales. Apparently, short term and small regional questions require high precision predictions, while long term and basin-wide questions may be satisfied with simpler models with larger uncertainties that can be corrected by data-assimilation. A perfect model that can accurately describe all observations is unreachable. An acceptable model should at least be physically convincing, be based on sound assumptions, and provide predictions that can capture most of the observations.

An inevitable question is then: what are the right models at various scales? Hibler (1977) provided an inspiring idea which could shed some light. In that study, it was shown through simple mathematics, that a stochastic average of any plasticity law for ice dynamics would result in a viscous law. The scale of validity at which this smoothing worked depended on the statistical characteristics of the deformation field. Although his study was not about wave-ice interaction, such philosophical thinking is general.

So far, a path to rigorously determine the "proper" model for wave-ice interactions at various scales has not been clearly identified. At present, we are still at the stage of making observations and identifying mechanisms that may contribute to the wave-ice interaction. We have not determined which ones are important at which scale, let alone how to simplify the model when many processes need to be included simultaneously.

Although a perfect model is unreachable, this does not mean that we should stop striving for model improvements. There are two mutually helpful ways to go forward. We must investigate each identified mechanisms thoroughly, using both laboratory and field observations and theoretical studies. We also must use simplified models to evaluate their performances against each other and against observations. All mechanical systems are fundamentally governed by a set of very simple conservation laws. It is the number of variables and their interactions that create the complexity. We are often fooled by the simplicity of the individual mechanisms to be overly optimistic about what a model can and should do. This optimism makes us overlook the compounded effects in a multi-variable and multi-physics system. We look forward to findings from general dynamic systems studies that are above individual mechanisms and specific models to help us see an overall picture.

Acknowledgements This study is supported in part by the Office of Naval Research grant numbers #N00014-13-1-0294 and N00014-17-1-2862. The author would also like to thank Steve Ackley for his introduction to sea ice and to Vernon Squire, with whom continued research discussions over the past several decades have been a constant source of inspiration.

References

L.G. Bennetts, T.D. Williams, Water wave transmission by an array of floating discs. Proc. r. Soc. A **471**, 20140698 (2015). https://doi.org/10.1098/rspa.2014.0698

S. Cheng, A. Tsarau, K. U. Evers, H. Shen, Floe size effect on gravity wave propagation through ice covers. JGR–Oceans **124**, 320–334 (2019). https://doi.org/10.1029/2018JC014094

CICE Consortium. CICE Documentation. https://buildmedia.readthedocs.org/media/pdf/cice-consortium-cice/master/cice-consortium-cice.pdf. pp. 104 (2019)

G. Carolis De, D. Desiderio, Dispersion and attenuation of gravity waves in ice: a two-layer viscous fluid model with experimental data validation. Phys. Lett. A **305**(6), 399–412 (2002). https://doi.org/10.1016/S0375-9601(02)01503-7

M. J. Doble, M. D. Coon, P. Wadhams, Pancake ice formation in the Weddell Sea, JGR–Oceans. 108:3209 (2003). https://doi.org/10.1029/2002JC001373

M. J. Doble, P. Wadhams, Dynamical contrasts between pancake and pack ice, investigated with a drifting buoy array. JGR–Oceans. **111**, C11S24. (2006) https://doi.org/10.1029/2005JC003320

M.J. Doble, G. De Carolis, M.H. Meylan, J.-R. Bidlot, P. Wadhams, Relating wave attenuation to pancake ice thickness, using field measurements and model results. Geophys. Res. Lett. **42**, 4473–4481 (2015). https://doi.org/10.1002/2015GL063628

W. Flügge, Viscoelasticity. Springer (1975). https://doi.org/10.1007/978-3-662-02276-4,194pp

C. Fox, V. A. Squire, Reflection and transmission characteristics at the edge of shore fast sea ice. JGR–Oceans. **95**, 11629–11639. (1990). https://doi.org/10.1029/JC095iC07p11629

C. Fox, V. A. Squire, Strain in shore fast ice due to incoming ocean waves and swell. JGR – Oceans. **96**, 4531–4547 (1991). https://doi.org/10.1029/90JC02270

C. Fox, V.A. Squire, On the oblique reflexion and transmission of ocean waves at shore fast sea ice. Phil. Trans. Royal Soc. a. **347**(1682), 185–218 (1994). https://doi.org/10.1098/rsta.1994.0044

A.G. Greenhill, Wave motion in hydrodynamics. Am. J. Math. **9**(1), 62–96 (1886)

A. Herman, Wave-induced stress and breaking of sea ice in a coupled hydrodynamic discrete-element wave–ice model. Cryosphere. **11**, 2711–2725 (2017). https://doi.org/10.5194/tc-11-2711-2017

A. Herman, S. Cheng, H. H. Shen, Wave energy attenuation in field of colliding ice floes. Part 1: discrete-element modelling of dissipation due to ice–water drag. Cryosphere. **13**, 2887–2900, (2019a) https://doi.org/10.5194/tc-13-2887-2019

A. Herman, S. Cheng, H. H. Shen, Wave energy attenuation in field of colliding ice floes. Part 2: a laboratory case study. Cryosphere. 13, 2901–2914, (2019b). https://doi.org/10.5194/tc-13-2901-2019

W. D. Hibler, A viscous sea ice law as a stochastic average of plasticity. JGR – Oceans. **82**, 3932–3938. (1977).https://doi.org/10.1029/JC082i027p03932

J.B. Keller, M. Weitz, Reflection and transmission coefficients for waves entering or leaving an icefield. Commun. Pure Appl. Math. **6**, 415–417 (1953)

J.B. Keller, Gravity waves on ice covered water. JGR–Oceans **103**(C4), 7663–7669 (1998). https://doi.org/10.1029/97JC02966

A.L. Kohout, M.H. Meylan, S. Sakai, K. Hanai, P. Leman, D. Brossard, Linear water wave propagation through multiple floating elastic plates of variable properties. J. Fluids and Structures **23**(4), 649–663 (2007). https://doi.org/10.1016/j.jfluidstructs.2006.10.012

A. L. Kohout, M. H. Meyla, An elastic plate model for wave attenuation and ice floe breaking in the marginal ice zone. JGR–Oceans, **113**(C9) (2008). https://doi.org/10.1029/2007JC004434

A. Kohout, M. Williams, S. Dean et al., Storm-induced sea-ice breakup and the implications for ice extent. Nature **509**, 604–607 (2014). https://doi.org/10.1038/nature13262

R. Kwok, Declassified high-resolution visible imagery for Arctic sea ice investigations: an overview, Remote Sens. Environ. **142**, 44–56 (2014). https://doi.org/10.1016/j.rse.2013.11.015

M.A. Lange, S.F. Ackley, P. Wadhams, G.S. Dieckmann, H. Eicken, Development of sea ice in the Weddell sea. Annals Glacio. **12**, 92–96 (1989). https://doi.org/10.3189/S0260305500007023

C. M. Lee, J. Wilkinson, J. Thomson, T. Maksym (eds), Special feature: marginal ice zone processes in the summertime arctic, Elementa. University of California Press, E-ISSN:2325–1026, (2016)

J. Li, A.L. Kohout, M.J. Dobel, P. Wadhams, C. Guan, H.H. Shen, Rollover of apparent wave attenuation in ice covered seas. JGR–Oceans **12**, 8557–8566 (2017). https://doi.org/10.1002/2017JC012978.

Liu and Mollo-Christensen, 1988. A. K. Liu, E. Mollo-Christensen, Wave propagation in a solid ice pack. J. Phys. Oceanogr. **18**(11), 1702–1712, (1988) https://doi.org/10.1175/1520-0485(1988)018<1702:WPIASI>2.0.CO;2.

L. E. Malvern. Introduction to the mechanics of a continuous medium. Prentice-Hall, ISBN 13–487603–2, pp. 711, (1969)

Medea, Scientific utility of naval environmental data – A MEDEA Special Task Force Report. a report to the navy meteorology and oceanography command. Va.: Medea Program Office McLean. 1995.

M.H. Meylan, L.G. Bennetts, Three-dimensional time-domain scattering of waves in the marginal ice zone. Philos. Transactions. Ser. A. **376**(2129), 20170334 (2018). https://doi.org/10.1098/rsta.2017.0334.

M. H. Meylan, L. G. Bennetts, J. E. M Mosig, W. E. Rogers, M. J. Doble, M. A. Peter, Dispersion relations, power laws, and energy loss for waves in the marginal ice zone. JGR–Oceans. **123**, 3322–3335. (2018). https://doi.org/10.1002/2018JC013776.

L. Mitnik, V. Dubina, E. Khazanova, New ice formation in the Okhotsk sea and the Japan Sea from C- and L-band satellite SARs. 2016 IEEE International Geoscience and Remote Sensing Symposium (IGARSS), Beijing, pp. 4853–4856. (2016). https://doi.org/10.1109/IGARSS.2016.7730266.

F. Montiel, F. Bonnefoy, P. Ferrant, L. G. Bennetts, V. A. Squire, P. Marsault (2013) Hydroelastic response of floating elastic discs to regular waves. Part 1 wave basin experiments. J. Fluid Mech. 723 604 628 https://doi.org/10.1017/jfm.2013.123

F. Montiel, L.G. Bennetts, V.A. Squire, F. Bonnefoy, P. Ferrant, Hydroelastic response of floating elastic discs to regular waves. Part 2. modal analysis. J. Fluid Mech. J. Fluid Mech. **723**, 629–652 (2013). https://doi.org/10.1017/jfm.2013.124

J. E. Mosig, M. F. Montiel, V.A. Squire, Comparison of viscoelastic-type models for ocean wave attenuation in ice-covered seas, JGR–Oceans. **120**, 6072–6090, (2015). https://doi.org/10.1002/2015JC010881

K. Newyear, S. Martin, Comparison of laboratory data with a viscous two-layer model of wave propagation in grease ice. JGR–Oceans. **104**(C4):7837–7840. (1999). https://doi.org/10.1029/1999JC900002.

W. Perrie, Y. Hu, Air–ice–ocean momentum exchange. part ii: ice drift. J. Phys. Oceanogr. **27**, 1976–1996 (1997). https://doi.org/10.1175/1520-0485(1997)027%3c1976:AIOMEP%3e2.0.CO;2

J. Rabault, G. Sutherland, A. Jensen, K.H. Christensen, A. Marchenko, Experiments on wave propagation in grease ice: Combined wave gauges and particle image velocimetry measurements. J. Fluid Mech. **864**, 876–898 (2019). https://doi.org/10.1017/jfm.2019.16

L. A. Roach, M. M. Smith, S. M. Dean, Quantifying growth of pancake sea ice floes using images from drifting buoys. JGR–Oceans. **123**, 2851–2866. (2018a) . org/https://doi.org/10.1002/2017JC013693

L. A. Roach, C. Horvat, S. M. Dean, C. M. Bitz, An emergent sea ice floe size distribution in a global coupled ocean–sea ice model. JGR – Oceans. **123**, 4322–4337, (2018b). https://doi.org/10.1029/2017JC013692

W. E. Rogers, S. Zieger, New wave-ice interaction physics in WAVEWATCH III®. Proc. 22nd IAHR Int. Symp. Ice, IAHR, ISBN: 978–981–09–0750–1, p. 1252, (2014). https://doi.org/10.3850/978-981-09-0750-1

F. Santi De, P. Olla, Effect of small floating disks on the propagation of gravity waves Fluid Dynam. Res. **49**(2), (2017). https://doi.org/10.1088/1873-7005/aa59e1

S. Sakai, K. Hanai, Empirical formula of dispersion relation of waves in sea ice. Proceedings of 16th IAHR international symposium on Ice in the environment, December 2–6, 2002, Dunedin, New Zealand

H. H. Shen, S. F. Ackley, Y. Yuan, Limiting diameter of pancake ice, JGR–Oceans. 109(C12035) (2004). https://doi.org/10.1029/2003JC002123

M. Smith, J. Thomsonm, Pancake sea ice kinematics and dynamics using shipboard stereo video. Ann. Glacio. 1–11, (2019) https://doi.org/10.1017/aog.2019.35

M. Smith, S. Stammerjohn, O. Persson, L. Rainville, G. Liu, W. Perrie, et al. Episodic reversal of autumn ice advance caused by release of ocean heat in the beaufort sea. JGR–Oceans. **123**, 3164–3185, (2018). https://doi.org/10.1002/2018JC013764

V.A. Squire, A. Allan, Propagation of flexural gravity waves in sea ice, in *Sea Ice Processes and Models*. ed. by R. Pritchard (Univ. of Wash. Press, Seattle, Wash, 1980), pp. 327–338

V.A. Squire, S.C. Moore, Direct measurement of the attenuation of ocean waves by pack ice. Nature **283**, 365–368 (1980). https://doi.org/10.1038/283365a0

V.A. Squire, A comparison of the mass-loading and elastic plate models of an ice field. Cold Reg. Sci. Tech. **21**, 219–229 (1993). https://doi.org/10.1016/0165-232X(93)90066-H

V.A. Squire, W. Robinson, M.H. Meylan, T.G. Haskell, Observations of flexural waves on the Erebus Ice Tongue, McMurdo Sound, Antarctica, and nearby sea ice. J. Glacio **40**(135), 377–385 (1994). https://doi.org/10.3189/S0022143000007462

V.A. Squire, Of ocean waves and sea-ice revisited. Cold Reg. Sci. Tech. **49**(2), 110–133 (2007). https://doi.org/10.1016/j.coldregions.2007.04.007

V.A. Squire, A fresh look at how ocean waves and sea ice interact. Phil. Trans. r. Soc. a. **376**, 20170342 (2018). https://doi.org/10.1098/rsta.2017.0342

V.A. Squire, Ocean wave interactions with sea ice: a reappraisal. Ann. Rev. Fluild Mech. **52**, 37–60 (2020). https://doi.org/10.1146/annurev-fluid-010719-060301

D. K. K. Sree, A.W. K. Law, H.H. Shen, An experimental study on gravity waves through a floating viscoelastic cover, Cold Reg. Sci. Tech. (2018). https://doi.org/10.1016/j.coldregions.2018.08.013

The WAVEWATCH III® development group. 2019. User manual and system documentation of WAVEWATCH III® version 6.07. Tech. Note 333, NOAA/NWS/NCEP/MMAB, College Park, MD, USA, 465 pp. + Appendices.

J. Thomson, S. Ackley, H. H. Shen, W. E. Rogers, The balance of ice, waves, and winds in the arctic autumn Eos 98 (2017). https://doi.org/10.1029/2017EO066029

J. Thomson, S. F. Ackley, F. Girard-Ardhuin, F. Ardhuin, A. Babanin, G. Boutin, et al. Overview of the arctic sea state and boundary layer physics program, JGR–Oceans. **123**, 8674–8687, (2018). https://doi.org/10.1002/2018JC013766.

Voermans et al., 2019. J. J. Voermans, A. V. Babanin, J. Thomson, M. M. Smith, H. H. Shen, Wave attenuation by sea ice turbulence. Geophys. Res. Lett. **46**, (2019). https://doi.org/10.1029/2019GL082945

P. Wadhams, The effect of a sea ice cover on ocean surface waves, Ph.D. dissertation, University of Cambridge, pp. 223, (1973)

P. Wadhams, A mechanism for the formation of ice edge bands, JR –Oceans. **88**, 2813–2818, (1983), https://doi.org/10.1029/JC088iC05p02813

P. Wadhams, Ice In the ocean, CRC Press, 364 pp. ISBN- **10**, 97 (2000)

P. Wadhams, V.A. Squire, J.A. Ewing, R.W. Pascal, The Effect of the Marginal Ice Zone on the Directional Wave Spectrum of the Ocean. J. Phys. Oceanogr. **16**, 358–376 (1986). https://doi.org/10.1175/1520-0485(1986)016%3c0358:TEOTMI%3e2.0.CO;2

P. Wadhams, V.A. Squire, D.J. Goodman, A.M. Cowan, Moore S.C, The attenuation rates of ocean waves in the marginal ice zone, JGR – Oceans. 93:6799– 6818 (1988), doi:https://doi.org/10.1029/JC093iC06p06799.

P. Wadhams, V.A. Squire, P. Rottier, A.K. Liu, J. Dugan, P. Czipott, H.H. Shen, Workshop on wave-ice interaction. Eos. Trans. Am. Geophys. Union, **73**(35):375–378 (1992). doi: https://doi.org/10.1029/91EO00288.

R. Wang, H.H. Shen, Gravity waves propagating into ice-covered ocean: a visco-elastic model. JGR – Oceans. 115(C06024) (2010a). doi:https://doi.org/10.1029/2009JC005591

R. Wang, H.H. Shen, Experimental study on surface wave propagating through a grease-pancake ice mixture. Cold Reg. Sci. Tech (2010b). doi:https://doi.org/10.1016/j.coldregions.2010.01.011

J.E. Weber, Wave Attenuation and Wave Drift in the Marginal Ice Zone. J. Phys. Oceanogr. **17**(12), 2351–2361 (1987). https://doi.org/10.1175/1520-0485(1987)017%3c2351:WAAWDI%3e2.0.CO;2

W. Weeks, A. Assur, The mechanical properties of sea ice, US Army Cold Regions Research and Engineering Monograph DA Project 1VO25001A130, pp. 94 (1967)

M. Weitz, J.B. Keller, Reflection of water waves from floating ice in water of finite depth. Commun. Pure Appl. Math. **3**, 305–318 (1950)

H. Wiebe, G. Heygster, T. Markus, Comparison of the ASI Ice Concentration Algorithm With Landsat-7 ETM+ and SAR Imagery. IEEE Trans. Geosci. Remote Sensing **47**(9), 3008–3015 (2009). https://doi.org/10.1109/TGRS.2009.2026367

T.D. Williams, L.G. Bennetts, V.A. Squire, D. Dumont, L. Bertino, Wave–ice interactions in the marginal ice zone. Part 1: Theoretical foundations. Ocean Model. **71**, 81–91 (2013). https://doi.org/10.1016/j.ocemod.2013.05.010

T.D. Williams, P. Rampal, S. Bouillon, Wave–ice interactions in the neXtSIM sea-ice model. Cryosphere **11**, 2117–2135 (2017). https://doi.org/10.5194/tc-11-2117-2017

L.J. Yiew, L.G. Bennetts, M.H. Meylan, B.J. French, G.A. Thomas, Hydrodynamic responses of a thin floating disk to regular waves. Ocean Model. **97**, 52–64 (2016). https://doi.org/10.1016/j.ocemod.2015.11.008

S. Zhang, J. Zhao, M. Li et al., An improved dual-polarized ratio algorithm for sea ice concentration retrieval from passive microwave satellite data and inter-comparison with ASI, ABA and NT2. J. Ocean. Limnol. **36**, 1494–1508 (2018). https://doi.org/10.1007/s00343-018-7077-x

Zhao, X., and Shen, H.H. 2015. Wave propagation in frazil/pancake, pancake, and fragmented ice covers, *Cold Reg. Sci. Tech.*, doi:https://doi.org/10.1016/j.coldregions.2015.02.007

X. Zhao, S. Cheng, H.H. Shen, Nature of wave modes in coupled viscoelastic layer over water. J. Eng. Mech. **143**(10), 04017114 (2017). https://doi.org/10.1061/(ASCE)EM.1943-7889.0001332

X. Zhao, H.H. Shen, A three-layer viscoelastic model with eddy viscosity effect for flexural-gravity wave propagation through ice covers. Ocean Model. **151**, 15–23 (2018). https://doi.org/10.1016/j.ocemod.2018.08.007

X. Zhao, H. Su, A. Stein, X. Pang, Comparison between AMSR-E ASI sea-ice concentration product, MODIS and pseudo-ship observations of the Antarctic sea-ice edge. Ann. Glacio. **56**(69), 45–52 (2015). https://doi.org/10.3189/2015AoG69A588

Part IV
Thermodynamics

Chapter 10
Thermo-Hydrodynamics of Sea Ice Rubble

A. Marchenko

Abstract Results of the field work performed in the end of April (2017–2019) on drifting ice in the Barents Sea region extended between the island Hopen and Bear Island are discussed. The field investigations included measurements of ice rubble sizes and shapes, vertical profiles of ice temperature and salinity inside ice rubble, uniaxial compressive strength of ice cores taken from ice rubble, and vertical permeability of ice rubble. The ocean heat fluxes below the drift ice were measured in several expeditions in the Barents Sea since 2005 including the filed works in 2017–2019. A mathematical model was formulated, and numerical simulations were performed to explain the formation of completely consolidated ice rubble. It is shown that the ocean heat flux, the initial macro-porosity and the initial draft of the rubble are the main parameters influencing the consolidation process when the heat fluxes from the ice into the air are small. Numerical simulations showed that complete consolidation of ice rubble may occur in one year or even several months when the ocean heat flux is of about 20 W/m^2 and the initial draft of ice rubble is smaller 10 m.

List of Symbols

P	Pore pressure of brine in sea ice, Pa
P_w	Hydrostatic pressure of water, Pa
P_a	Atmosphere pressure, Pa
q	Flux of liquid brine in sea ice, m/s
$\mu = 1.5$ mPa·s	The dynamic viscosity of water
κ	Permeability of sea ice by liquid brine, m^2
$\rho_{sw} = 1020$ kg/m^3	The mean sea water density
ρ	Density of sea water brine at the freezing point, kg/m^3
ρ_{si}	Sea ice density, kg/m^3
$c = 4.19$ kJ/kg·C	The specific heat capacity of water

A. Marchenko (✉)
The University Centre in Svalbard, Longyearbyen, Norway
e-mail: aleksey.marchenko@unis.no

J. Tuhkuri and A. Polojärvi (eds.), *IUTAM Symposium on Physics and Mechanics of Sea Ice*, IUTAM Bookseries 39, https://doi.org/10.1007/978-3-030-80439-8_10

T	Water temperature, C
T_{si}	Temperature of sea ice, C
σ	Sea water brine salinity
σ_{si}	Sea ice salinity
ν_b	Liquid brine content of sea ice
Q_w	The vertical ocean heat flux, W/m^2
E	Cumulative vertical ocean heat flux, MJ/m^2
h_r	Ice rubble draft, m
h_{cl}	Consolidated layer thickness, m
h_{ur}	The thickness of unconsolidated ice rubble, m
k	Thermal conductivity of sea ice, W/m·C
S_{ur}	The amount of salts in a vertical column of unconsolidated rubble, kg/m^2

10.1 Introduction

Compression of sea ice influences stress concentrations at the edges of interacting floes and separation of ice blocks from them. Ice blocks pushed by converging floes in the water and on the surface of level ice form ice rubble. Ice rubble extended along specific direction and having smaller dimensions in the transversal directions is called ice ridge. Usually, ice ridges are extended along floes boundaries, but in case of relatively small diameters of interacting floes the rubble can occupy regions of arbitrary shape. Above water and submerged parts of ice ridges are named sail and keel. Usually, vertical dimension of keels is greater than vertical dimension of sails in 3–5 times (Timco and Burden 1997).

Internal structure of ice rubble is characterized by macro-porosity which is the volume of voids filled with water within unit volume of the rubble. Macro-porosity is calculated from the results of vertical drilling. The speed of drill-bit is high when it passes the voids and much lower when it passes solid ice. The speed of drill-bit is analyzed depending on the depth of the drilling, and depth intervals with high speeds of the drilling are recorded in a table. Macro-porosity in each drilling location is calculated as a ratio of total length of intervals with high drilling speed to total length of the drilling well. The drilling can be performed with mechanical augers, thermo-electric drill or hot water drill. Interpretation of the drilling results depends on the drilling rig and personal experience of the operator. Sometimes drilling speed changes gradually from high speed drops related to a pass of the drill-bit through voids filled by water to small speeds related to a pass of the drill-bit through solid ice. Intermediate speeds of the drilling can be registered in slush and broken ice. They also can be explained by a sliding of a drill-bit along inclined surfaces of submerged floes.

Although vertical sizes of ice rubble exceed the thickness of level ice in several times the initial strength of ice rubble is not high since submerged blocks of ice are not frozen with each other. Thermodynamic effects influence the formation of slush and freeze bonds between submerged blocks of ice and make ice rubble stronger. This process is called thermodynamic consolidation. The thermodynamic consolidation occurs due to atmosphere cooling (Lepparanta et al. 1995; Hoyland 2002), release of cold reserves from submerged ice blocks (Hoyland and Liferov 2005; Marchenko 2008) and penetration of fresher water inside ice rubble keel (Shestov and Marchenko 2016). Usually, the shape of consolidated region inside rubble keel is similar a layer. Therefore, it is named consolidated layer.

Penetration of sea water inside ice rubble was registered by direct measurement of water velocity between submerged ice blocks inside ice ridge keels in the Barents Sea (Gorbatsky and Marchenko 2007; Marchenko and Hoyland 2008). The measurements were performed with Acoustic Doppler Velocimeter (ADV) mounted on vertical stick and placed inside the keels between submerged blocks of ice. It was possible when gaps between the ice blocks were wider 1 m. The mean speed of water streams inside the keels was measured up to 10 cm/s, and the kinetic energy of fluctuations of water streams was similar to the kinetic energy of the mean flow. The last property means high turbulence of water streams inside ice ridge keels. Shestov and Marchenko (2016) measured water temperature at different depths inside ice ridge keels in the Barents Sea and investigated the dependence of water temperature fluctuations from the depth. It was found that the fluctuation amplitudes inside the keels increased sharply at depth of about 5 m and deeper and became similar to the amplitudes of water temperature fluctuations below the level ice near the ice rubble.

The latent heat spent for the freezing of the unit mass of ice rubble with macro-porosity p equals pL_{sb}, where L_{sb} is the latent heat spent for the melting of submerged ice blocks. According to the Stefan equation (Stefan 1891) the thickness of growing level ice is proportional in quasi-static approximation to $1/\sqrt{L_{si}}$, where L_{si} is the latent heat of sea ice. The latent heat of saline ice equals $L_{si} = (1 - \sigma - \sigma/S)L_i$, where L_i is the latent heat of fresh ice, σ is the salinity, and S is fractional salt content of the ice (Schwerdtfeger 1963). Since the salinity of submerged ice blocks can be greater than the salinity of sea ice (Kovacs 1983) then $L_{sb} \leq L_{si}$. The ratio of level ice thickness to the consolidated layer thickness is proportional to $\sqrt{pL_{sb}/L_{si}} \leq \sqrt{p}$. Measured values of the macro-porosities of ice rubble are shown in Table 10.1. A comprehensive review of field studies of ice rubble morphology performed from 1975 to 2011 (Strub-Klein and Sudom 2012) gives the mean value of the macro-porosity of 0.2. The macro-porosity of unconsolidated rubble changes between 0.3 and 0.4 for the first-year ice rubble (Table 10.1). Therefore, the ratio of level ice thickness to the consolidated layer thickness is estimated smaller than 0.5. It is confirmed by numerous field investigations (Lepparanta et al. 1995; Hoyland and Loset 1999; Kharitonov 2008, 2019; Ervik et al. 2018; Guzenko et al. 2019).

Multiyear ice rubble can be completely consolidated (Kovacs 1983). Hoyland et al. (2008), Strub-Klein et al (2009), and Shestov et al (2012) discovered that the second-year ice ridges in the Fram Strait are also completely consolidated. Marchenko (2018) elaborated a model describing consolidation of rubble ice due to the melting of the keel under the influence of ocean heat flux. Melt water with lower salinity and lower density penetrates inside the keel and freezes there since the ice temperature inside the

Table 10.1 Ice rubble porosity measured in field experiments; M—mechanical drilling, ET—electro thermal drilling and HW—hot water drilling. UR is unconsolidated rubble below consolidated layer

	Zubakin et. al. (2005), Barents Sea, 2003. M	Hoyland (2007), Barents Sea, 2002–2005. M	Lepparanta et al. (1995), Baltic Sea, 1990. M	Gudoshnikov et al. (2003), Pechora Sea, 1996–1999, 2001,2003. M
Sail, mean/max	0.08/0.16	0.15–0.29		0.13 (mean)
Keel, mean/max	0.06/0.15	0.1–0.46 (UR)	0.322–0.289 (UR)	0.38 (mean)
Total, mean/max	0.07/0.15		0.281–0.175	0.29/0.62
	Kharitonov (2008), Sakhalin, 1998. ET	Kharitonov (2008), Pechora Sea, 1999. ET	Kharitonov (2008), Caspian Sea, 2003. ET	Kharitonov (2012), Central Arctic, 2011. ET
Sail, mean/max	0.06–0.08	0.17 (mean)	0.11/0.14	0.086/0.1
Keel, mean/max	0.22–0.24	0.17 (mean)	0.18/0.24	0.11/0.18
Total, mean/max		0.17 (mean)	0.17/0.23	0.11/0.16
	Shestov et al (2012), Fram Strait, M	Kharitonov (2019), Shokal'skogo Strait, 2016. HW	Guzenko et al (2019), Kara and Laptev Seas, 2013–2017. HW	Naumov et al. (2019), Ob' Bay, 2005–2015. M
Sail, mean/max	0	0.19/0.24		0.13/0.16
Keel, mean/max	0/0.07	0.13/0.15 0.27/0.3 (UR)		0.2/0.37 0.38/0.74 (UR)
Total, mean/max	0	0.13/0.15	0.17/0.21	0.2/0.34

keel is lower than the freezing point of the melt water. In the present paper the results of the field investigations of completely consolidated first-year ice rubble discovered in the cruises of Polarsyssel in the Barents Sea in 2017–2019 are described and discussed. Numerical simulations are performed to explain the fast consolidation of ice rubble.

10.2 Field Investigations of Ice Rubble

In this section the field investigation on the drift ice in the region extended between Spitsbergen, Hopen Island and Bear Island are described. The field works were performed in the end of April in 2017–2019. The field works locations are shown by black squares in Fig. 10.1. Main goal of the field works was to investigate characteristics of drift ice and under ice boundary layer in the region. Offshore supply ship Polarsyssel was moored to selected floes in 2017 and 2018, and we had access to the floes directly from the ship (Fig. 10.2). In 2019 the floe had very complicated shape

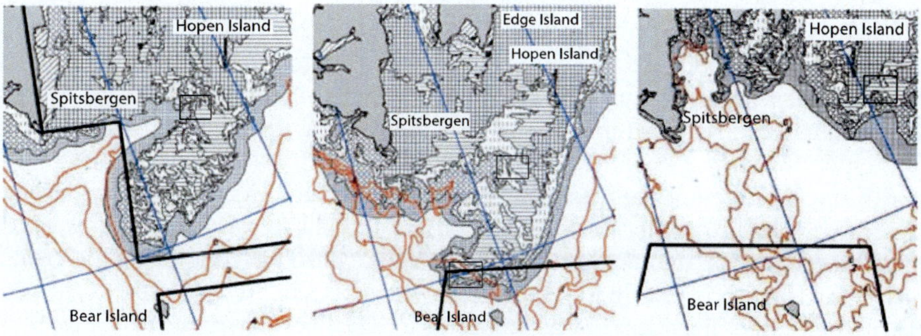

Fig. 10.1 Locations of field works of drift ice on April 28, 2017 (left panel), April 27, 2018 (middle panel) and April 26, 2019 (right panel) are shown by black rectangles

Fig. 10.2 Field works on the drift floe in 2017 (photograph of S. Sicora)

and the field works on the floe were organized by plastic boats Polarcirkel based on Polarsyssel (Fig. 10.3). Many similar floes were observed around the selected floes during the field investigations.

Fig. 10.3 Field works on the drift floe in 2019 (photograph of N. Marchenko)

10.2.1 Morphology and Structure of Drifting Ice Rubble

The floes shapes were analyzed by laser scanning and drilling (Marchenko 2019). In 2017 and 2018 the floes had smooth shapes of their sails and keels, maximal thickness of the floes reached 5–6 m, and they were completely consolidated (Fig. 10.4, left panel). In 2019 the floe sail consisted of numerous ice blocks (Fig. 10.3). Drilling studies performed in selected locations on the floe surface showed that the floe draft exceded 8.5 m and the keel was completely consolidated (Fig. 10.5). Visual observations performed with ROV equipped with StarOddi pressure sensors showed that in some points the floe keel was deeper 8.5 m. The floes shapes shown in Fig. 10.4 (left panel) are similar ice rubble shape since the floes bottoms and surfaces are not flat and relatively smooth. The floe in Fig. 10.3 looks also as a fragment of ice rubble. Right photograph in Fig. 10.4 shows big ice rubble field with vertical dimension of 30–40 m and free board of above 4 m discovered in the region of land fast near the Stone-breen glacier (Marchenko and Marchenko 2017). Fragments of the ice rubble can drift to the South and appear on Spitsbergen Bank.

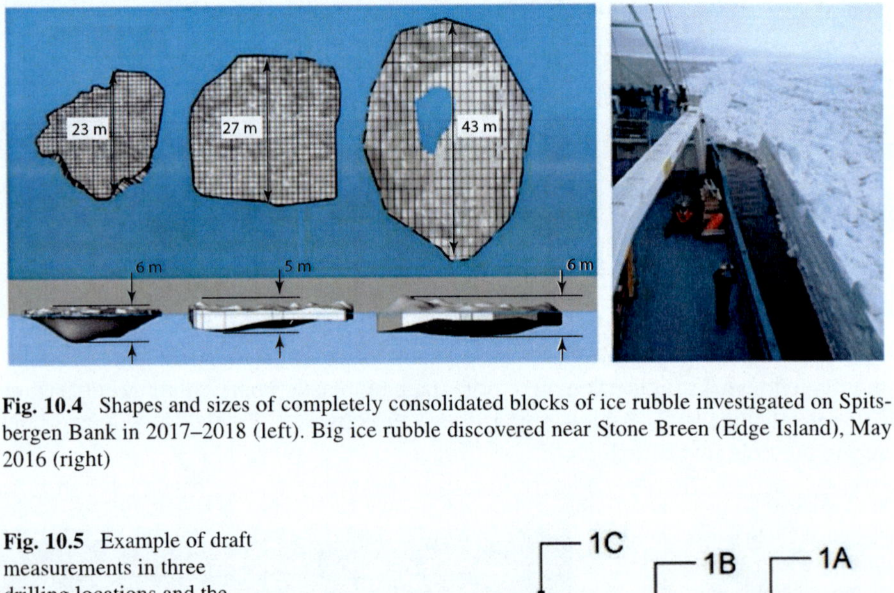

Fig. 10.4 Shapes and sizes of completely consolidated blocks of ice rubble investigated on Spitsbergen Bank in 2017–2018 (left). Big ice rubble discovered near Stone Breen (Edge Island), May 2016 (right)

Fig. 10.5 Example of draft measurements in three drilling locations and the shape of ice rubble sail by the data of laser scanner, Barents Sea—2019

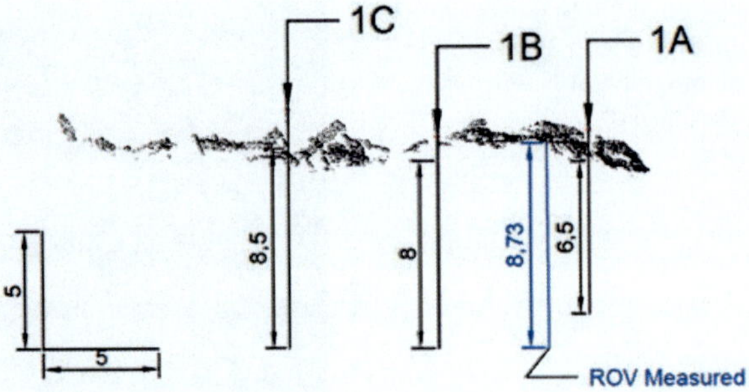

In October 2017 thin sections were made in the UNIS cold laboratory from the ice core taken from the top layer of the floe located at the distance smaller than 1 m from the sea level, and from the ice core taken from the middle layer of the floe located at depth of around 2 m from the sea level. The photographs of the thin sections are shown in Figs. 10.6 and 10.7. Gray strips show length scale of 5 cm. One can see that in the top layer the ice has almost granular structure with small elongation of grains in the vertical direction. In the middle layer the columnar structure is better visible. The sizes of grains shown in Figs. 10.6 and 10.7 are slightly smaller than grain sizes of ice cores taken from the consolidated layer of ice ridges located near the island Hopen in 2002–2005 (Hoyland 2007). Hoyland (2007) also discovered that the ice had less columnar structure in deeper layers of the consolidated layer of the ice ridges in comparison with their surface layer.

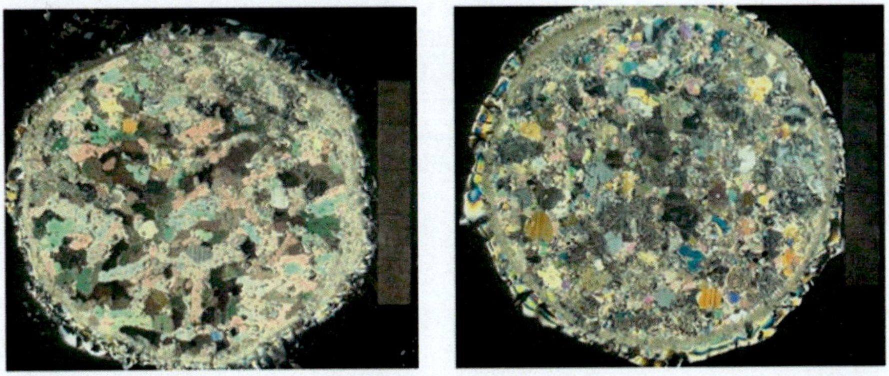

Fig. 10.6 Horizontal thin sections from top (left panel) and middle (right panel) layers of the floe. Barents Sea—2017. Scale of 5 cm is shown to the right

Fig. 10.7 Vertical thin sections from the top (left panel) and middle (right panel) layers of the floe. Barents Sea—2017. Scale of 5 cm is shown to the right

10.2.2 Physical Properties of Ice Rubble

The ice cores were taken from different depths of the floes in 2017 and 2018 to measure the ice temperature and salinity. The temperature was measured every 10 cm directly after taking the cores out. Some parts of the cores were placed in the boxes to measure their salinity after the melting. The results are shown in Figs. 10.8 and 10.9. In 2017 the ice salinity was around 2 ppt and the ice temperature was around the freezing point within an ice layer of 2 m thickness near the keel bottom. Above of this depth the salinity was around 6 ppt, and the ice temperature varied between −2 and −9 °C. Low values of the ice salinity could be explained by a leak of the brine out of the ice cores before they were placed in the boxes. In 2018 the salinity of ice cores taken from the upper layer of the floe was measured of about 4 ppt. The salinity measurements and visual observations showed that the floes consisted of the first-year sea ice with possible inclusions of glacier ice. Potential origin of the ice ridge is near the glacier Stone-Breen on the East coast of Edge Island (Fig. 10.4, right panel) (Marchenko and Marchenko 2017).

Tests on uniaxial compression strength of ice cores taken from the floes were performed onboard of Polarsyssel within few hours after the ice cores were taken out of the floes. The temperature and the salinity of the ice cores were measured after the tests. The compression rig (Kompis) and the testing procedures were described

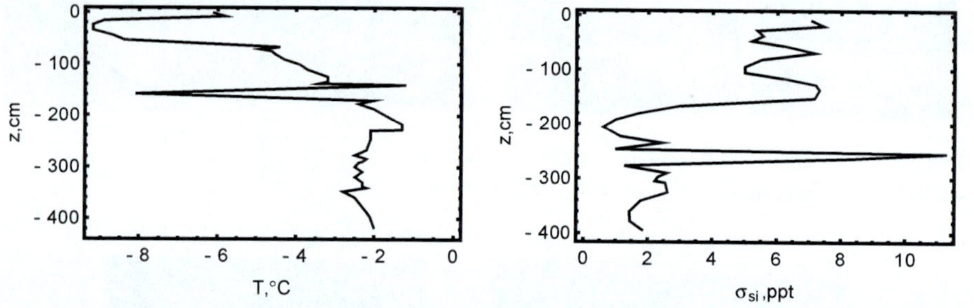

Fig. 10.8 Ice rubble salinity and temperature, Barents Sea—2017

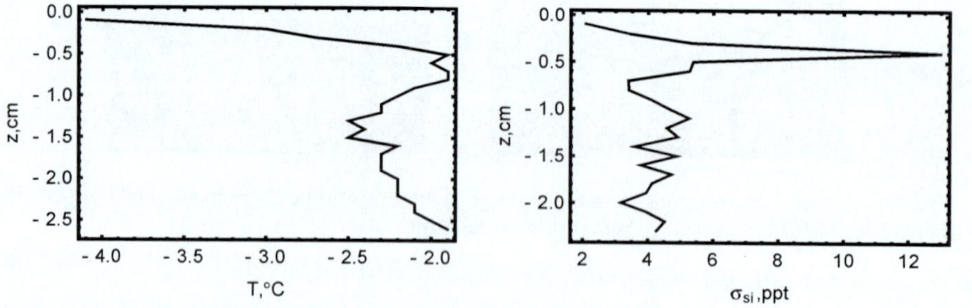

Fig. 10.9 Ice rubble salinity and temperature, Barents Sea—2018

by Moslet (2007). The strength is defined as the maximal force divided by the area of the horizontal cross-section of the sample. The piston speed corresponded to a nominal strain rate of 10^{-4} to 10^{-3} s^{-1}. Results of the tests on compressive strength are shown in Fig. 10.10 versus the liquid brine content calculated by the formula of Frankenstein and Garner (1967).

$$\nu_b = \sigma_{si}\left(\frac{49.185}{|T_{si}|} + 0.532\right), \tag{10.1}$$

where T_{si} and σ_{si} are sea ice temperature and salinity measured after the tests. Higher values of the compression strength in 2019 are explained by taking of several vertical and all horizontal ice cores from the ice rubble above sea surface.

Ice failure in compression tests was either ductile either brittle in 2017, and only ductile in 2019. Types of ice failure were not registered in compression tests in 2018. Obtained values of compressive strength of the ice cores varied in the range from 0.3 to 2.5 MPa, when their liquid brine content decreased from 14 to 1% excluding ice cores taken from the ice blocks above the sea level (Fig. 10.10). Measured in the field compressive strength of ice cores taken from consolidated layer of several ice ridges near Hopen Island in 2004 (Hoyland 2007) varied from 2.17 to 2.9 MPa when the ice porosity decreased from 10–15% to 0–5%. Their porosity was calculated using the data on ice temperature, salinity and density. It can be greater than the liquid brine content because of the gas content. Compressive strength of ice cores measured in the field tests in 2017–2019 is smaller than compressive strengths of ice cores measured in the field tests in 2004, and it is similar to the strength of ice cores taken from the first-year ice ridges in the Fram Strait (Ervik et al. 2018).

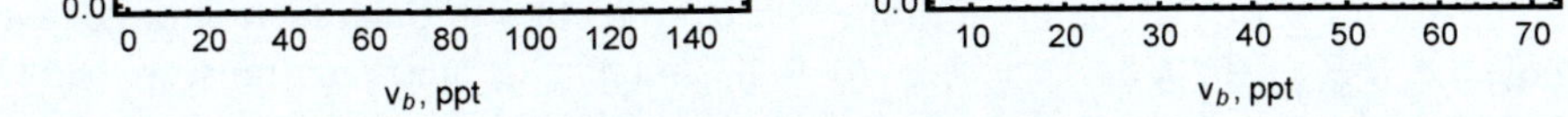

Fig. 10.10 Uniaxial compressive strength of vertical (left) and horizontal (right) ice cores taken from the floes in 2017 (circles), 2018 (squares) and 2019 (rhombuses)

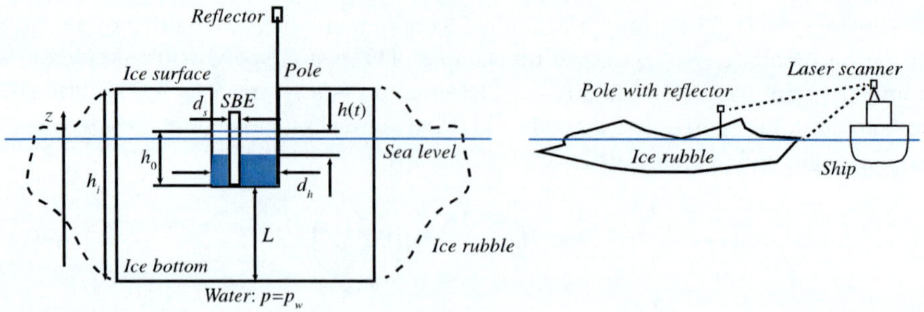

Fig. 10.11 Test on vertical permeability of floating ice (left panel). Laser scanning of the ice rubble from ship (right panel)

10.2.3 Permeability of Ice Rubble

In 2019 four tests on vertical permeability of ice rubble keel were performed during the field works. The scheme of the permeability tests is shown in Fig. 10.11 (left panel). Non-through vertical hole with diameter of 10 cm was drilled in the ice by ice core drill. Pressure recorder was placed at the bottom of the hole immediately after the core was taken out of the hole and recorded water pressure near the bottom of the hole. Water level in the hole increased due to the migration of liquid brine through the bottom and walls of the hole. Pressure and temperature recorders SBE-39 were used in the tests. The records were performed with sampling interval of 1–2 s.

The momentum balance of liquid brine inside the ice is described by the Darcy's equation

$$\nabla P + \frac{\mu}{\kappa}\mathbf{q} - \rho\mathbf{g} = 0, \tag{10.2}$$

where P is the pore pressure of the brine, $\mathbf{q}$ is the brine flux through the unit area of ice surface, $\mathbf{g} = (0, 0, -g)$ is the gravity acceleration, μ is the dynamic viscosity of water, κ is the permeability, and ρ is the brine density.

It is assumed that vertical pressure gradient and columnar structure of ice influence dominant migration of the brine in the vertical direction through the bottom of the hole, i.e. $\nabla P = (0, 0, \partial P/\partial z)$, $\mathbf{q} = (0, 0, q)$ and $q \geq 0$. The water pressure below the ice rubble equals $P_w + P_a$, where P_w is the hydrostatic pressure of water below the ice rubble, and P_a is the atmosphere pressure. The water pressure at the bottom of the hole is $\rho g(h_0 - h) + P_a$, where h_0 and h are shown in Fig. 10.11 (left panel); h_0 is a maximal value of the water level in the hole. Thus, the pressure gradient in Eq. (10.2) equals

$$\frac{\partial P}{\partial z} = -\frac{P_w - \rho g(h_0 - h)}{L} - \frac{\delta_z P}{L}, \tag{10.3}$$

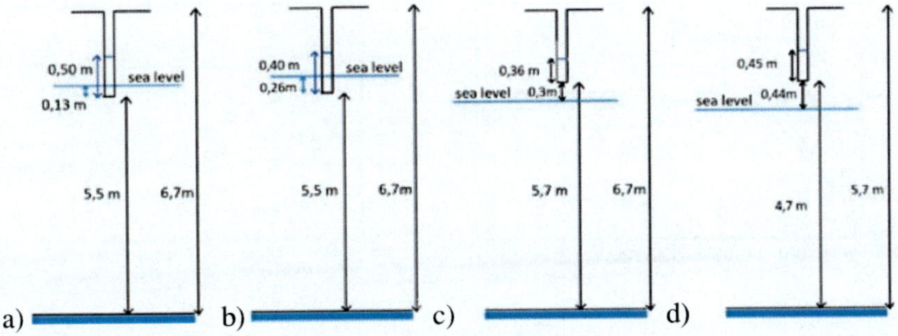

Fig. 10.12 Four drilled holes used in the permeability experiment in 2019

where L is the distance from the bottom of the hole to the bottom of the ice rubble (Fig. 10.11, left panel), and $\delta_z P/L$ is additional pressure gradient which can appear due to lateral compression of the ice $\delta_z P$. The lateral compression of the ice can exist because of spatial variations of ice rubble density related to spatial variations of ice rubble temperature and salinity.

The value of $\delta_z P$ is calculated from (10.2) and (10.3) by $q = 0$ and $h = 0$ as follows

$$\delta_z P = g(\rho(L + h_0) - P_w). \tag{10.4}$$

The mass balance of the brine in the hole is expressed by the equation

$$\frac{dh}{dt} = -q\left(1 - \frac{A}{A_s}\right), \tag{10.5}$$

where $A = \pi d_h^2/4$ is the area of the hole bottom, and $A_s = \pi d_s^2/4$ is the area of the horizontal cross-section of the housing of the pressure recorder. The values $d_h = 10$ cm and $d_s = 4.82$ cm were used for the data processing.

From (10.2)–(10.5) it follows

$$\frac{dh}{dt} = -\gamma h, \gamma = \frac{\rho g \kappa}{\mu L}. \tag{10.6}$$

The solution of (10.6) can be written in the form

$$\frac{h_0 - h(t_1)}{h_0 - h(t_2)} = \frac{1 - e^{-\gamma t_1}}{1 - e^{-\gamma t_2}}, \tag{10.7}$$

where t_1 and t_2 are two arbitrary times belonging to the computational domain.

Figure 10.12 shows the characteristics of four drilling holes used in the experiment on ice rubble permeability in 2019. The pressure records are shown in Fig. 10.13

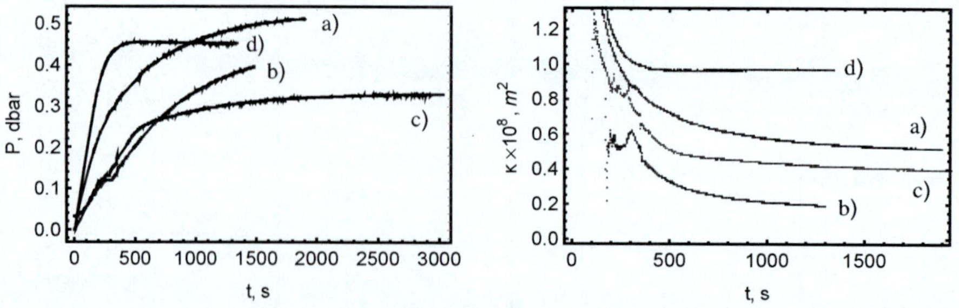

Fig. 10.13 Records of water pressure at the bottom of the drilled holes versus the time (left panel). Ice rubble permeability calculated versus the time (right panel)

(left panel). The permeability of ice rubble calculated from Eq. (10.7) is shown in Fig. 10.13 (right panel) versus the time. One can see that the permeability changes from $0.2 \cdot 10^{-8}$ to 10^{-8} m^2. These values are greater than the permeability of level ice estimated lower than 10^{-9} m^2 (Golden et al. 2007), and correspond well to the permeability of summer Arctic ice (Freitag and Eicken 2003).

Water levels in the drilled holes were compared with sea level using laser scanning from the ship (Fig. 10.11, right panel). Four poles with reflectors on the ends were placed in the drilled holes (Fig. 10.11, left panel), and their vertical coordinates were compared with lowest value of the vertical coordinates of the ice rubble sail. Vertical distances between sea level and the bottoms of the drilled holes were calculated since the lengths of the poles were known and water levels in the holes h_0 were measured. Figure 10.12 shows that brine level in all drilling holes was higher than the sea level. Lateral compression of the ice $\delta_z P$ was estimated from formula (10.4) as 1.3 kPa (a), 2.6 kPa (b), 6.6 kPa (c), and 8.9 kPa (d).

10.2.4 Ocean Heat Flux

Ocean heat flux was calculated using the records of water velocity fluctuations and water temperature fluctuations near ice rubble keels during the field works in 2017, 2018 and 2019. Velocity measurements were performed with acoustic doppler velocimeter (ADV) SonTek ocean probe 5 MHz with sampling frequency of 10 Hz. Water temperature was measured with the recorder SBE-39Plus with sampling frequency of 2 Hz. The deployment scheme is shown in Fig. 10.14. Black points show the locations of measurements of the water velocity and water temperature. The mounted sensors were mounted and left on the ice rubble for several hours. The ship was disconnected from the rubble and drifted on the distance of several hundred meters from the rubble during the measurements.

The vertical ocean heat flux was calculated with the formula

$$Q_w = \rho_{sw} c \langle v_z' T' \rangle, \tag{10.8}$$

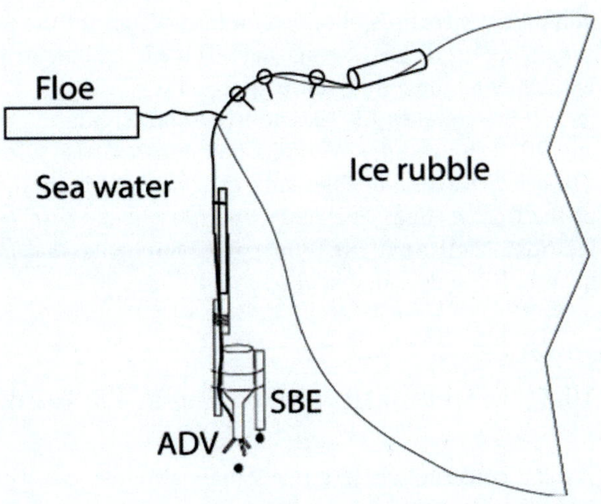

Fig. 10.14 The deployment scheme of sensors ADV SonTek and SBE

where ρ_{sw}=1020 kg/m^3 is the mean sea water density, $c = 4.19$ kJ/kg·C is the specific heat capacity of water, v_z' is the fluctuation of vertical velocity of the water, and T' is the fluctuation of the water temperature. Symbol $\langle\rangle$ means the averaging over the specific interval of the measurements (burst interval), which was programmed to 20 min. The fluctuations were calculated by the formulas

$$v\prime_z = v_z - v_z,\ T\prime = T - \langle T \rangle, \tag{10.9}$$

where v_z and T are measured values of the water velocity and temperature.

Figure 10.15 shows an example of measured in 2017 sea water temperature near the floating rubble, and cumulative heat flux ($E = \int_{t_i}^{t} Q_w dt$) versus the time, where i = I, II, III, and the times t_i corresponds to the beginning of the measurements.

Fig. 10.15 Sea water temperature (left) and cumulative heat flux (energy) (right) versus the time constructed from the data of the field measurements in 2017

Temperature records obtained on three stages of the measurements I, II, and III are shown in Fig. 10.15. During stage III the ice rubble drifted in the region with warmer water having the temperature up to −1.2 °C. The averaged heat fluxes on stages I, II and III are equal to 13, 19, and 91 W/m^2. Similar measurements performed in 2018 and 2019 showed the averaged heat fluxes of 15 W/m^2 and 27 W/m^2 respectively. These heat fluxes are of the same order as ocean heat fluxes measured in other regions of the Barents Sea. Gorbatsky and Marchenko (2007) calculated the heat fluxes to ice ridge keels up to 30 W/m^2, and Marchenko and Hoyland (2008) calculated the heat fluxes to ice ridge keel up to 24 W/m^2.

10.3 Mathematical Modeling of Ice Rubble Consolidation

For the construction of a mathematical model describing thermodynamic consolidation of ice rubble the following assumptions are made. Ice rubble consists of a consolidated layer in the upper part of the rubble and unconsolidated rubble below it (Fig. 10.16). Volumetric portion of water in the rubble is characterized by the macro-porosity p. Macro-porosity of consolidated layer is equal to zero. The thicknesses of consolidated layer and unconsolidated rubble, and the thickness of ice rubble keel are denoted as h_{cl}, h_{ur}, and h_r respectively. Thermal inertia is neglected, and the temperature profile in consolidated layer is assumed to be linear. The temperature of unconsolidated rubble equals to sea water temperature which is equal to the freezing point.

The heat flux directed from the ice into the air influences water freezing between submerged ice blocks. This process leads to the formation of consolidated layer extended from the air–water interface to the depth h_{cl}. The consolidated layer thickness h_{cl} increases with the time depending on the surface heat flux according to the Stefan equation

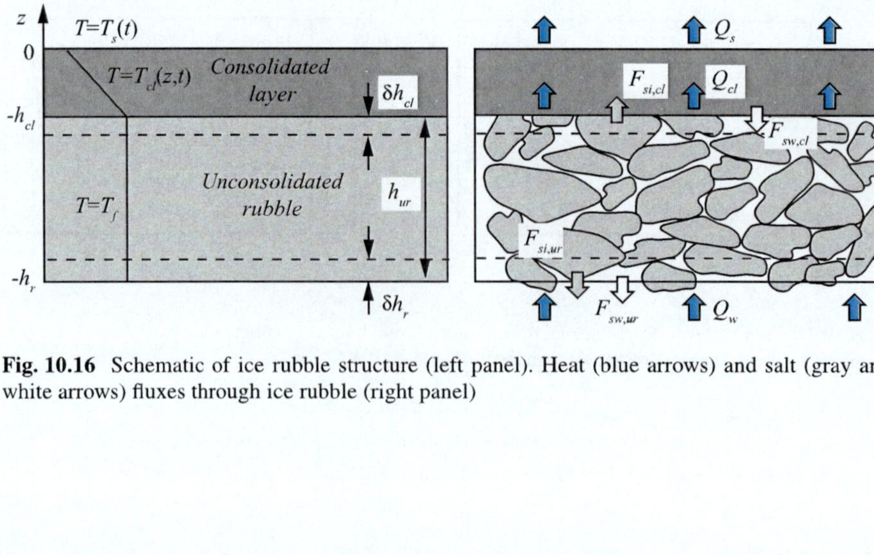

Fig. 10.16 Schematic of ice rubble structure (left panel). Heat (blue arrows) and salt (gray and white arrows) fluxes through ice rubble (right panel)

$$\rho_{si} p L_{si} \frac{dh_{cl}}{dt} = Q_{cl}, \tag{10.10}$$

where ρ_{si} is the sea density, L_{si} is the latent heat of sea ice, and Q_{cl} is the heat flux in the consolidated layer. The heat flux is proportional to the vertical temperature gradient

$$Q_{cl} = k \frac{\Delta T}{h_{cl}}, \tag{10.11}$$

where ΔT is the difference between the temperatures at the bottom and the surface of consolidated layer, and k is the thermal conductivity of ice.

Ice rubble melts from below due to the action of the ocean heat flux, and melt water penetrates inside the rubble. As a result, its thickness and macro-porosity decrease. The Stefan boundary condition at the bottom of unconsolidated rubble is written as follows

$$\rho_{si}(1-p) L_{si} \frac{dh_r}{dt} = -Q_w, \, h_r = h_{cl} + h_{ur}. \tag{10.12}$$

The amount of salts in a vertical column of unconsolidated rubble with the horizontal cross-section of unit area equals

$$S_{ur} = h_{ur}[\sigma \rho p + \sigma_{si} \rho_{si}(1-p)], \tag{10.13}$$

where ρ and σ are the density and the salinity of sea water brine at the freezing point, and S_{ur} is given in kg/m^2. The amount of salts changes due to the changes of macro-porosity p and the thickness h_{ur} of unconsolidated rubble.

Salt fluxes at the boundaries of unconsolidated rubbles influence the water salinity inside it and may cause the formation of new ice or melting of existing ice. The salt flux at the bottom of consolidated layer consists of two part. The first part ($F_{sw,cl} > 0$) describes salt expulsion into the water due to the growth of consolidated layer

$$F_{sw,cl} = p(\sigma \rho - \sigma_{si} \rho_{si}) \frac{dh_{cl}}{dt}. \tag{10.14}$$

The second part describes salt deflux ($F_{si,cl} < 0$) from unconsolidated rubble due to the incorporation of ice blocks into consolidated layer

$$F_{si,cl} = -\sigma_{si} \rho_{si}(1-p) \frac{dh_{cl}}{dt}. \tag{10.15}$$

The salt flux at the bottom of unconsolidated rubble consists also of two parts. The first part ($F_{si,p} < 0$) is related to the extrusion of sea water trapped inside the rubble by the melt water

$$F_{si,ur} = \rho_{si}(1-p)(\sigma-\sigma_{si})\frac{dh_r}{dt}. \tag{10.16}$$

The second part ($F_{sw,r} < 0$) is related to salt deflux due to the discharge of water trapped inside unconsolidated rubble

$$F_{sw,ur} = \sigma\rho p\frac{dh_r}{dt}. \tag{10.17}$$

The salt balance is expressed by the equation

$$\frac{dS_{ur}}{dt} = F_{sw,cl} + F_{si,cl} + F_{si,ur} + F_{sw,ur}. \tag{10.18}$$

The directions of the heat and salt fluxes through unconsolidated rubble are shown in Fig. 10.16 (right panel).

Substitution of formulas (10.13)–(10.17) into Eq. (10.18) leads to the equation

$$\rho_{si}L_{si}(h_r-h_{cl})(\sigma\rho-\sigma_{si}\rho_{si})\frac{dp}{dt} = (2\sigma\rho-\sigma_{si}\rho_{si})Q_{cl} - \rho_{si}(\sigma-2\sigma_{si})Q_w. \tag{10.19}$$

Three Eqs. (10.10), (10.12) and (10.19) describe temporal evolution of three functions $h_{cl}(t)$, $h_r(t)$ and $p(t)$. Further these equations are investigated in the dimensionless variables

$$\tau = \frac{t}{t^*}, \quad \eta_{cl} = \frac{h_{cl}}{h_0}, \quad \eta_r = \frac{h_r}{h_0}. \tag{10.20}$$

The representative time t^* is determined by the formulas

$$t^* = h_0\rho_{si}L_{si}/Q^*, \quad Q^* = k\cdot 1\,°\text{C}/h_0, \quad h_0 = 1\cdot\text{m}, \tag{10.21}$$

where Q^* is the heat flux over the ice layer with the thickness of 1 m caused by 1 °C difference of the ice temperature at the boundaries of the layer. Since $k \approx 2$ W/(m·C) then $Q^* \approx 2$ W/m^2. Assuming $\rho_{si} = 920$ kg/m^3 and $L_{si} = 300$ kJ/kg we find that $t^* \approx 1597$ days.

Equations (10.10), (10.12) and (10.19) are written in the dimensionless variables in the form

$$\frac{d\eta_{cl}}{d\tau} = \frac{\Delta_T}{\eta_{cl}p}, \quad \frac{d\eta_r}{d\tau} = -\frac{c}{1-p}, \quad \frac{dp}{d\tau} = \frac{a\Delta_T}{\eta_{cl}(\eta_r-\eta_{cl})}\left(1-\frac{bc}{\Delta_T}\eta_{cl}\right), \tag{10.22}$$

$$a = \frac{2\sigma\rho-\rho_{si}\sigma_{si}}{\sigma\rho-\rho_{si}\sigma_{si}}, \ b = \frac{\rho_{si}(\sigma-2\sigma_{si})}{2\sigma\rho-\rho_{si}\sigma_{si}}, \ c = \frac{Q_w}{Q^*}, \ \Delta_T = \Delta T/1\,°\text{C}.$$

Densities of sea water brine at the freezing point, ice and sea ice are determined by the formulas (Schwerdtfeger 1963)

$$\rho = 999(1+\alpha T)\,\mathrm{kg/m^3},\ \rho_i = \frac{916}{1+1.54\cdot 10^{-4}T}\,\mathrm{kg/m^3},$$
$$\rho_{si} = \frac{\alpha\rho\rho_i T}{\sigma_{si}(\rho_i-\rho)+\alpha\rho T(1-\sigma_{si})}\,\mathrm{kg/m^3}, \tag{10.23}$$

where the temperature is given in Celsius degrees, and $\alpha = -0.0182\,\mathrm{C}^{-1}$. Sea water brine salinity equals $\sigma = \alpha T/(1+\alpha T)$. Numerical estimates show that $a \approx 2.24$ and $b \approx 0.4$ when the ice and the water temperatures are $T = -1.9$ °C, and the ice salinity is $\sigma_{si} = 7$ ppt.

The consolidation is now investigated when $\Delta_T = 0$, i.e. the heat flux into the air is absent and the thickness of consolidated layer is zero. The second and the third Eqs. (10.22) are written in the form

$$\frac{d\eta_r}{d\tau} = -\frac{c}{1-p_{lp}},\qquad \frac{dp}{d\tau} = -\frac{abc}{\eta_r}. \tag{10.24}$$

Integration of Eqs. (10.24) leads to the formula

$$\left(\eta_{r,0}/\eta_r\right)^{ab} = \frac{1-p}{1-p_0}. \tag{10.25}$$

Substitution of formula (10.25) into the second Eq. (10.24) and integration give the dimensionless time during which the macro-porosity of unconsolidated rubble changes from p_0 to p

$$\Delta\tau = \frac{\eta_{r,0}}{c}F(p,p_0),\qquad F = \frac{(1-p_0)^{1/ab}\left((1-p)^{1-1/ab}+(1-p_0)^{1-1/ab}\right)}{ab-1}, \tag{10.26}$$

where $\eta_{r,0}$ is the initial value of the ice rubble draft.

Figure 10.17 shows the dependence of F from p constructed with two values of the initial macro-porosity: $p_0 = 0.2$ and $p_0 = 0.3$. Assuming the ocean heat flux $Q_w = 20$ W/m^2 and $\eta_{r,0} = 10$ (the dimensional draft is 10 m) we find that $\frac{\eta_{r,0}}{c} = 1$. In this case the dimensionless time to complete consolidation of the rubble ($p = 0$) is $\Delta\tau = 0.2$ when $p_0 = 0.2$, and $\Delta\tau = 0.27$ when $p_0 = 0.3$. The dimensional times equal 319 days and 431 days respectively. The dimensional times of the consolidation of the rubble to the porosity $p = 0.1$ are 159.7 days and 303 days respectively. These times decrease when the initial draft of the rubble decreases and the ocean heat flux increases.

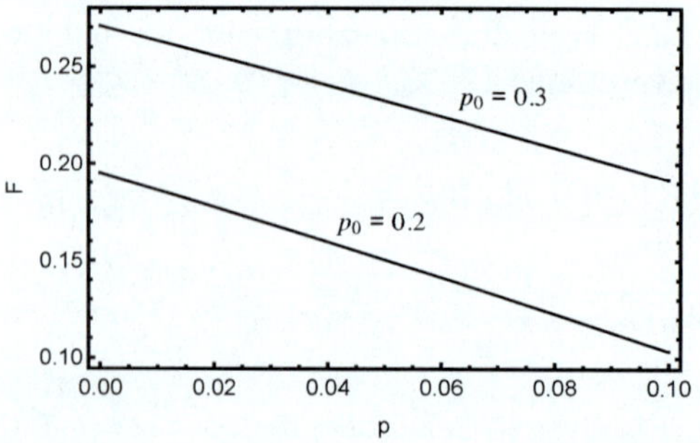

Fig. 10.17 Graphs of the function $F(p, p_0)$ constructed with $p_0 = 0.2$ and $p_0 = 0.3$

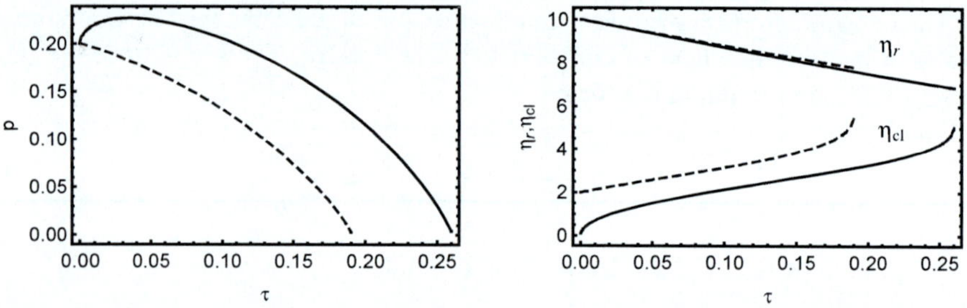

Fig. 10.18 Macro-porosity of unconsolidated rubble (left panel), thickness of consolidated layer and ice rubble draft (right panel) versus the dimensionless time calculated with $Q_w = 20$ W/m^2 and $\Delta_T = 5$. Initial thickness of consolidated layer is 0.1 m (solid lines) and 2 m (dashed lines)

Figure 10.18 shows results of numerical simulations of Eqs. (10.22) with $Q_w = 20$ W/m^2 and $\Delta_T = 5$. The initial macro-porosity equals 0.2. The simulations were performed with two values of the initial thickness of consolidated layer: $\eta_{cl,0} = 0.1$ and $\eta_{cl,0} = 2$. In the dimensional variables it means 0.1 m and 2 m respectively. One can see that the times to complete consolidation are greater when $\Delta_T = 5$ than by $\Delta_T = 0$. It is explained by the influence of brine rejection into unconsolidated rubble due to the growth of consolidated layer. Solid line in the left panel of Fig. 10.17 shows that the brine rejection influences an increase the macro-porosity by $\tau < 0.03$. The complete consolidation occurs over 415 days when $\eta_{cl,0} = 0.1$ and over 303 days when $\eta_{cl,0} = 2$. The consolidation to macro-porosity of 0.1 occurs over 335 days when $\eta_{cl,0} = 0.1$ and over 207 days when $\eta_{cl,0} = 2$.

10.4 Discussion and Conclusions

Field investigations performed in the end of April (2017–2019) in the Barents Sea region extended between Spitsbergen, Hopen Island and Bear Island where the sea surface was covered by broken ice. Among flat floes with thickness of about 30 cm ice floes with drafts 5–10 m and diameters up to 30 m were discovered. Drilling studies showed that they are completely consolidated. Masses of these ice formations were estimated from one to five thousand tons. Visual observations and salinity measurements support that thermodynamic consolidation can explain their formation from the first-year ice rubble.

Temperature of ice rubble keels was at the freezing point of surrounded sea water. Uniaxial compressive strength of ice cores taken from the floes was smaller than the compressive strength of ice cores taken from the consolidated layer of drifting ice ridges in the Hopen area in 2004 (Hoyland 2007), and similar to the strength of ice cores taken from the first-year ice ridges in the Fram Strait (Ervik et al. 2018). The permeability of the ice rubble keels corresponds well to the permeability of summer Arctic ice (Freitag and Eicken 2003).

The mean ocean heat flux registered below the ice rubble in 2017–2019 ($\sim$20 W/m^2) was similar the heat fluxes registered below level ice, ice ridges and inside ice ridges during field works on larger fields of drift ice in the Barents Sea (Gorbatsky and Marchenko 2007; Marchenko and Hoyland 2008). In 2017 the heat flux rose up to 91 W/m^2, when the ice rubble drifted into the warm water with temperature $-1.5°$ to -1.2 °C. The warm Atlantic waters interact in the region with the cold Arctic waters coming from the North by the East Spitsbergen current, and this interaction increases the ocean heat flux. Similar effects were observed in the Nansen Basin where the cold Arctic waters interact with the warm Atlantic waters (Peterson et al. 2017; Shestov et al. 2018.

A mathematical model of thermodynamic consolidation of ice rubble was elaborated to investigate the influence of melt water on the consolidation of ice rubble. The melt water is formed at the bottom of ice rubble due to the ice melt under the influence of the ocean heat flux. This effect is usually ignored in the models describing thermodynamic consolidation of ice rubble, where the growth of consolidated layer occurs due to the atmosphere cooling. In this approach the thickness of consolidated layer is approximately two times greater than the thickness of level ice subjected to the same atmospheric cooling. For the Barents Sea it means that the thickness of consolidated layer of the first-year ice rubble is smaller 2 m. Therefore, the atmosphere cooling can't explain the formation of observed completely consolidated ice rubble with drafts of 5–10 m.

In the considered model the macro-porosity of unconsolidated rubble changes under the influence of salt fluxes at the boundaries with the consolidated layer and ocean. The macro-porosity of ice rubble adjusts to the changes of water salinity inside the rubble by the formation of new ice or melting of existing ice. The temperature of unconsolidated rubble and water salinity inside the rubble are equal to the temperature and salinity of sea water below the rubble. The ocean heat flux, the initial macro-porosity and the initial draft of the rubble controls the consolidation process if the heat flux into the atmosphere is small. Numerical simulations showed that the time of the consolidation of ice rubble with 10 m draft and initial macro-porosity of 0.2 to the macro-porosity of 0.1 takes of about 6–7 months (smaller than 210 days) when the ocean heat flux is 20 W/m^2. The consolidation time of ice rubbles with smaller drafts is shorter and it is inversely proportional to the ocean heat flux.

Acknowledgements The author wishes to acknowledge the support of the Research Council of Norway through the SFI SAMCoT and AOCEC (IntPart), and help of students of AT-211 course at UNIS.

References

A. Ervik, K.V. Hoyland, A. Shestov, T.S. Nord, On the decay of first year ice ridges: Measurements and evolution of rubble macroporosity, ridge drilling resistance and consolidated layer strength. Cold. Reg. Sci. Techn. **151**, 196–207 (2018)

G.E. Frankenstein, R. Garner, Equations for determining the brine volume of sea ice from −0.5 to −22.9°C. J. Glaciol. **6**(48), 943–944 (1967)

J. Freitag, H. Eicken, Meltwater circulation and permeability of Arctic summer sea ice derived from hydrological field experiments. J. Glaciology **49**(166), 349–358 (2003)

K.M. Golden, H. Eicken, A.L. Heaton, J. Miner, D.J. Pringle, J. Zhu, Thermal evolution of permeability and microstructure in Sea ice. Geoph. Res. Lett. **34**, L16501 (2007)

V.V. Gorbatsky, A.V. Marchenko, On the influence of turbulence in ice adjacent layer on water-ice drag forces and heat fluxes in the Barents Sea. Recent development of Offshore Engineering in Cold Regions, in *POAC-07*, Dalian, China, June 27–30, 2007, ed by Yue (Dalian University Press, Dalian, 2007), pp. 648–659.

Y.P. Gudoshnikov, G.K. Zubakin, A.K. Naumov, Morphological characteristics of ice formations of Pechora Sea by multiyear data of the field works, in *Proceedings of the RAO-03* (St'Petersburg, Russia, 2003), pp. 295–299

R.B. Guzenko, Y.U. Mironov, R.I. May, V.S. Porubayev, V.V. Kharitonov, S.V. Khotchenkov, K.A. Kornishin, Y.O. Efimov, P.A. Tarasov, Morphometry and internal structure of ice ridges in the Kara and Laptev Seas, in *Proceeding of the Twenty-Ninth International Ocean And Polar Engineering Conference*, Honolulu, Hawaii, USA (2019), pp. 647–654

K.V. Hoyland, S. Loset, Measurements of temperature distribution, consolidation and morphology of a first-year sea ice ridge. Cold. Reg. Sci. Techn. **29**, 59–74 (1999)

K.V. Hoyland, Consolidation of first-year sea ice ridges. J. Geophys. Res. **107**(C6), 3062 (2002)

K.V. Hoyland, P. Liferov, On the initial phase of consolidation. Cold. Reg. Sci. Techn. **41**, 49–59 (2005)

K.V. Hoyland, Morphology and small-scale strength of ridges in the North-western Barents Sea. Cold. Reg. Sci. Techn. **48**, 169–187 (2007)

K.V. Hoyland, S. Barrault, S. Gerland, H. Goodwin, M. Nicolaus, O.M. Olsen, E. Rinne, The consolidation in second- and multi-year sea ice ridges Part I: Measurements in early winter, in *Proceedings of 19th IAHR International Symposium on Ice*, Vancouver, BC, Canada (2008), pp. 1439–1449

V.V. Kharitonov, Internal structure of ice ridges and stamukhas based on thermal drilling data. Cold. Reg. Sci. Techn. **52**, 302–325 (2008)

V.V. Kharitonov, Internal structure and porosity of ice ridges investigated at "North Pole-38" drifting station. Cold. Reg. Sci. Techn. **82**, 144–152 (2012)

V.V. Kharitonov, On the results of studying ice ridges in the Shokal'skogo Strait, part II: porosity. Cold. Reg. Sci. Techn., 166, 102842 (2019)

V Kovacs, Characteristics of multi-year pressure ridges, in *Proceedings of POAC-83*, Helsinki, Finland, vol. 2 (1983), 173–182

M. Lepparanta, M. Lensu, P. Koslof, B. Veitch, The life story of a first-year sea ice ridge. Cold. Reg. Sci. Techn. **23**, 279–290 (1995)

A. Marchenko, K. Hoyland, Properties of sea currents around ridged ice in the Barents Sea, in *Proceedings of 19th IAHR International Symposium on Ice*, Vancouver, BC, Canada (2008), pp. 1251–1262

A. Marchenko, Thermodynamic consolidation and melting of sea ice ridges. Cold Reg. Sci. Techn. **52**, 278–301 (2008)

N. Marchenko, A. Marchenko, Investigation of large ice rubble field in the Barents Sea. POAC17-097 (2017)

A. Marchenko, Influence of the water temperature on thermodynamic consolidation of ice rubble, in *Proceedings of the 24th IAHR International Symposium on Ice*, Vladivostok, Russia (2018), p. 18-005.

N. Marchenko, Ice at its southern limit in the Barents Sea: field investigation near Bear Island in April 2017–2018. POAC19-58 (2019)

P.O. Moslet, Field testing of uniaxial compression strength of columnar sea ice. Cold Reg. Sci. Techn. **48**, 1–14 (2007)

A.K. Naumov, E.A. Skutina, N.V. Golovin, N.V. Kubyshkin, I.V. Buzin, Y.P. Gudoshnikov, A.A. Skutin, in *Proceeding of the Twenty-Ninth International Ocean and Polar Engineering Conference*, Honolulu, Hawaii, USA (2019), pp. 684–690

A.K. Peterson, I. Fer, M.G. McPhee, A. Randelhoff, Turbulent heat and momentum fluxes in the upper ocean under Arctic sea ice. J. Geophys. Res Ocean. **122**, 1439–1456 (2017)

P. Schwerdtfeger, The thermal properties of sea ice. J. Claciol. **4**(36), 789–807 (1963)

A. Shestov, K.V. Høyland, O.C. Ekeberg, Morphology and physical properties of old sea ice in the Fram Strait 2006–2011, in *Proceedings of the 21th International Symposium on Ice (IAHR)*, Dalian, China. Paper P052 (2012)

A.S. Shestov, A.V. Marchenko, Thermodynamic consolidation of ice ridge keels in water at varying freezing points. Cold Reg. Sci. Techn. **121**, 1–10 (2016)

A.S. Shestov, K.V. Høyland, Å. Ervik, Decay phase thermodynamic of ice ridges in the Arctic Ocean. Cold Reg. Sci. Technol. **152**, 23–34 (2018)

J. Stefan, Uber die theorie der eisbildung, insbesondere uber die eisbildung in polararmeere. Ann. Phys. **42**(2), 269–286 (1891)

L. Strub-Klein, S. Barrault, H. Goodwin, S. Gerland, Physical properties and comparison of first- and second-year sea ice ridges. POAC09-117 (2009)

L. Strub-Klein, D. Sudom, A comprehensive analysis of the morphology of first-year sea ice ridges. Cold Reg. Sci. Techn. **82**, 94–109 (2012)

G.M. Timco, R.P. Burden, An analysis of the shapes of sea ice ridges. Cold Reg. Sci. Techn. **25**, 65–77 (1997)

G.K. Zubakin, Y.P. Gudoshnikov, A.K. Naumov, I.V. Stepanov, N.V. Kubyshkin, Peculiarities of the structure and properties of ice ridges in the Eastern Barents Sea based on the 2003 expedition data. Int. J. Offshore Polar Eng. **15**(1), 28–33 (2005)

Part V
Computational Ice Mechanics

Chapter 11
Ridge Load on the Monopile—A Comparison Between FEM-CEL–Simulations and ISO 19906

Jaakko Heinonen

The shallow coastline and the consistent wind conditions in the Gulf of Bothnia provide a good environment for wind energy production. However, the sea freezes annually, introducing the most significant uncertainties in the support structure design for offshore wind turbines. Ice ridges are common ice features in the Gulf of Bothnia. Therefore, the interaction with an ice ridge must be considered in the structural design for support structures of offshore wind turbines. A special emphasis was made on narrow structures.

This presentation focused to ridge keel loads on the cylindrical bottom-fixed monopiles (diameter of 6.0 m). The ridge-structure interaction was studied by FEM-CEL-simulations. The ice rubble was modelled as a continuum with a shear-cap failure criterion, which describes both the shear failure and compaction. Mechanical properties of the ridge keel are essential input parameters to predict ridge loads. The rubble parameters, e.g. cohesion and friction angle, were selected based on ridge keel punch shear tests in the Gulf of Bothnia. The shape of ridge keel was chosen according to guidelines in ISO 19960 standard with the maximum thickness of 8.0 m. In the simulations, the ridge was drifting with a constant velocity against the structure.

Numerical simulations resulted in the ridge keel forces on the structure. I addition, the simulations created more knowledge about various failure modes and how the failure progression affects the keel load.

A comparison between the simulated maximum keel loads and the modified Dolgolov's analytical model, which is recommended by the ISO 19906 standard, was carried out. The reasons behind much higher ridge load estimate based on the ISO standard were discussed.

J. Heinonen (✉)
VTT Technical Research Centre of Finland Ltd., Tietotie 1A2, P.O.Box 1000, FIN-02044 Espoo, Finland
e-mail: Jaakko.Heinonen@vtt.fi

J. Tuhkuri and A. Polojärvi (eds.), *IUTAM Symposium on Physics and Mechanics of Sea Ice*, IUTAM Bookseries 39, https://doi.org/10.1007/978-3-030-80439-8_11

11.1 Introduction

Ice ridges are deformed sea ice features. Ridges are commonly found in Northern seas and they usually introduce a major load scenario of offshore structures. Ice ridges consist of a sail above the water and a keel beneath the water line. In first-year ridges the keel consists of an upper refrozen or consolidated layer, and unconsolidated rubble below. The strength of the consolidated layer is about same than as in the level ice, whereas ice rubble is significantly weaker. The sail is usually ignored in the ridge load predictions, because its load contribution is minimal compared to the consolidated layer and keel.

Even though the ridges introduce an important design loading case, the guidance is fairly poor. Serré and Liferov (2010) and Croasdale (2011) both concluded that regarding the ridge or rubble loads the ISO 19906 (2010) may lead non-conservative solutions. Serré and Liferov (2010) demonstrated based on model-scale studies that the ISO guideline for the ridge load on conical structures (diameter 14 m in full-scale) is underconservative. The load underestimation was also observed for a narrow cylindrical structure (diameter 1 m in full-scale), but it depended much on the surcharge effect, weather the load underestimation was considered or not. Croasdale (2011) studied a wide sloping structure (~hundred meters wide) in shallow water, in which the rubbling of broken ice plays an important role in ice-interaction, but poorly considered in the ISO standard causing load underestimation. It is however important to notice that this conclusion regarding the underestimation in the ISO standard is limited to selected case studies with certain structural dimensions and ridge size. Also, the conclusions are based on either model-scale or numerical studies. Therefore, there is a crucial need for further knowledge of ridge-structure interaction.

The sail and keel in the ridge are composed of large number of ice blocks, which are considered to be partly frozen together. When the ridge interacts with an offshore structure, the process can be considered to be relatively slow. In that case the load applied to the ice rubble is transmitted by contact forces developed between adjacent blocks, along a skeleton type of structure. The deformations in the rubble are mainly governed by the interaction between individual ice blocks. The ice rubble undergoes large deformations during the ridge-structure interaction. The main rubble deformation mechanisms are caused by the freeze-bond failure, by individual ice block rotation and displacement in contact with neighbouring blocks and by individual ice block failure. Due to complicated nature of ice rubble, the modelling of deformation and failure processes is demanding. The failure criterion has often been modelled based on the mechanics of granular media, e.g. using Mohr–Coulomb criterion by Heinonen (1999) and Drucker-Prager-Cap models (DPC) by Serré and Liferov (2010) and by Serré (2011a, b). A specific variation of DPC, a shear-cap model, was introduced by Heinonen (2004).

Recently, Kulyakhtin and Høyland (2015) has utilized Cam-Clay model, which was shown to offer better control of how the volumetric changes influence to the shear strength compared to the Mohr–Coulomb criterion.

The Coupled Eulerian–Lagrangian (CEL) framework based on the continuum approach enables robust modelling of large deformations due to material failure. In a Lagrangian analysis nodes are fixed within the material, and elements deform as the material deforms and each element is 100% full of a material. In an Eulerian analysis nodes are fixed in space and the loading cause the ice rubble to flow through the element mesh that do not deform. Therefore, typical mesh distortion problems presented by the Lagrangian framework can be avoided. Eulerian elements may be partially or completely filled of material or void (Simulia Abaqus 6.14 user's guide). The user may consider Eulerian material (i.e. ice rubble) to interact with Lagrangian elements (i.e. offshore structure) through Eulerian–Lagrangian contact, which is a typical application of the CEL method.

Previous studies of CEL have been focused to model-scale studies: punch test and oedometer studies (Heinonen and Høyland 2013; Serré 2011a; Heinonen 2016), ridge actions on a conical structure (Serré and Liferov 2010) and ridge keel action on subsea structures (Serré 2011b). Furthermore, Patil et al. (2015) have applied smooth particle hydrodynamics (SPH) method to simulate full-scale punch tests. As in the CEL, an advantage of using SPH is to avoid undesired mesh distortion caused by large deformation. Despite using a continuum approach for ice rubble, SPH formulation can be used to simulate discrete nature of rubble. However, Patil et al. (2015) reported about restrictions within SPH method, namely the friction between particles cannot be introduced.

The discrete element method (DEM) has been recently studied widely to model ice failure and interaction processes in the ice rubble within various applications with promising results (e.g. Tuhkuri and Polojärvi 2018). The main idea in DEM is to model each individual ice block in the rubble as a single particle and by modelling the interaction between the blocks by contact laws. The latter is often applied by the force–displacement relationship, which can be defined separately in the contact normal and tangential direction. This makes the modelling of ice rubble more realistic compared to the continuum approaches. Once the DEM models the internal rubble texture more in details, more information is required. Even though the nature of discrete modelling corresponds well to complicated rubble texture and topology, one major challenge is to determine correctly the shape of ice blocks in rubble. The individual particle shape, the shape and size distributions are not well known for real ridges and therefore difficult to replicate in the numerical model. The effect of internal ice block structure to the ice failure process is also not well known. Ice block interaction—the freeze bonds—are also challenging to model because so far the bond strength and corresponding contact area as well as their relation to the ridge consolidation are not well known. Computational resources for the DEM simulations are typically very high, but recent developments in the super computing and DEM software development based on e.g. a non-smooth discrete element modelling decrease the computation time drastically (van den Berg et al. 2018). Despite these challenges, there exist several good attempts to model the rubble failure processes with the DEM e.g. Polojärvi and Tuhkuri (2013).

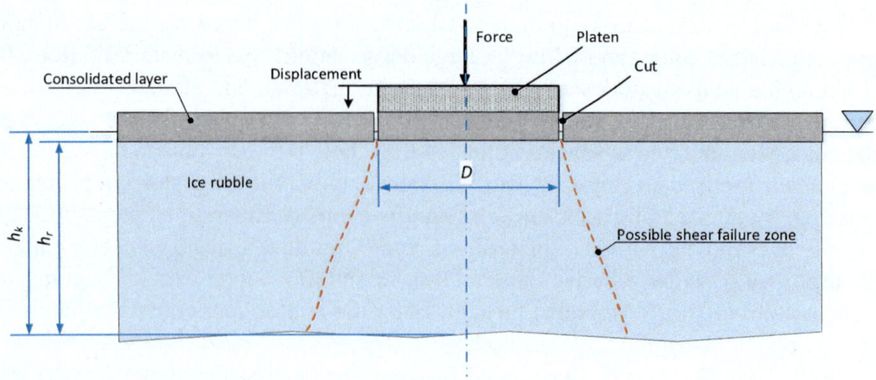

Fig. 11.1 Principal sketch of the punch shear test

11.2 Punch Shear Tests

In-situ full-scale loading tests were conducted during five winters (1998–2003) in the Northern Gulf of Bothnia in order to study the failure mechanisms of the keel and to measure its mechanical properties because they are essential input parameters to predict ridge loads. The rubble parameters, e.g. cohesion and friction angle, were selected based on ridge keel punch shear tests in the Gulf of Bothnia. In a punch test, a circular cut was first made along the perimeter of the platen. The cut was extended through the consolidated layer. In the test, the platen was pushed downwards to until the keel underneath broke totally (Fig. 11.1). In various tests the diameter varied between 2.5 and 4.7 m, the keel depth between 3.0 and 6.4 m and the maximum stroke was 0.7 m. The force and displacement of the platen was measured. Also, at selected locations the displacements both inside and at the bottom of the keel were measured. The measurements were combined together with visual observations of the keel failure and deformations. Experimental studies and the evaluation of the material parameters based on finite element simulations have been presented in detail by Heinonen (2004). Also, some individual tests have been studied by the discrete element simulation (DEM) (Sorsimo and Heinonen 2019; Polojärvi and Tuhkuri 2013).

11.3 Material Model for Ice Rubble

For modeling a failure process in the ice rubble, a shear-cap failure criterion was earlier developed by the author (Heinonen 2004). This model is suitable for numerical continuum based finite element simulations. The main failure mechanisms modelled by the shear-cap model are the shear failure and compaction of ice rubble. As in the other frictional-cohesive material models, the main parameters in the shear-cap

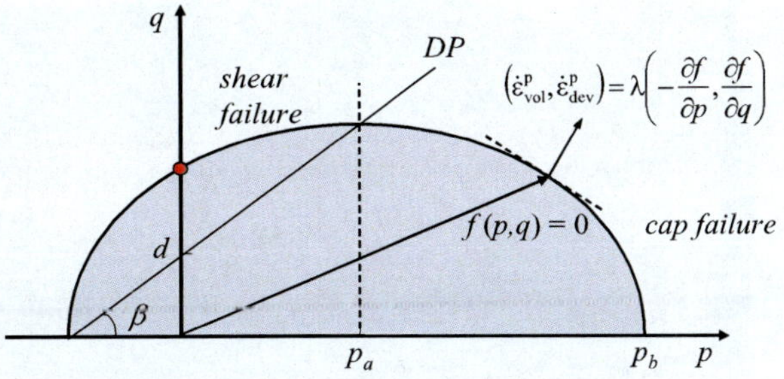

Fig. 11.2 Shear-cap yield function in the meridian plane. p is the hydrostatic stress and q is the second deviatoric stress invariant (von Mises stress). The straight line describes the corresponding Drucker-Prager failure surface

model to describe the shear failure are the cohesion and friction angle. The post failure behaviour is modelled by the cohesive softening. Due to the porous nature of ice rubble, the modelling of volumetric behaviour is essential. Volumetric change in the rubble is modelled using the cap hardening feature. Deformations in the failure process are modelled with an associative flow rule which describes dilatation during the shear failure and compaction during the cap failure.

The failure criterion combines two elliptical curves in the meridian plane: One for the shear failure f_s and one for the cap (compaction) failure f_c:

$$f_s = \sqrt{[(p - p_a)\tan\beta]^2 + q^2} - (d + p_a \tan\beta); p \le p_a : \text{shear failure}$$

$$f_c = \sqrt{(p - p_a)^2 + (Rq)^2} - R(d + p_a \tan\beta); p > p_a : \text{compaction failure} \tag{11.1}$$

in which d and β are the corresponding Drucker-Prager parameters for the cohesion and friction angle. Parameter R defines the cap shape and p_a describes the hydrostatic stress that divides the shear and cap failure parts, as shown in Fig. 11.2. More details about the material model are described in Heinonen (2004).

11.4 Numerical Model for Monopile Interaction with Ridge Keel

The finite element model for a monopile interaction with the ridge keel is shown in Fig. 11.3. A commercial software Abaqus/Explicit version 6.12 was applied for numerical simulations using Coupled Eulerian–Lagrangian (CEL) framework. The shear-cap material model was implemented via user subroutine (VUMAT in Abaqus).

The monopile structure, diameter of 6.0 m, was modelled with rigid element. The ridge-structure interaction process was considered quasi-static, so the dynamic behavior of the structure could be ignored. Therefore, from the structural point of view, only the shape (cylindrical) and the size (diameter) were important factors in

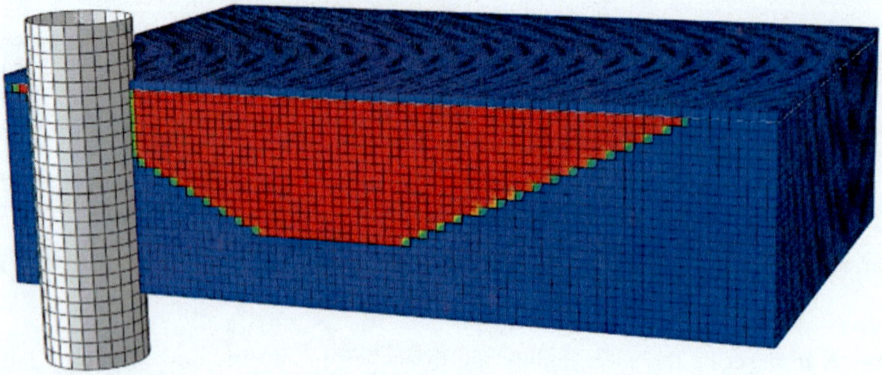

Fig. 11.3 Coupled Eulerian–Lagrangian simulation model of the monopile interaction with the ridge keel. Half of 3D geometry was modelled with symmetric boundary conditions. Modelled domain consists of two regions: Red colour indicates the region of ice rubble, blue one is empty (without ice)

the model. The interaction between the structure and ice was introduced by a general contact algorithm based on Coulomb friction model (friction coefficient 0.1).

The shape of ridge keel was chosen according to guidelines in ISO 19960 standard as shown in Fig. 11.4. Common values from the Gulf of Bothnia were chosen for geometrical dimensions. By utilizing symmetricity, only half of 3D ridge geometry was modelled. In the CEL-modelling the material flows through the finite element mesh. Therefore, the user needs to define a computational domain large enough for the material itself plus necessary additional space for the deformations to avoid boundary effects or material losses from the model as illustrated in Fig. 11.3. In the CEL-model the red colour indicates the region of ice rubble and the blue colour the "empty volume" (without ice).

The ridge width in the horizontal direction perpendicular to ridge motion was modelled large enough to avoid any boundary effects. Eulerian boundary conditions at that far-end surface was given so that theoretically the rubble material can flow in

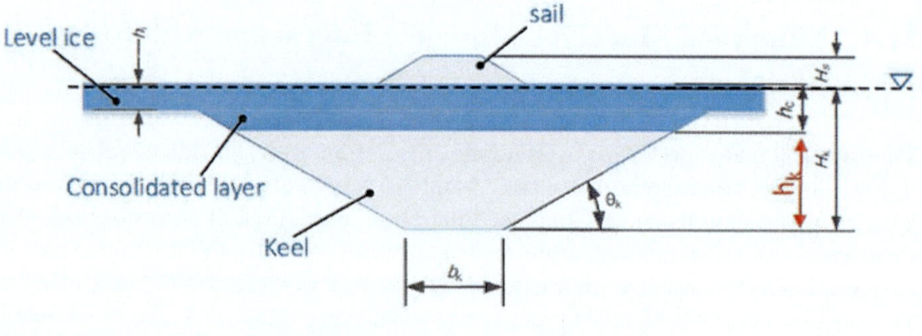

Fig. 11.4 Cross-section view of a ridge according ISO 19906 standard

Table 11.1 Main dimensions in the numerical model

Variable	Symbol	Value (m)
Diameter of the structure	D	6.0
Thickness of the keel	h_k	8.0
Ridge width at the keel top	b_t	40.8
Ridge width at the keel bottom	b_k	8.0
Ridge width in horizontal direction perpendicular to ridge motion		60 (30 m half-model)

or out of the Eulerian domain freely. To implement the far-end boundary in this way mitigate the stress waves and prevents the wave mirroring from the model boundary back to the active ice failure zone.

The consolidated layer was introduced as a horizontal Lagrangian contact plane to restrict the up-flow of rubble due to the buoyancy and interaction with the structure. The contact plane was placed at the top of the rubble. The implementation for the consolidated layer was done in this way to introduce realistic boundary condition at the top of keel rubble without modelling the complicated ice failure processes in the consolidated layer. Therefore, only the rubble part in the keel and its interaction with the monopile was modelled. The sail was ignored, because its load contribution is small compared to the keel load.

The simulation was made in two analysis steps; the first one was used to apply the internal stress state in keel due to the buoyancy. In the second step the monopile structure penetrated with a constant motion through the ridge. The main output quantities were local contact forces and global resultant forces on the monopile structure, kinematic quantities: displacements, velocities and accelerations of each element, stresses and strains of each elements as well as information about the failure mode and void ratio in each ice rubble element.

The main dimensions of ridge keel and material parameters describing the properties of ice rubble are shown in Tables 11.1 and 11.2.

11.5 Analyses

Main aspects from the simulations were to analyse the resultant forces on the structure and to analyse the failure progression in the keel. And more in details to understand the failure mode and how the failure progression affects the keel load.

The failure progression in the keel is shown in by snap-shot pictures in Fig. 11.5. The shear failure is introduced by the equivalent deviatoric plastic strain. The volumetric changes and relocation of broken ice rubble is introduced by a deformed state (elements with red colour). The time history plots of global ice load is given in Fig. 11.6. The initial peak is caused by a dynamic impact when the structure hits the ice edge first time at the start of the simulation. Thereafter, almost immediately

Table 11.2 Mechanical properties of ice rubble (Heinonen 2004)

Variable	Symbol	Value	Unit
density of ice	ρ_{i}	910	kg/m^3
density of water	ρ_{w}	1000	kg/m^3
porosity of rubble	η	0.3	[-]
density of rubble[a]	ρ_{r}	937	kg/m^3
Elastic modulus	E	1.1	GPa
Poisson value	υ	0.3	[-]
Cohesion	d	5	kPa
Friction angle	β	30	deg
Cap shape factor	R	0.5	[-]
Hydrostatic pressure strength	p_O	8.55d	Pa

[a]Submerged rubble: $\rho_r = \rho_i(1-\eta) + \rho_w\eta$

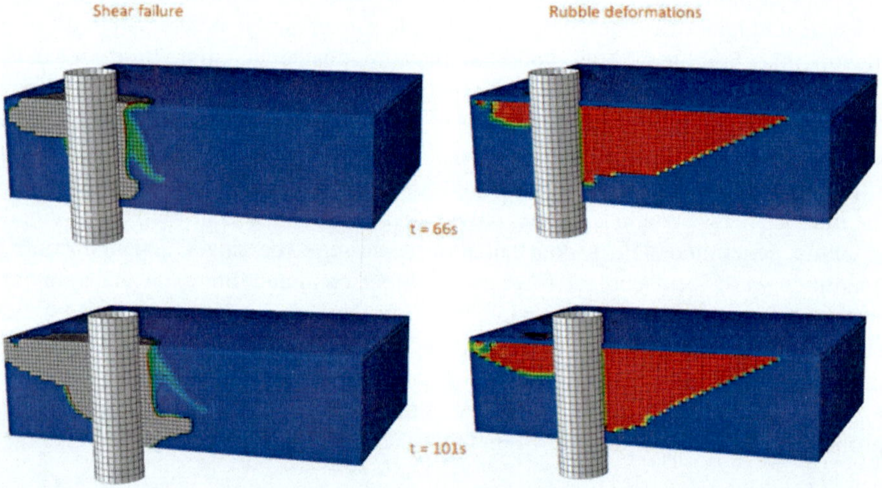

Fig. 11.5 Snap-shot pictures of the failure mechanisms in the keel. Left: The equivalent deviatoric plastic strain describing the shear failure. Two time instants—66 s and 101 s—are selected to visualize the local collapse of the keel (indicative contour distribution without scale). Right: Deformed rubble in the keel. Red colour indicates the region of ice rubble, blue one is empty (without ice)

when the penetration begins, with small rubble deformations, the ice rubble starts to fail. When the penetration continued, the load in the ice drift direction started to fluctuate mostly due to local wedge-like failures in the keel. As seen in Fig. 11.5 an inclined shear failure zone proceeds from top towards the keel bottom. Observation from the numerical simulation is in line with previous observations made by other theoretical models as shown in various references (i.e. Palmer and Croasdale (2012), see Fig. 11.7). The wedge-like failure mode happened right after a local force peak,

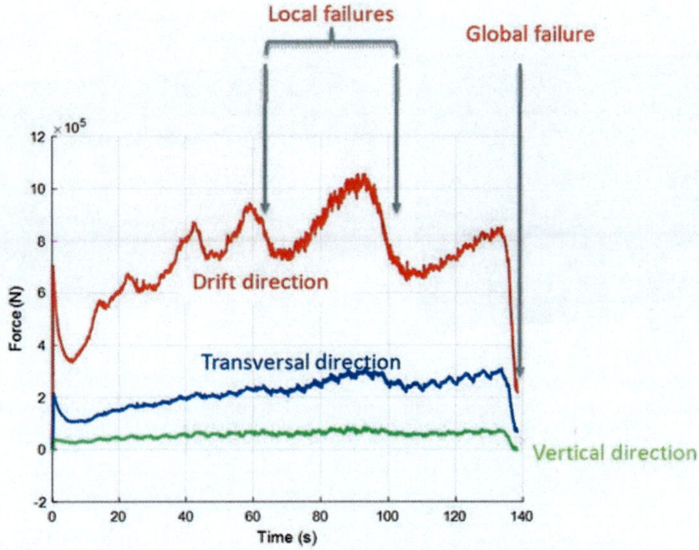

Fig. 11.6 Simulated force time history plot of the global keel load on the structure

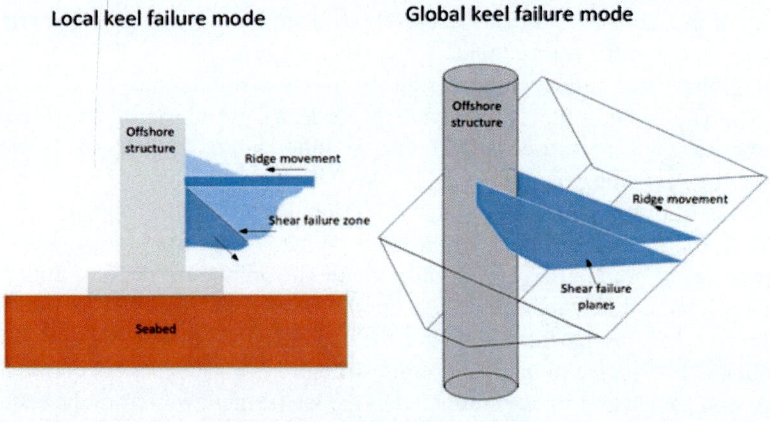

Fig. 11.7 Local and global failure modes in the keel

as shown in Fig. 11.6. This process was repeated after each load peak, when the keel started to soften due to the shear failure process. As the penetration progressed, the keel became thicker due to its triangular shape. Therefore the following local peak forces became higher and higher until the structure reached the position when the keel was thickest. At this point the global keel failure took place and thereafter, the force peak did not increase anymore compared to preceding peaks. Even though the structure is relatively slender (the diameter is less than the keel depth), some ice accumulation in the front of the structure was found at the keel bottom. That was partly

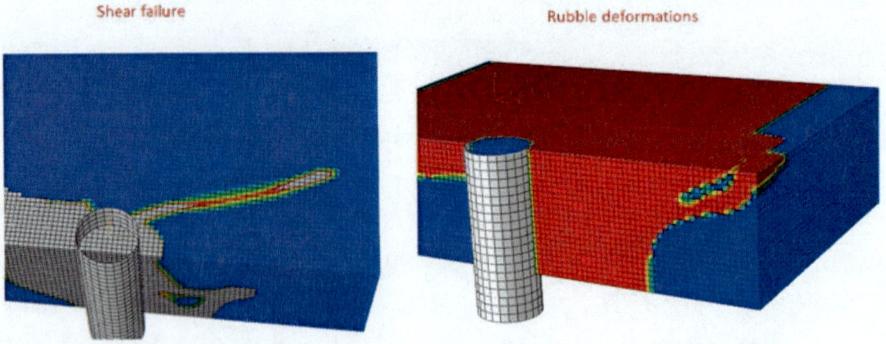

Fig. 11.8 Different snap-shot views of the failure mechanisms when the global keel failure occurs. Left: The equivalent deviatoric plastic strain describing the shear failure (indicative contour distribution without scale). Right: Deformed rubble in the keel. Red colour indicates the region of ice rubble, blue one is empty (without ice)

caused by the rubble dilatation (expansion) due to the shear failure and relocation of broken ice material. One may conclude here that due to the rubble accumulation the ice contact area on the structure increases. That can be considered in the ridge load models by adding so called surcharge factor. However, the keel thickness increase in the front of the structure is mostly caused by broken ice mass having only frictional resistance, because the freeze bonds are already broken.

The global keel failure progression from CEL-simulations is visualized in Fig. 11.8. The global plug failure takes place in a later phase of the ridge penetration because the energy needed to break the ridge in that way becomes lower than breaking the keel along the local wedge-like failures. In the case of single ridge, as modelled here by triangular like shape, the global plug failure happens only once, because it causes a total collapse of the ridge. When seeing from top, a shear failure zone proceeded through the keel slightly inclined from the ice drifting direction, as seen in the left hand side in Fig. 11.8. An important observation is that the failure zone does not go along the shortest way in the drifting direction because there is additional accumulated broken ice in the front of the structure. This ice mass becomes thicker and it is compacted and strong compared to the ice further away from the centerline. The energy needed to break the ridge along a longer inclined path is therefore the smallest. At same time a portion of ice rubble mass—bounded by the structure and shear failure zone—moves in the front of the structure similarly as introduced by Palmer and Croasdale (2012). After the global keel failure takes place, the keel load decreased drastically (Fig. 11.6).

According the ISO 19906 the maximum keel load on the monopile structure can be predicted using Dolgolov's analytical model with some modifications.

$$F_k = \mu_\varphi h_k w \left(\frac{h_k \mu_\varphi \gamma_e}{2} + 2c \right) \left(1 + \frac{h_k}{6w} \right) \tag{11.2}$$

$$\mu_\varphi = \tan\left(45^\circ + \frac{\varphi}{2}\right) \tag{11.3}$$

$$\gamma_e = (1 - e)(\rho_w - \rho_i)g \tag{11.4}$$

where w is the width of the structure, μ_φ is a passive pressure described as a function of internal friction angle φ. c is rubble cohesion and γ_e is the effective buoyancy of ice rubble. One should note that evaluation of the cohesion and internal friction in the Dolgopolov's model is based on the Mohr–Coulomb model and is therefore not equivalent with the shear-cap model. That makes a straight comparison between the analytical and numerical models challenging. A red dot along the vertical axis in Fig. 11.2 shows the selected failure state, which is used for the evaluation of corresponding Mohr–Coulomb parameters. Based on commonly known matching between the Mohr–Coulomb and the Drucker-Prager models in the shear failure state, one finds following relationships:

$$\tan\beta = \sqrt{3}\,\sin\varphi; \qquad d = \sqrt{3}\,c\cos\varphi \tag{11.5}$$

where the values for d and β are found from Fig. 11.2 (at the red dot) by a procedure described in Heinonen (2004). Finally, this results in the values for the Mohr–Coulomb cohesion 9.1 kPa and friction angle 24.8°, which were used in the analytical model (Eqs. (11.2)–(11.4)).

To compare the maximum keel loads between the numerical simulation and analytical solution, one observes the force equal to 1.05 and 2.0 MN, respectively. Several reasons for such a difference can be noted. Firstly, Dolgopolov's model is defined for two dimensional case: only ice drift and vertical directions are considered. Any influence of cylindrical shape of the structure is ignored. Therefore is it not clear how well the Dolgopolov's model is valid for narrow structures. Secondly, Dolgopolov's model assumes simultaneous failure progression (shear band) through the keel. This seems to be very conservative estimation as in the simulations the failure progresses as a function of the structure penetration into the ridge. Also, how to match the material parameters from the numerical model to the analytical is not unambiguous. Thirdly, Dolgopolov's model does not consider the ridge keel shape anyhow. Only the keel thickness is a variable. The simulations show that for the maximum load the keel shape is important factor. As a conclusion, a proper matching of material parameters should be highlighted. Due to highly varying stress state in the keel (shear, compression and tension) and non-linear numerical failure model, the evaluation of material parameters used in the analytical models is a key issue.

11.6 Conclusions

Monopiles are typically used for offshore wind turbine sub-structures. To use monopiles in ice-infested sea areas is challenging because of high ice loads. This presentation concludes observations about the numerical FEM-CEL simulation to determine ridge loads on the typical monopile structure with diameter of 6.0 m.

The shear-cap material model to describe the constitutive behavior of ice rubble during the interaction with a monopile structure was presented. FEM-CEL simulations of ridge-structure interaction visualize local failure processes in the keel and how the wedge-like failure mode affects the global ridge load evaluation. Even though the ridge interaction process is quasi-static, some load fluctuations were observed and connected to local wedge formation in the keel. Observations from the numerical simulation are in line with previously reported observations. The wedge-like failure patterns were repeated as the structure penetrated deeper into the ridge and the following local peak forces became higher and higher until the structure reached the position when the keel was thickest. Even though the structure is relatively slender, some ice accumulation was found in the front of the structure. When the global ridge failure took place, the ice rubble accumulation caused the shear failure zone to incline from the ice drifting direction.

Comparison between the FEM-CEL simulations and ISO 19906 standard showed that with corresponding parameters the standard predicts about double high loads. As the Dolgopolov's model is based on two-dimensional theory and includes an assumption of simultaneous failure progression with simplified keel geometry, one should be careful if using it in the case of monopiles to avoid conservative design.

Acknowledgements The author wishes to acknowledge the Strategic Research Council in Finland for funding the SmartSea project (Strategic research programme [grant numbers 292985 and 314225]), Mr. Juha Kurkela for the numerical work related to the case study and prof. Knut V. Høyland for fruitful discussions about the research idea.

References

M. Berg, R. Lubbad, S. Løset, An implicit time-stepping scheme and an improved contact model for ice-structure interaction simulations. Cold Reg. Sci. Technol. **155**, 193–213 (2018). https://doi.org/10.1016/j.coldregions.2018.07.001

K.R. Croasdale, Platform shape and ice interaction: a review, in *Proceedings of the 21st International Conference on Port and Ocean Engineering under Arctic Conditions*, July 10–14, 2011, Montréal, Canada (2011)

J. Heinonen, Simulating ridge keel failure by finite element method, in *The 15th International Conference on Port and Ocean Engineering Under Arctic Conditions (POAC)*, Helsinki, Finland, vol 3 (1999), pp. 956–963

J. Heinonen, Constitutive Modeling of Ice Rubble in First-Year Ridge Keel, VTT Publications 536, Espoo 2004, 142 p., Doctoral thesis. ISBN 951-38-6930-5 (sort back ed.) (2004). http://www.vtt.fi/inf/pdf/publications/2004/P536.pdf

J. Heinonen, CEL-analysis of punch shear tests to evaluate mechanical properties of ice rubble, in *23nd IAHR International Symposium on Ice*, Ann Arbor, Michigan USA, May 31 to June 3, 2016 (2016)

J. Heinonen, K.V. Høyland, Strength and Failure Mechanisms in Scale-Model Ridge Keel Punch through Tests - FE-Analysis, Proceedings of the 22nd International Conference on Port and Ocean Engineering under Arctic Conditions, June 9-13, 2013, Espoo, Finland (2013)

ISO/FDIS/19906, Petroleum and natural gas industries-arctic offshore structures. Tech.rep., International Standard, International Standardization organization, Geneva, Switzerland (2010), 434 p.

S. Kulyakhtin, K.V. Høyland, Ice rubble frictional resistance by critical state theories. Cold Reg. Sci. Technol. **119**(2015), 145–150 (2015)

A. Patil, B. Sand, L. Fransson, Smoothed particle hydrodynamics and continuous surface cap model to simulate ice rubble in punch through test, in *Proceedings of the 23nd International Conference on Port and Ocean Engineering under Arctic Conditions*, June 14–18, 2015, Trondheim, Norway (2015)

A.C. Palmer, K. Croasdale, *Arctic Offshore Engineering* (2012)

A. Polojärvi, J. Tuhkuri, On modelling cohesive ridge keel punch through tests with a combined finite-discrete element model. Cold Reg. Sci. Technol. **85**(2013), 191–205 (2013). https://doi.org/10.1016/j.coldregions.2012.09.013

N. Serré, P. Liferov, Loads from ice ridge keels-experimental vs. numerical vs. analytical, in *Proceeding of the 20 Int. Symposium on Ice (IAHR)*, 2010, Lahti, Finland. Paper # 92 (2010)

N. Serré, Mechanical properties of model ice ridge keels. Cold Reg. Sci. Technol. **67**(2011), 89–106 (2011). https://doi.org/10.1016/j.coldregions.2011.02.007

N. Serré, Numerical modelling of ice ridge keel action on subsea structures. Cold Reg. Sci. Technol. **67**(2011), 107–119 (2011)

Simulia Abaqus 6.14 user's guide (2014), © Dassault Systèmes

A. Sorsimo, J. Heinonen, Modelling of ice rubble in the punch shear tests with cohesive 3D discrete element method. Eng. Comput. (2019). https://doi.org/10.1108/EC-11-2017-0436

J. Tuhkuri, A. Polojärvi, A review of discrete element simulation of ice–structure interaction. Phil. Trans. R. Soc. A **376**, 20170335 (2018). https://doi.org/10.1098/rsta.2017.0335

Chapter 12
Safer Operations in Changing Ice-Covered Seas: Approaches and Perspectives

Yevgeny Aksenov, Stefanie Rynders, Danny L. Feltham, Lucia Hosekova, Robert Marsh, Nikolaos Skliris, Laurent Bertino, Timothy D. Williams, Ekaterina Popova, Andrew Yool, A. J. George Nurser, Andrew Coward, Lucy Bricheno, Meric Srokosz, and Harold Heorton

Abstract The last decades witnessed an increase in Arctic offshore operations, partly driven by rising energy needs and partly due to easing of sea ice conditions and improved accessibility of shipping routes. The study examines changes in sea ice and ocean conditions in the Arctic with their implications for off-shore safety. The objective of the research is to develop a basis for forecasting technologies for maritime operations. We assess loads on off-shore structures from sea ice and ocean in centennial climate future projections and implications for the accessibility and future Arctic shipping. As a test case, we calculate loads on a tubular structure of 100-m wide and 20-m tall, similar to installations in the Beaufort Sea in the 1980s. With sea ice retreating, loads are predicted to increase from ~0.1 × 10^6 N (MN) at present to ~50–200 MN in the 2090s, primarily due to wave loads. This study asserts

---

Y. Aksenov (✉) · S. Rynders · E. Popova · A. Yool · A. J. G. Nurser · A. Coward · M. Srokosz
Marine Systems Modelling Group, National Oceanography Centre, European Way, Southampton, UK
e-mail: yka@noc.ac.uk

D. L. Feltham · L. Hosekova
Centre for Polar Observation and Modelling, Department of Meteorology, University of Reading, Reading, UK

L. Hosekova
Applied Physics Laboratory, University of Washington, Seattle, WA, USA

R. Marsh · N. Skliris
School of Ocean and Earth Science, University of Southampton, Southampton, UK

L. Bertino · T. D. Williams
Nansen Environmental and Remote Sensing Centre and Bjerknes Centre for Climate Research, Bergen, Norway

L. Bricheno
Marine Systems Modelling Group, National Oceanography Centre, Joseph Proudman Building, Liverpool, UK

H. Heorton
Department of Earth Science, University College London, London, UK

J. Tuhkuri and A. Polojärvi (eds.), *IUTAM Symposium on Physics and Mechanics of Sea Ice*, IUTAM Bookseries 39, https://doi.org/10.1007/978-3-030-80439-8_12

the need for new approaches in forecasting to make marine operations in the Arctic safer.

12.1 Introduction

The Arctic has become a prominent highlight in climate change news and discussions. Recurrent summer sea ice records apparently support the view that Arctic sea ice is on a long-term decline trajectory, with ice-free summers projected to occur as early as in the 2030s (Stroeve et al. 2012). More evidence from the observational record on the unprecedented changes in the Arctic system has recently come to light, including those in the ocean, sea state, atmosphere, glaciers, subsea permafrost, ocean biology and ecosystems and also on land (Polyakov et al. 2017; Yool et al. 2015; Shakhova et al. 2017). The thinning of the sea ice cover and the appearance of large areas of open water in the summer in the Arctic generate more waves, breaking-up pack ice and creating an area of fragmented sea ice, known as Marginal Ice Zone (MIZ) (Stopa et al. 2016). This can potentially lead to the further decline of the pack ice and increase of the MIZ area as a proportion of the total ice cover (Fig. 12.1), although the evidence for this from the satellite records remains inconclusive and depends on details of the data processing and the definitions of thresholds.

Understanding these changes can improve our ability to forecast how the Arctic system is evolving, but can also give us valuable insights into climate change elsewhere. These new insights will allow us to build more rigorous climate predictions

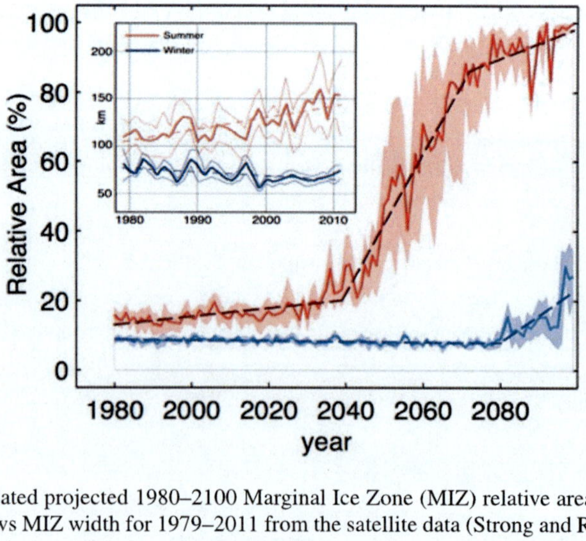

Fig. 12.1 Simulated projected 1980–2100 Marginal Ice Zone (MIZ) relative area (Aksenov et al. 2017). Inset shows MIZ width for 1979–2011 from the satellite data (Strong and Rigor 2013). MIZ is defined as sea ice with fraction of 0.15–0.80. Blue lines are winter MIZ (December–January–February) MIZ and red lines are summer MIZ (June–July–August). The shading and thin lines marks one standard deviation. Dashed lines show fitted linear trends

for the next decades to a century. The effects of global warming are translated into a multitude of socio-economics impacts in the Arctic. The last decades have witnessed an increase in Arctic offshore operations, partly driven by increasing energy needs and partly due to easing of sea ice conditions and improved accessibility of the shipping routes (Aksenov et al. 2017; Melia et al. 2016). A comprehensive assessment of the changes in the environment will allow industries, governing and regulatory bodies and local communities to plan for a variety of economics and societal development scenarios. The study presents an analysis of the environmental risks relevant to future Arctic offshore operations and shipping. The aim is twofold: (i) to examine changes in sea ice and oceanic conditions in the Arctic and (ii) to assess their relevance to off-shore shipping and operations both now and in the future. We use a suite of high-resolution Ocean General Circulation Model (OGCM) simulations to examine key environmental parameters of the current and future climates as far as the end of the century, along with output from the ocean waves model WaveWatch^{TM}III, and apply these to analyse operational risks. The paper is structured as follows: Sect. 12.2 introduces analysis methods and describes the OGCM simulations, new environmental variables are presented in Sect. 12.3; the present-day climate assessment of structural loads is presented in Sect. 12.4; Sects. 12.5 and 12.6 examine future climate scenarios; Sect. 12.7 discusses the results and Sect. 12.8 presents summary of the study.

12.2 Methods

a. Models

For the present-day climate analysis, we use high-resolution global Ocean General Circulation Model (OGCM) NEMO (Nucleus for European Modelling of the Ocean) coupled to Los Alamos sea ice model CICE (Madec et al. 2017; Hunke et al. 2015). NEMO is a Boussinesq hydrostatic model and uses finite differences on the global tripolar orthogonal mesh with Arakawa C-grid discretization (Storkey et al. 2018). To avoid singularity at the North Pole, the mesh has two poles in the Siberia and Canada with the third mesh pole at the South Pole. In the vertical, there are 75 levels with resolution of 1-m at the surface, ~2 m in the top 50 m and ~4 m in the top 100 m. The high model resolution and partial-step model bottom topography improves simulations of the ocean currents on the continental shelf and shelf slope. CICE is dynamics-thermodynamics model, shares the same tripolar mesh but is discretized on Arakawa B-grid.

CICE thermodynamics is energy-conserving, with four layers of ice and one layer of snow to model vertical heat conduction. The balance of the fluxes controls sea ice and snow melting from the top. Surface melt ponds are simulated from a topographic melt pond model. The bottom ice growth and melt are governed by the heat conduction through ice and oceanic heat flux to ice base. Ice age tracer allows to keep track of first-year and multi-year level and ridged ice. The dynamical part of CICE includes continuum Elastic-Viscous-Plastic rheology (EVP), combining non-linear viscous-plastic (VP) rheology with elastic term for regularization of VP for strain rates approaching zero (Hunke et al. 2015). Sea ice is driven by winds and ocean currents and it resists deformation with a compressive strength that depends on ice thickness and concentration. The momentum balance accounts for the atmosphere–ice and ice-ocean stresses, Coriolis force, slope of the sea surface and ice internal stresses (Hunke et al. 2015). The model calculates ice thickness distribution in each model cell from ice thermal evolution and mechanical redistribution, we use five ice thickness categories. Details of the NEMO-CICE and validation are presented elsewhere (Storkey et al. 2018). NEMO-CICE is employed in forecasting and climate research by the UK Meteorological Office (UKMO), is a part of the Global Monitoring for Environment and Security (Copernicus) and of the Inter-governmental Panel on Climate Change (IPCC) assessments.

For this study we have updated CICE model with collisional rheology to represent fragmented ice dynamics in MIZ (Feltham 2005; Rynders et al. 2021). To simulate sea ice break-up by waves, floe size distribution (FSD) evolution and wave attenuation by sea ice we developed the Waves-in-ice interaction Module (WIM), based on the framework by Williams et al. (2013). The updates included up-stream scheme for wave advection in sea ice, FSD advection using linear remapping and its evolution following lateral melt of ice floes. We have included wave mixing in the Generic Length Scale (GLS) turbulent closure (Rynders et al. 2021; Bateson et al. 2020). We have conducted simulations of the coupled NEMO-CICE-WIM at a 1/4° horizontal resolution (28 km globally, 9–14 km in the Arctic) for 1958–2015, forced with 6-hourly atmospheric DRAKKAR reanalysis (DFS5.2) and waves data from the European Center for Medium Range Weather Forecasting (ECMWF) (Brodeau et al. 2010). For the analysis of the future projections output is taken from the NEMO simulations completed by the authors of this study under the *Regional Ocean Acidification Modelling* project (ROAM) forced with the Representative Concentration Pathway 2.6 and 8.5 (*RCP2.6* and *RCP8.5*) scenarios from IPCC AR5 (Yool et al. 2015). These scenarios feature low and high carbon dioxide (CO_2) emissions with moderate and strong climate warming by the end of the twenty-first century respectively. To examine the future wave field, we use WaveWatch^{TM}III spectral wave model (hereafter, WWIII) simulations for the *RCP8.5* scenario completed by the authors under the *Coordinated Ocean Wave Climate Project (COWCLIP)* integrations (Bricheno and Wolf 2018). The model has the resolution of 0.70° × 0.46° in longitudinal and latitudinal directions with a global domain extending from 80°S to 83°N. The simulations were forced with 3-hourly atmospheric 10-m wind and daily sea ice concentration taken from the EC-EARTH model runs. The latter is the 1° NEMO-LIM2 sea-ice-ocean model coupled to 1.125° ECMWF Integrated Forecasting System), integrated for 1970–2100 (Morim et al. 2020).

b. Model validation

To gain confidence in OGCMs skills to simulate present-day climate we have compared the NEMO-CICE-WIM and NEMO-ROAM simulations with available observations, focusing on sea ice metrics (concentration, thickness and drift) and ocean fields (temperature, salinity, mixed layer depth, sea surface heights and currents). UKMO Hadley Centre Sea Ice and Sea Surface Temperature (HadISST1) and World Ocean Atlas datasets, sea ice thickness from the Pan-Arctic Ice-Ocean Modeling and Assimilation System reanalysis (PIOMAS) and satellite dynamical topography and sea ice drift from CERSAT have been employed (Yool et al. 2015; Aksenov et al. 2017; Rynders et al. 2021). We concluded that the models are good agreement with observations and fit for the study. Comparison of the future NEMO-ROAM projections with the *Coupled Model Intercomparison Project 5* (CMIP5) ensembles shows that NEMO-ROAM sea ice extent, area and concentration fields are very close to the CMIP5 ensemble mean (Yool et al. 2015; Aksenov et al. 2017). This gives us confidence in the model skills to predict a plausible state of the Arctic sea ice for the CO_2 emissions and climate warming scenarios. The WWIII model has been extensively validated for the open ocean in COWCLIP (Morim et al. 2020), although, observational uncertainty is still large in ice covered areas (Stopa et al. 2016; Heorton et al. 2021). We have used technique from the University College London and extracted significant wave heights $\boldsymbol{H_s}$ in MIZ from CryoSat-2 (Heorton et al. 2021). Comparison of 2002–2015 modelled and observed seasonal averages in the Arctic Ocean and North Atlantic for Winter (December–January–February, henceforth DJF) and Summer (June–July–August, henceforth JJA) shows agreement within 10% for means and standard deviations, giving us confidence in the model simulations (Table 12.1).

c. Morison's Equation

Here we describe the use of environmental information to calculate the load maps for the off-shore structures and assess risk for ships. The method follows Morison's equation (henceforth ME) to estimate the total hydrodynamic (waves plus currents) forces. MEis composed of Froude-Krylov force and accelerated fluid force "inertia" terms and a boundary layer influence through the drag term (Morison 1950; Chakrabarti 1987). Without repeating derivation details given in the literature we shall treat the inertia—drag dependent total hydrodynamic load on horizontal cross-section of a thickness $\boldsymbol{dz}$ of a "fixed" cylindrical structure due to ocean waves and spatially and temporarily varying currents as:

Table 12.1 Mean significant wave heights H_s(m) and standard deviations for 2002–2015 from model and Cryosat-2 (Heorton et al. 2021)

Region	Arctic (>66°N)	Arctic (>66°N)	Arctic and N.Atlantic >60°N	Arctic and N.Atlantic >60°N	N.Atlantic 60°–66°N	N.Atlantic 60°–66°N
Season	Winter (DJF)	Summer (JJA)	Winter (DJF)	Summer (JJA)	Winter (DJF)	Summer (JJA)
Model $\boldsymbol{H_s}$(m)	**2.68** ± 0.79	**1.06** ± 0.37	**3.02** ± 0.94	**1.19** ± 0.46	**3.35** ± 0.96	**1.31** ± 0.46
Cryosat-2 $\boldsymbol{H_s}$(m)	**2.70** ± 1.03	**0.97** ± 0.29	**2.47** ± 1.21	**1.07** ± 0.38	**3.04** ± 1.22	**1.29** ± 0.46

$$dF(x, z, t) = \left[\rho AC_M \dot{u} + \frac{\rho}{2} DC_D U \left| \vec{U} \right| \right] \cdot dz \quad (12.1)$$

Here, $\vec{u}\,(x, z, t)$ and $\dot{u}(x, z, t)$ are the wave-induced velocity and modulus of its time derivative in the location of the structure at a given time t, and $\vec{U} = \vec{U}\,(x, t)$ is the velocity of the ocean current, assumed to be constant with depth and equal to the ocean surface velocity; x is coordinate in the direction of wave propagation, with $x = 0$ aligned with the vertical axis of the cylindrical structure; z is the vertical coordinate; C_M and C_D are inertia and drag coefficients; $\rho = 1025\left(kg/m^3\right)$ is seawater density; D is the structure diameter, and $A = \pi D^2/4$ is the cross-section area. For the fixed structures the full derivative of the relative velocity between the structure and the ambient water is neglected in (12.1). The fixed structure condition can be easily relaxed by adding relative displacement of the structure to this equation, although this is out of the scope of the present study. Here we also neglect the spatial variation in the ambient water flow near the cylinder, assuming undisturbed flow in the immediate vicinity of the cylinder at the scale of the cylinder diameter D is about the same at any given time. For linear waves propagating in x-direction, wave-induced velocity and acceleration at location $x = 0$ (aligned with the vertical axis of the structure), are given by:

$$u(x = 0, z, t) = \frac{agk}{\omega} \frac{\cosh(k(h + z))}{\cosh(kh)} \cos(\omega t) \quad (2a)$$

$$\dot{u}(x = 0, z, t) = agk \frac{\cosh(k(h + z))}{\cosh(kh)} \sin(-\omega t) \quad (2b)$$

Notations here are as follows: ω is wave angular velocity; a is maximum wave amplitude; k is the wave number; g is acceleration of gravity, h is water column total depth, z is depth, and t is time. The drag term in (12.1) depends on the velocity, whereas the inertia term depends on the acceleration. Hence, the occurrence of the maximum drag force and the maximum inertia force are lagged by a phase shift of 90° and the maximum force is calculated as the maximum value over a wave period. Both the C_M and C_D are functions of Keulegan-Carpenter number (Keulegan and Carpenter 1958; Clauss et al. 1992), a measure for the ratio between the wave height and the cylinder diameter, and Reynolds number (*Re*). In addition, C_D increases with increasing local surface roughness of the structure, whereas C_M decreases with increasing roughness. We use typical values of $C_M = 0.3$ and $C_D = 0.45$ (Halse, 1997). To obtain the total load on the structure F_{tot} we integrate (12.1) by dz for the whole height of the cylindrical structure. Since the structure displacement is neglected, we can drop the dependency on the coordinate x:

$$F_{tot}(t) = \int_0^Z dF(z, t) \cdot dz \quad (12.3)$$

Here, $\boldsymbol{Z}$ is the height of the cylindrical structure. The integration of (12.3) is done numerically using the Simpson's method. The input variables for the above calculations are as follows. We use the wave peak frequency $\boldsymbol{f_p}$ from the WWIII wave model to calculate angular peak frequency as $\boldsymbol{\omega} = 2\pi\boldsymbol{f}$ and significant wave height $\boldsymbol{H_s}$ to obtain maximum wave amplitude: $\boldsymbol{a} = 1.8 \cdot \boldsymbol{H_s}$. The wave fields from WWIII and ocean currents from NEMO are at hourly frequency. ME in the above form can be applied to tubular (cylindrical) columns of varying diameters that represents several types of offshore structures typically used in the offshore oil and gas and offshore wind industries: fixed jackets, fixed monopiles, floating (spar type) monopiles and artificial islands (e.g., Skliris et al. 2021). To build a demonstration case, here we use the method for a tubular structure 100-m wide and 20-m tall, the loading changes with diameter, the choice based on structures employed in the North Sea and the Arctic (Sanderson 1988).

d. Ice loads

We incorporate forces arising from the ice floes collisions with the structure in the presence of the wave field, considering both frictional and collisional (dynamical) sea ice loads and associating collisional loading with the turbulent velocity of sea ice floes. Using an approach to account for the rapid turbulent velocities of the individual ice floes in sea ice rheology (Feltham 2005; Haff 1983; Shen et al. 1987), we split the ice velocity $\boldsymbol{U_{ice}}$ into mean velocity $\boldsymbol{u_{ice}}$ for the model grid cell, area-averaged of all ice floes in the model cell, and a randomly-oriented rapidly fluctuating turbulent velocity $\boldsymbol{u}'$ as: $\overrightarrow{\boldsymbol{U}}_{\boldsymbol{ice}} = \overrightarrow{\boldsymbol{u}}_{\boldsymbol{ice}} + \overrightarrow{\boldsymbol{u}}'$. Following this decomposition, the internal sea ice stress tensor $\boldsymbol{\sigma}$ is expressed as a sum of the frictional $\boldsymbol{\sigma_{fr}}$ and collisional $\boldsymbol{\sigma_{col}}$ parts $\boldsymbol{\sigma} = \boldsymbol{\sigma_{fr}} + \boldsymbol{\sigma_{col}}$, with the sea ice internal force $\boldsymbol{F_{ice}^{total}} = \nabla \cdot \boldsymbol{\sigma}$ being a sum of frictional and collisional forces: $\boldsymbol{F_{ice}}^{\boldsymbol{total}} = \boldsymbol{F_{icefrictional}} + \boldsymbol{F_{icecollisional}}$ (Feltham 2005).

Assuming that a not moving (fixed) structure is imbedded in the drifting sea ice, we adapt the model for sea ice forces on icebergs (Lichey and Hellmer 2001; Martin and Adcroft 2010) to calculate the frictional loads as:

$$F_{icefrictional}(t) = \frac{\rho_{ice}h_{ice}}{2} \cdot D \cdot C_d^{ice} \cdot \left|\overrightarrow{u}_{ice}\right|^2 \tag{12.4}$$

Here, $\boldsymbol{\rho_{ice}} = 917(\boldsymbol{kg/m^3})$ and $\boldsymbol{h_{ice}}$ are the sea ice density and thickness; $\overrightarrow{\boldsymbol{u_{ice}}}$ is ice mean velocity; $\boldsymbol{D}$ is the structure diameter; $\boldsymbol{C_d^{ice}} = 1.0$ is the non-dimensional ice drag coefficient; we chose the highest of the suggested values (Lichey and Hellmer 2001; Martin and Adcroft 2010; Timco and Weeks 2010). The key feature arising from the structure immobility in Eq. (12.4) is the non-zero load for sea ice moving against the structure for all the non-zero ranges of sea ice concentrations, this renders our calculations of the loads to their maximum values. In contrast, Eq. (12.5) in Lichey and Hellmer (2001) leads to the same zero loads for the sea ice concentration less than 15% (loose ice) or exceeding 90% (pack ice), which is not physical and erroneous from the observational data. Hence, the step function applied for sea ice concentration in Lichey and Hellmer (2001) is not applicable to stationary structures. We note, that our Eq. (12.4) is in the same form as Equation (A2c) given by Martin and Adcroft

(2010) and this also justifies our approach due to the lack of literature suggesting a better way.

Most of the sea ice models do no calculate turbulent ice velocities, but only mean velocity for a given model time step, this prevents calculating collisional loads, which can be substantial. Using the both mean and turbulent sea ice velocities, we are able to fill this gap and calculate the collisional impact following:

$$\delta F_{icecollisional}(t) = h_{ice} \cdot \rho_{ice} \cdot G_t \pi \left(\frac{L_f}{2}\right)^2 \tag{12.5}$$

with $\boldsymbol{L_f}$ being mean floe size and $\boldsymbol{G_t}$ granular temperature of ice drift, explained below. By definition, the average of the turbulent velocity $\boldsymbol{u}'$ is zero over a model grid cell, however the associated mass-specific kinetic energy $\boldsymbol{G_t} = \boldsymbol{u}'^2/2$, hereafter "granular temperature of ice drift" (Feltham 2005), is not zero. Granular temperature is a model prognostic parameter, calculated from the evolution equation, accounting for sources and sinks in the turbulent ice drift due to air and water turbulence, waves and floe-to-floe collisions (Rynders et al. 2021). To obtain the total collisional load we need to add collisions of all floes with structure during the given time period. We use "raindrop model" to calculate probability of floes collisions with structures occurring over the model timestep (one hour in our simulations) (see Onishchenko (2009) for discussion and further references to the raindrop model). The impact probability of collision of sea ice floe with the structure is given by:

$$P_{collisions}(t) = n_o \cdot (L_f + D) \cdot \left|\overrightarrow{U_{ice}}\right| \tag{12.6}$$

In (12.6) $\boldsymbol{n_0}$ is the area density of the floes in the model grid cell where the structure is located and is defined as below, where $\boldsymbol{a_{ice}}$ is the ice area in the model cell (with area of $\boldsymbol{a_{cell}}$) and $\boldsymbol{a_{floe}}$ is the floe area:

$$n_0 = \left(\frac{a_{ice}}{a_{floe}}\right) \cdot \left(\frac{1}{a_{cell}}\right) = \frac{A_{ice}}{\left(\pi/4 \cdot L_f^2\right)} \tag{12.7}$$

The total collisional load on the structure over period $\boldsymbol{\Delta T}$ is:

$$F_{icecollisional} = \delta F_{icecollisional} \cdot P_{collisions} \cdot \frac{\Delta T}{L_f} \tag{12.8}$$

with $\boldsymbol{\Delta T} = 3600(s)$ being the model time step.

The ice loads calculations require modelling dynamics of the fragmented ice cover and floe collisions. (Simulations of the ice floes dynamics are presented in the next section.) Following the methodology detailed here, we calculate loads from waves, currents and sea ice using OGCM simulations of the present climate and future projections to compute 2-D spatial maps of the total loads and examine both current hazards in the Arctic environment and changes in the future.

e. Shipping risks

To assess future shipping risks in the Arctic we use approach taken by the Arctic Ice Regime Shipping System (AIRSS) to determine sea ice thickness (age) thresholds for ships of different ice classes being able safely enter and navigate sea ice. AIRSS defines the concept of Ice Numerals (IN) as a sum of Ice Multipliers (IM) for different ice thickness bins $\left[\boldsymbol{h}_{ice}^{i}, \boldsymbol{h}_{ice}^{i+1}\right]$, $\boldsymbol{i} = [1, \boldsymbol{N}]$ weighted by their partial fraction: $\boldsymbol{IN} = \sum_{i=1}^{i=N} \boldsymbol{A}^{i}{}_{it} \times \boldsymbol{IM}_{i}$ (Aksenov et al. 2017). $\boldsymbol{IM}$ are obtained empirically for a range of ship classes, the values are given in Aksenov et al. (2017), Table A1. For positive $\boldsymbol{IN}$, risk from the ice conditions risks is low, ship can sail in sea ice with safe speed assesses from the $\boldsymbol{IN}$ values. If $\boldsymbol{IN}$ is zero or negative, the sailing is unsafe. From $\boldsymbol{IN}$ access maps are being produced.

12.3 Simulated Environmental Parameters

To calculate ocean and ice loads on structures the method requires ocean currents, wave heights, ice thickness and fragmentation (floe sizes) and the turbulent velocity of the ice drift (Rynders et al. 2021). In the MIZ, sea ice cover is broken by ocean waves and consists of mobile ice floes. Sea ice drift in the MIZ is subject to large variations due to wind and water turbulence, wave surge and internal ice stresses which are transmitted through floe-to-floe collisions (Rynders et al. 2021). Here we analyse the NEMO-CICE-WIM simulations 2000–09. The results show that sea ice presence rapidly attenuates wave energy within ~50–100 km distance of the ice edge. In the ice-free regions of the Norwegian, Greenland and Barents seas, wave height can reach 3 m in winter.

In summer ice floes sizes can decrease to <200 m in the Arctic Ocean and the seasonal ice retreat opens Arctic shelf seas to waves up to 2-m high (Fig. 12.2). The inner boundary of the summer MIZ (defined as an ice fraction of 0.80 (Strong and Rigor 2013) moves northward, beyond the continental shelf slope, and the southern periphery of the pack ice is transformed into MIZ (Fig. 12.2). Seasonally, floe sizes in the MIZ increase in mode values from 20–30 m in winter to 30–40 m in summer due to the melting of the small ice floes (Fig. 12.3).

12.4 Loads on Off-Shore Structures

To estimate present-day loads on off-shore structures, and the associated risks, we calculate combined loads from the ocean (waves and currents) and frictional and collisional loads from sea ice as set out in Sect. 12.2.

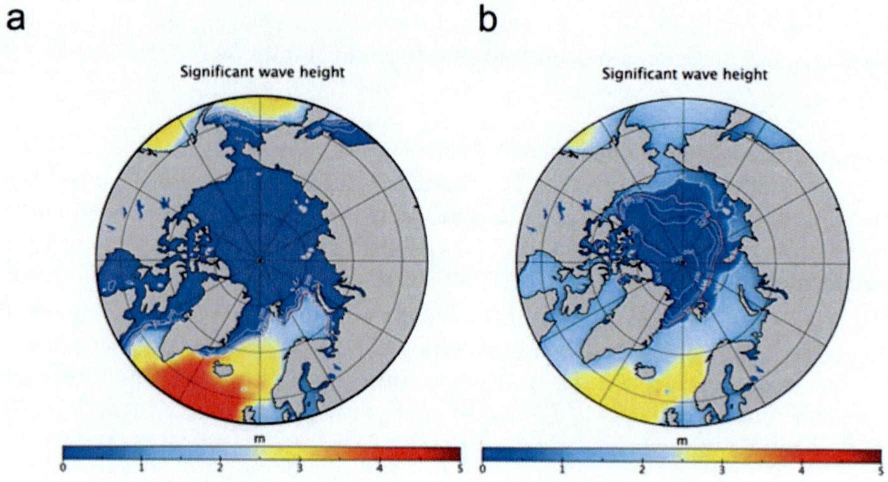

Fig. 12.2 Winter (DJF) (**a**) and summer (JJA) (**b**) 2000–2009 seasonal variations in wave heights (colour), with floe sizes (white contours). Magenta marks MIZ inner boundary (0.8 ice fraction). NEMO-CICE-WIM simulations

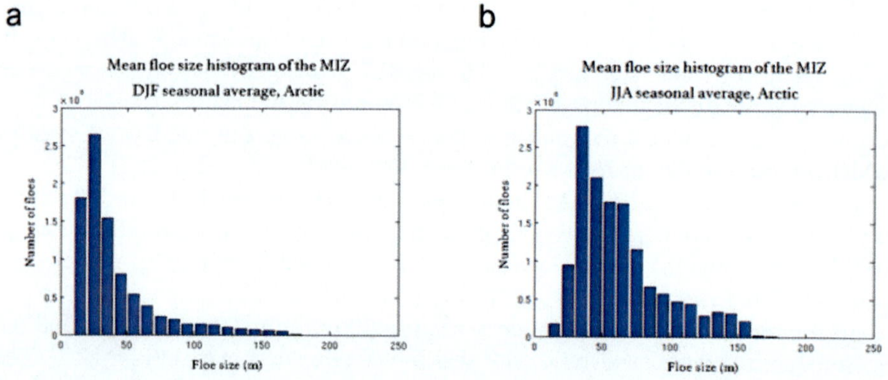

Fig. 12.3 Floe size distribution in the Arctic Marginal Ice Zone (MIZ) in winter (**a**) and summer (**b**) 2000—2009 from NEMO-CICE-WIM simulations

We choose January 2014 as a typical winter period with a strong storm in the North Atlantic and extensive Arctic ice cover to assess relative contributions from different environmental factors in these conditions. The analysis can be easily extended to synoptic and seasonal scales; however, this is beyond the scope of this study.

In the ice-free North Atlantic, the ME model predicts that wave loads reach 50–200 $\times$ 10^6 N (MN) and dominate the total load. Here the waves are 4–5 m in height and peak loading maps well onto the areas with maximum wave heights (Fig. 12.4a, b—open water and ice covered areas are marked). In the ice-covered Arctic Ocean

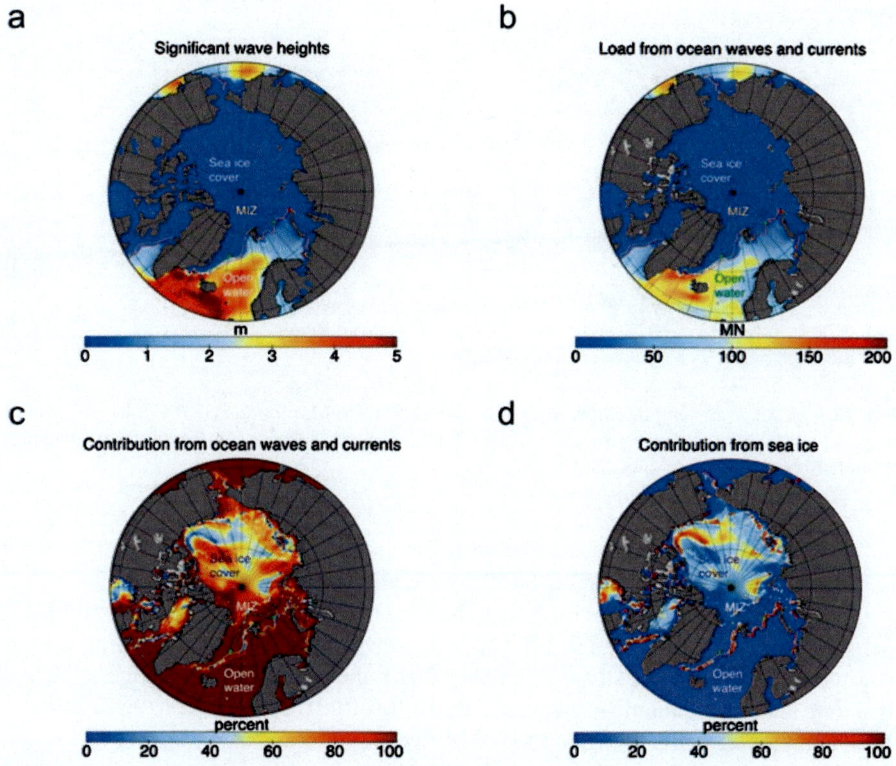

Fig. 12.4 Significant wave height (**a**) and loads on structures from waves and currents in January 2014 (**b**). **c** and **d** show the contributions of waves and currents and of sea ice to the total loads. Red and yellow lines mark MIZ boundaries. NEMO-CICE-WIM simulation

the contribution to the total load from the waves and currents are substantial, at 60–80% in some areas (Fig. 12.4c), although the overall loads from waves, current and sea ice are of $o(0.10\ \mathrm{MN})$, much smaller than in the open ocean. Ice loads are of the same magnitude as the wave loads and the spatial distribution of the ice load complements that of the wave load, with the ice load fraction increasing up to 80% in the areas of ice convergence on the Laptev, East-Siberian, Chukchi and Beaufort seas (cf. Fig. 12.4c, d). The ice loads in the central Arctic Ocean result almost entirely from frictional loading, with the collisional load of contributing up to 60–90% of the total load in MIZ in the Icelandic, Greenland and Bering seas, in the Baffin Bay and Hudson Bay and in the Laptev and Beaufort seas due to the high turbulent energy of ice motion (granular temperature) (Figs. 12.4d and 12.5a). Spatial variability of the collisional ice load is influenced by variation in the floe sizes (mass factor) and granular temperature of ice. Floe sizes decrease in the MIZ towards the ice edge due to break-up, with ice granular temperature higher in the MIZ (Figs. 12.2 and 12.5a). As a result of these two competing tendencies, the area of the maximum ice

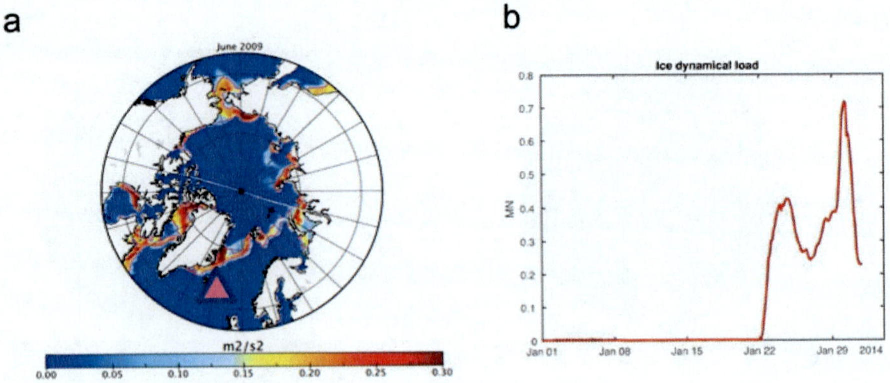

Fig. 12.5 **a** Sea ice granular temperature in NEMO-CICE-WIM in January 2014. **b** Ice collisional (dynamical) load on a virtual structure on the Greenland Shelf, location marked with the magenta triangle in (**a**). Magenta shows MIZ boundaries

collisional loads in the MIZ coincides with areas with a sea ice fraction of ~0.15–0.80 (a band of the higher ice load between the red and yellow lines in Fig. 12.4d).

It should be noted that the load contributions from different factors vary substantially on the synoptic scale (5–7 days), with wave and ice loads increasing during storm events. During the storm in January 2014, the collisional loads on the virtual structure in the Greenland Sea increased about 100-fold (to 0.4 MN) on the 22 January 2014 and then a further twofold (to ~0.7 MN) on the 30 January 2014 (Fig. 12.5a, b). The contribution of the ocean currents to the loads is more moderate. Average values in the Arctic Ocean are of $o(0.01$ MN), with the loads peaking at of ~0.08–0.20 MN where there are strong surface currents. For example, in the Bering Strait inflow; north of Barrow; off the Norwegian coast; in the East and West Greenland current systems; and east of Baffin Island and Labrador (not shown).

12.5 Future Changes in the Ocean and Sea Ice

Arctic sea ice decline in all seasons during the last four decades has been documented in satellite data and OGCMs with a variable but overall reasonable degree of robustness (Stroeve et al. 2012; Morim et al. 2020). Future projections of the Arctic marine environment present scenarios for sea ice changes, although changes in the ocean are not often examined in detail (Aksenov et al. 2017). Here we combining outputs from the OGCM and WWIII simulation (Yool et al. 2015; Bricheno and Wolf 2018) and use these in Sect. 12.6 to assess projected loads on off-shore structures and risks for shipping.

The most striking result of the forward OGCM simulations is the extension of the ice-free season later into the year, leaving Arctic virtually ice-free until December in

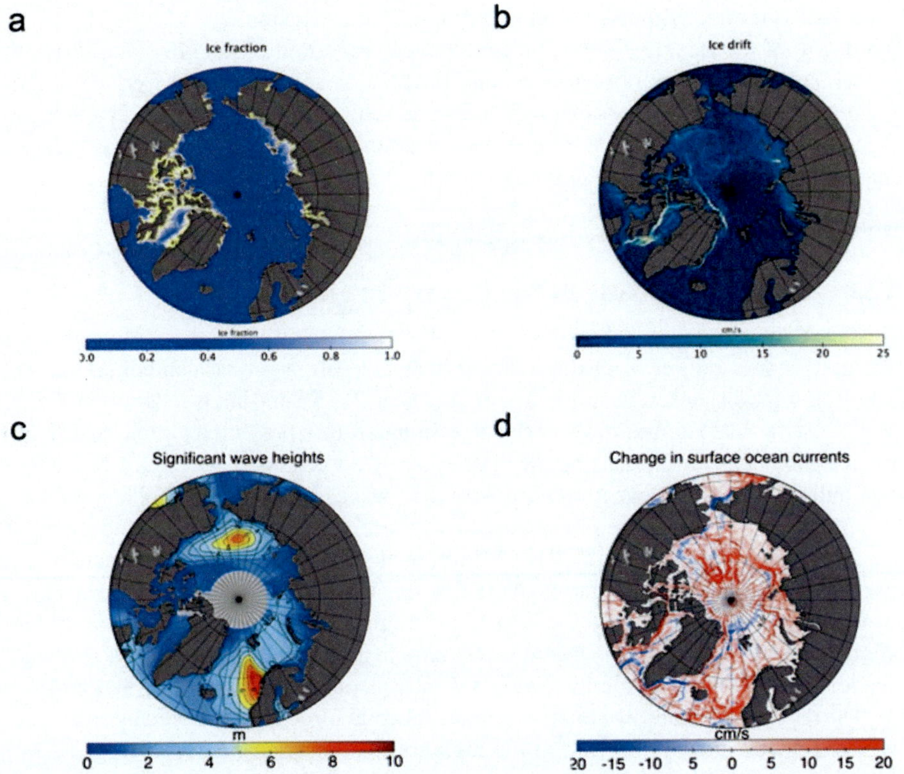

Fig. 12.6 Future Arctic in Jan 2090 (*RCP8.5*): **a** sea ice fraction (colour) and thickness (contours); **b** ice drift and **c** wave heights; **d** surface currents change 1990s to 2090s. **a**, **b**, **d**—from the high-resolution forward NEMO-ROAM simulations; **c**—WWIII. The light grey shading around the North Pole in **c** shows area outside WWIII model domain

the 2090s (Fig. 12.6a). Following the open ocean conditions, Fig. 12.6 shows development of strong storms with waves reaching in excess of 5 m in the Norwegian and Beaufort seas (maximum wave height reaches 9 m off the Norwegian coast). These are analogues of the 2012 storm (Stopa et al. 2016), but with stronger wind and waves. The lower ice conditions increase momentum transfer to the ocean from the atmosphere, accelerating spin-up of the ocean currents and faster drift of the fragmented sea ice (Fig. 12.6b, d). This drift can form "ice jets" (Rynders et al. 2021) with a large spatial shear in ice deformations in the Beaufort Sea, on the North and East Greenland continental shelves, in the Canadian Archipelago and Baffin Bay (Fig. 12.6b). Significantly, these areas are mineral-rich and are considered prospective for offshore mining. The patterns of ocean currents also change in these simulations, with the Fram Strait branch of the Atlantic inflow diminishing, and an accelerated Arctic Boundary Current along the Siberian shelf slope (Fig. 12.6d). The Beaufort Gyre has been predicted to shift from its current position in the west Beaufort Sea to the shallow Chukchi Cap, spinning-up strong eddies off-shore, and affecting ocean currents in the immediate vicinity of the shallows and in the Canada Basin (Fig. 12.6d). The westward shift of the Beaufort Gyre has already been recorded in satellite data, showing agreement between the observations and models. In conclusion, the sea ice retreat significantly affects upper ocean dynamics and sea state (waves). These changes are expected to impact future environmental extremes and to affect hazards ranking for marine operations discussed in the next sections.

12.6 Risks and Loads in the Future Projections

We used model output from the forward high-resolution forced centennial (2000–2099) simulations described previously (Yool et al. 2015), and analysed low (*RCP2.6*) and high (*RCP8.5*) emission scenarios to estimate future projected global loads from sea ice and ocean currents using the methodology described in Sect. 12.2. Since it is computationally expensive to run fully-coupled wave-sea ice-ocean models, we used a stand-alone WWIII wave model forced with projected atmospheric model output (to 2099) (Bricheno and Wolf 2018; Morim et al. 2020) to obtain an estimate of the wave loads in the future scenarios. Note that this approach is limited by its omission of wave-ocean feedbacks. To the best of the authors' knowledge there are only few analyses published on the projected wave climate in the Arctic, and these do not take sea ice-waves-ocean interaction processes into account either (Khon et al. 2014).

The results find that wave loads dominate the total forces experienced by structures with magnitudes of *o*(100 MN) in the high emission scenario *RCP8.5* (Fig. 12.7a). In these simulation total loads from waves reach very high values during the storm events in the Norwegian Seas and also in the most of the Arctic Ocean (Fig. 12.7a).

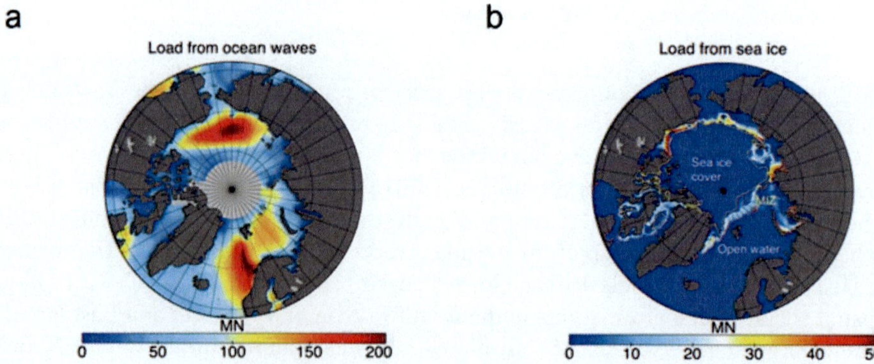

Fig. 12.7 Wave loads on virtual structure in mid-January 2090 calculated from WWIII run with *RCP8.5* forcing (**a**) and total sea ice load in July 2030–39 in the same *RCP8.5* scenario from the NEMO-ROAM simulations (**b**). The light grey shading around the North Pole in **a** shows area outside WWIII domain. Red and yellow lines in **a** mark outer and inner MIZ boundaries (0.15 and 0.80 of ice fraction)

In the *RCP8.5* simulations Arctic summer sea ice disappears by 2060s (Aksenov et al. 2017). By 2090, winter sea ice is largely absent in the Arctic Ocean except for thin ice on the Siberian shelves (0.2 m) and the Canadian Archipelago (0.6 m). Total ice loads in the high scenarios become negligible by the mid 21-st century due to low ice cover. However, in the 2030s, ice loads are still the second highest, of o(10 MN) (Fig. 12.7b).

Total loads from ocean currents are lower still, order of o(0.1 MN), in both low and high scenarios (results not shown). Meanwhile, in the *RCP2.6* simulations total ice loads remain significant until the end of the century, contributing of 10–15% to the total loads in the 2030s in the Central Arctic Ocean and up to 70–100% of total forces (results not shown). Sporadic high ice loading (reaching up to 100% of total loads) can also occur in the both emission scenarios in the mid-century in areas of ice drift convergence—the Arctic Transpolar Drift and north of Greenland. These are primarily from loading due to floe collisions with structures.

Using accessibility maps based on AIRSS model (Aksenov et al. 2017) we compared high-load risk areas from different environmental factors to illustrate regional differences for the marine operational hazards. From the *RCP2.6* forward simulations we have assessed accessibility in summer (JJA) 2030–39. The accessibility analysis has been based on the safe speed criteria for ship classes Type B and CAC4 (Aksenov et al. 2017) and the accessibility has been presented as a "traffic light" factor based on the Ice Numerals method (Aksenov et al. 2017). Figure 12.8 shows accessible areas in green (Ice Numeral > 0, safe to sail in) and inaccessible areas in red (Ice Numeral <= 0, not safe to sail in) for the two ship classes, general cargo Type B vessel and CAC4 ice-capable vessel, for the projected average summer (JJA) sea ice conditions in 2030–39 from the *RCP2.6* high-resolution forward simulations. In this decade the projected summer sea ice extent is similar to the present day, with the ice fraction is about 0.8–1.0, however ice is thinner, only ~0.5–1.5 m thick in the central Artic Ocean and less than 0.5-m thick in the Siberian shelf seas.

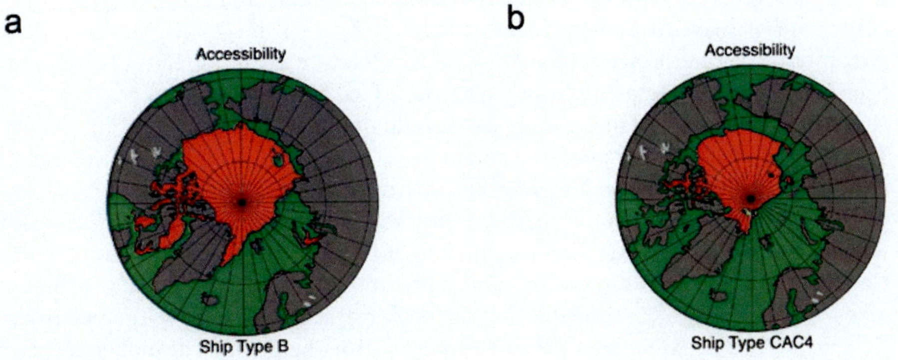

Fig. 12.8 Risk areas as "traffic lights". **a**, **b** accessibility based on AIRSS model (Aksenov et al. 2017) in summer (June–July–August) 2030–39 for ship classes Type B and CAC4, *RCP2.6* forward NEMO-ROAM simulations

These allows CAC4-type ice-capable vessels to sail unsupported by North Sea Route (NSR), whereas Type B general vessels cannot access the NSR without ice-breaker support. This analysis is in agreement with the Arctic navigability assessment from (Aksenov et al. 2017), although the approach taken by AIRSS includes only "static" risks from sea ice, aka areas covered by heavy ice, neither addresses risks from ice convergences nor from ice floes collisions with ship hull nor in calm seas nor in wavy conditions.

Comparison of the ship accessible areas in Fig. 12.8 with areas of the high ice loads in Fig. 12.7b demonstrates that the standard accessibility models can be underestimating the ice loads and risks in MIZ. The higher wave loads also broadly correspond to the areas of the higher ice loads in MIZ. Where sea ice is broken and piling up against the coast, the collisional loads are high in the presence of the high wave loads, e.g. the areas around the islands and north of the Cape Barrow, increasing combined risk from waves and ice dynamics. Keeping in mind that the time periods for the waves loads and ship accessibility analysis are for winter *vs.* summer, the overall conclusion is clear: the areas for the risks from waves are mostly on the Siberian shelf, whereas higher ice hazards are in the central Arctic Ocean, where there is more ice and less waves.

12.7 Discussion

In the analysis presented here, loads on Arctic structures are predicted to increase from ~0.1 MN at present to ~50–200 MN in the future, primarily due to increased wave loads. Although Arctic sea ice is simulated to decline into the future, ice loads will also increase due to dynamical impacts from floe-structure collisions, but remain of secondary importance at ~20–50 MN (cf. Figs. 12.4 and 12.7). These results challenge the view that the risks for off-shore operation in the Arctic will diminish with sea ice reduction (Melia et al. 2016). Our analysis also predicts loads from ocean currents to increase from the present peak level of ~0.01 to ~0.3 MN in the central Arctic in the future projections due to the acceleration of ocean currents, whereas frictional ice load decreasing from with reduced ice fraction and thickness.

The ME method used in the study has several limitations that can affect its results. The presented analysis focuses on mono-piles with large width. Since the inertia term in the Eq. (12.1) depends on D^2 and the drag term depends linearly on D, the width of the structure is more important for wave loads (as they quickly attenuate with water depth), whereas the structure height is more important for current loads. Consequently, different dimensions and types of structure will experience different loads. It is possible to scale the load for the tubular structures of various diameters and the analysis can be extended to the structures of various heights and a number of jacket legs. We have performed such an analysis for the fixed 100-m tall monopiles with diameters 1–10 m in the Northern Sea (Sanderson 1988). The results demonstrated that the wave force is at least an order of magnitude larger (~50 MN for 10-m diameter structure) than the current force. Relatively large current contribution is in the tidally-dominated areas (~1 MN for 10-m diameter structures) or in deeper regions, whereas large wave force is in shallow water. Our analysis excludes tides, these are important in the North Sea. In the Arctic tides may be crucial in regions, such as the Sea of Okhotsk.

Higher ice thickness in the ridges and higher strength of the consolidated ridges can result in higher loads. We used averaged ice velocities, in reality there is a fast stick-sleep loading of the structures. The above assumptions may explain lower ice frictional loads in our analysis as compared to the measured local loads on off-shore structures in the Arctic (Sanderson 1988; Jefferies et al. 2011). As a next step we plan to use the ice strength from the model to calculate deformation scale of the ice floes. This would allow to detect the different regimes of the floe-structure collision.

While our work has used what are currently high resolution models, even higher resolutions (<1 km) will further increase the speed of simulated currents and the resulting loads from them. Similarly, higher resolution will result in faster sea ice drift and higher ice stresses, making ice loads larger (Lemieux et al. 2020). Collisional ice loads have a strong sensitivity to ice floe sizes due to their dependency on floe mass: in our calculations the loads in-crease ~100-fold when mean floe size increases from 50 to 300 m. However, for ice–ship interactions, the AIRSS model does not include collisions with ice floes, underestimating risks. Linking floe sizes with their dynamical impact on ship hulls from scaled model ship basin experiments and Discrete Element Modelling (DEM) provides method of assessment (Huang et al. 2019). The environmental models used in this study do predict iceberg trajectories and their probabilistic distribution and risk of collision (Storkey et al. 2018). However, to estimate the loads and damage due to ship-icebergs collision the full external and internal mechanics should be considered (Liu and Amdahl 2010), this requires additional development linking the environmental modelling with ship mechanics. The above uncertainties are likely to result in the underestimation of loads and associated risks for other factors than sea ice, therefore supporting the main conclusion of the study that future sea ice retreat, while opening the Arctic to off-shore industries, may lead to higher risks from emerging environmental factors.

12.8 Summary

The reduction of the Arctic sea ice cover in the last decades has led to the easing of conditions for summer navigation along the NSR and an episodic opening of the Northwest Passage Route (NWP). This is often interpreted as overall improvement of marine access in the Arctic. However, the combined changes in sea ice cover and the ocean are not addressed in these assessments. In this study we have examined environmental factors which may increase risks to marine operations in the future. These environmental parameters should be considered in marine accessibility and risks planning for off-shore structures and navigation. The emerging future Arctic features more fragmented and active sea ice, as well as a more active ocean. These changes in the environment consequently require new approaches in monitoring and forecasting of the ocean and sea ice. New fundamental developments are also important for constraining future projections and improving forecasting skills, ultimately making navigation and off-shore marine industries in the Arctic safer for the environment and society.

Acknowledgements The authors express their gratitude to the International Union for Applied and Theoretical Mechanics for organizing and supporting the Symposium on Physics of Sea Ice in Helsinki in 2019 and providing an excellent forum for discussions and testing novel ideas. We also thank anonymous reviewers for the very useful feedbacks on the early version of the manuscript. For the research we acknowledge support from EU FP7 "Ships and waves reaching Polar Regions (SWARP)" (GA 607476), from the UK Innovation Safer Operations at Sea—Supported by Operational Simulations SOS-SOS (NE/N017099/1), "Towards a marginal Arctic sea ice cover" (NE/R000085/1), "Preconditioning the trigger for rapid Arctic ice melt" (NE/T000260/1) and funding from the European Union's Seventh Programme for Research, Technological Development, and Demonstration under grant agreement FP7-ENV-2013-Two-Stage-603396-RISES-AM. This work also has received financial support from the projects APEAR (NE/R012865/1, NE/R012865/2, #03V01461), ARISE (NE/P006000/1) and Arctic PRIZE (NE/P006078/1), as part of the Changing Arctic Ocean programme, jointly funded by the UKRI Natural Environment Research Council (NERC) and the German Federal Ministry of Education and Research (BMBF). The study was supported by the European Union's Horizon 2020 research and innovation programme under grant agreement no. 820989 (project COMFORT). The work reflects only the authors' view; the European Commission and their executive agency are not responsible for any use that may be made of the information the work contains. The sea ice-ocean-wave model simulations have been performed as part of the Regional Ocean Acidification (ROAM) (NE/H01732/1) and part-funded by the EU FP7 EURO-BASIN (FP7/2007-2013, ENV.2010.2.2.1-1; GA264933). This work also used the ARCHER UK National Supercomputing Service and JASMIN, the UK collaborative data analysis facility. Finally, we are thankful to Dr. Jean Bidlot (ECMWF) for providing model wave data and for his kind advice on the wave modelling.

References

Y. Aksenov, E. Popova, A. Yool, A.J.G. Nurser, L. Bertino, T.D. Williams, J. Bergh, On the future navigability of the arctic sea routes: high-resolution projections of the arctic ocean and sea ice decline. Mar. Policy **75**, 300–317 (2017). https://doi.org/10.1016/j.marpol.2015.12.027

A.W. Bateson, D.L. Feltham, D. Schröder, L. Hosekova, J.K. Ridley, Y. Aksenov, Impact of sea ice floe size distribution on seasonal fragmentation and melt of Arctic sea ice. The Cryosphere **14**(2), 403–428 (2020). https://doi.org/10.5194/tc-14-403-2020

L.M. Bricheno, J. Wolf, Future wave conditions of Europe, in response to high-end climate change scenarios. J. Geophys. Res. Oceans **123**(12), 8762–8791 (2018). https://doi.org/10.1029/2018JC013866

L. Brodeau, B. Barnier, A.M. Treguier, T. Penduff, S. Gulev, An ERA40-based atmospheric forcing for global ocean circulation models. Ocean Mod. **31**(3–4), 88–104 (2010). https://doi.org/10.1016/j.ocemod.2009.10.005

K. Chakrabarti, *Hydrodynamics of Offshore Structures*. pp. 440, Springer Berlin, (1987), ISBN 978-0-93121-516-2

G. Clauss, E. Lehmann, C. Östergaard, *Hydromechanical Analysis of Offshore Structures*. In Offshore Structures. Springer London, (1992). https://doi.org/10.1007/978-1-4471-3193-9_3

D.L. Feltham, Granular flow in the marginal ice zone. Phil. Trans. R. Soc. A. **363**, 1677–1700 (2005). https://doi.org/10.1098/rsta.2005.1601

P.K. Haff, Grain flow as a fluid-mechanical phenomenon. J. Fluid Mech. **134**, 401–430 (1983). https://doi.org/10.1017/S0022112083003419

K.H. Halse, *On vortex shedding and prediction of vortex-induced vibrations of circular cylinders.* Dr. Eng. Thesis, Norwegian University of Sci. Tech., pp. 270 (1997), ISBN 82-471-0076-2

H. Heorton, M. Tsamados, T. Armitage, A. Ridout, J. Landy, CryoSat-2 Significant Wave Height in Polar Oceans Derived Using a Semi-Analytical Model of Synthetic Aperture Radar 2011–2019. Remote Sens. (2021)

E.C. Hunke, W.H. Lipscomb, A.K. Turner, N. Jeffery, S. Elliott, CICE: the Los Alamos Sea Ice Model Documentation and Software User's Manual Version 5.1, LA-CC-06-012, Los Alamos Nat. Lab., NM 87545 (2015)

L. Huang, M. Li, B. Igrec, P. Cardiff, D. Stagonas, G. Thomas, Simulation of a ship advancing in floating ice floes. Port and Ocean Engineering under Arctic Conditions (POAC)., Simulation of a ship advancing in floating ice floes. In Proceedings of the International Conference on Port and Ocean Engineering Under Arctic Conditions, Paper no. POAC19-015 (2019). https://www.poac.com/Papers/2019/pdf/POAC19-015.pdf

M. Jefferies, B. Rogers, M. Hardy, B. Wright, Ice load measurement on Molikpaq: methodology and accuracy. In Proceedings of the International Conference on Port and Ocean Engineering Under Arctic Conditions, Paper no. POAC11–189, (2011). ISSN 0376-6756, https://www.poac.com/PapersOnline.html

H. Keulegan, L.H. Carpenter, Forces on cylinders and plates in an oscillating fluid. J. Res. Natianl Bureau of Standards **60**(5), 423–440 (1958). paper No 2857. http://resolver.tudelft.nl/uuid:172b871b-b5a1-4068-b8a6-96572059af52

V.C. Khon, I. I. Mokhov, F. A. Pogarskiy, A. Babanin, K. Dethloff, A. Rinke, H. Matthes, Wave heights in the 21st century Arctic Ocean simulated with a regional climate model. Geophys. Res. Lett. **41**(8), 2956–2961 (2014). https://doi.org/10.1002/2014GL059847

J.F. Lemieux, L.B. Tremblay, M. Plante, Toward a method for downscaling sea ice pressure for navigation purposes. The Cryosphere **14**(10), 3465–3478 (2020). https://doi.org/10.5194/tc-14-3465-2020

Z. Liu, J. Amdahl, A new formulation of the impact mechanics of ship collisions and its application to a ship–iceberg collision. Mar. Struct. **23**(3), 360–384 (2010). https://doi.org/10.1016/j.marstruc.2010.05.003

C. Lichey, H.H.Hellmer, Modeling giant-iceberg drift. J. Glaciol. **47**(158), 452–460 (2001). https://doi.org/10.3189/172756501781832133

T. Martin, A. Adcroft, Parameterizing the fresh-water flux from land ice to ocean with interactive icebergs in a coupled climate model. Ocean Mod. **34**(3–4), 111–124 (2010). https://doi.org/10.1016/j.ocemod.2010.05.001

G. Madec and NEMO System Team, NEMO ocean engine. Note du Pole de modelisation, Institut Pierre-Simon Laplace (IPSL), France, **27**, 1288–1619 (2017). https://doi.org/10.5281/zenodo.3248739

N. Melia, K. Haines, E. Hawkins, Sea ice decline and 21st century trans-Arctic shipping routes. Geophys. Res. Lett. **43**(18), 9720–9728 (2016). https://doi.org/10.1002/2016GL069315

J. Morim, C. Trenham, M. Hemer, X.L. Wang, N. Mori, M. Casas-Prat, ... , L. Erikson, A global ensemble of ocean wave climate projections from CMIP5-driven models. Scientific Data **7**(1), 1–10 (2020). https://doi.org/10.6084/m9.figshare.11940576

J.R. Morison, M.P. O'Brien, J.W. Johnson, S.A. Schaaf, The force exerted by surface waves on piles. Petrol. Trans., AIME **189**, (1950)

D.A. Onishchenko, Analytical approach to the calculation of the design values of the loads associated with discrete ice features. In Proceedings of the International Conference on Port and Ocean Engineering Under Arctic Conditions. Paper No. POAC09-126, (2009), ISSN: 0376-6756. https://www.poac.com/PapersOnline.html

I.V. Polyakov, A.V., Pnyushkov, M.B. Alkire, I.M. Ashik, T.M. Baumann, E.C. Carmack, ... A. Yulin, Greater role for Atlantic inflows on sea-ice loss in the Eurasian Basin of the Arctic Ocean. Sci. **356**(6335), 285–291 (2017). https://www.science.org/doi/10.1126/science.aai8204

S., Rynders, Y. Aksenov, D.L. Feltham, A.J. George Nurser, G. Madec, Impact of Granular Behaviour of Fragmented Sea Ice on Marginal Ice Zone Dynamics. In IUTAM Symposium on Physics and Mechanics of Sea Ice (2021), Springer Nature, https://doi.org/10.1007/978-3-030-80439-8_13

T.J.O. Sanderson, *Ice Mechanics*; *Risks to Offshore Structures*, pp. 253, Graham and Trotman pub. Boston (1988). ISBN: 978-0-86010-785-9

H.H. Shen, W.B. Hibler, M. Lepparanta, The role of floe collisions in sea ice rheology. J. Geophys. Res. Oceans **92**(C7), 7085–7096 (1987). https://doi.org/10.1029/JC092iC07p07085

N. Shakhova, I. Semiletov, O. Gustafsson, V. Sergienko, L. Lobkovsky, O. Dudarev, V. Tumskoy, M. Grigoriev, A. Mazurov, A. Salyuk, R. Ananiev, Current rates and mechanisms of subsea permafrost degradation in the East Siberian Arctic Shelf. Nature Comm. **8**(1), 1–13 (2017). https://doi.org/10.1038/ncomms15872

N. Skliris, R. Marsh, M. Srokosz, Y. Aksenov, S. Rynders, N. Fournier, Assessing extreme environmental loads on offshore structures in the North Sea from high-resolution ocean currents, waves and wind forecasting. J. Marine Sci. Eng. **9**(10), 1052 (2021). https://doi.org/10.3390/jmse9101052

C. Strong, I.G. Rigor, Arctic marginal ice zone trending wider in summer and narrower in winter. Geophys. Res. Lett. **40**(18), 4864–4868 (2013). https://doi.org/10.1002/grl.50928

D. Storkey, A.T. Blaker, P. Mathiot, A. Megann, Y. Aksenov, E.W. Blockley, D. Calvert, T. Graham, H.T. Hewitt, P. Hyder, T. Kuhlbrodt, UK Global Ocean GO6 and GO7: A traceable hierarchy of model resolutions. Geosci. Mod. Dev. **11**(8), 3187–3213 (2018). https://doi.org/10.5194/gmd-11-3187-2018

J.C. Stroeve, V. Kattsov, A. Barrett, M. Serreze, T. Pavlova, M. Holland, W.N. Meier, Trends in Arctic sea ice extent from CMIP5, CMIP3 and observations. Trends in Arctic. Geophys. Res. Lett. **39**(16), L16502 (2012). https://doi.org/10.1029/2012GL052676

J.E. Stopa, F. Ardhuin, F. Girard-Ardhuin, Wave climate in the Arctic 1992–2014: Seasonality and trends. The Cryosphere **10**(4), 1605–1629 (2016). https://doi.org/10.5194/tc-10-1605-2016

G. Timco, W. Weeks, A review of the engineering properties of sea ice. Cold Reg. Sci. Techn. **60**(2), 107–129 (2010). https://doi.org/10.1016/j.coldregions.2009.10.003

T.D. Williams, L.G., Bennetts, V.A., Squire, D., Dumont, L. Bertino, Wave–ice interactions in the marginal ice zone. Part 1: Theoretical foundations. Ocean Mod. **71**, 81–91 (2013). https://doi.org/10.1016/j.ocemod.2013.05.010

A.Yool, E. E. Popova, A.C. Coward, Future change in ocean productivity: Is the Arctic the new Atlantic? J. Geophys. Res. Oceans **120**(12), 7771–7790 (2015). https://doi.org/10.1002/2015JC011167

Chapter 13
Impact of Granular Behaviour of Fragmented Sea Ice on Marginal Ice Zone Dynamics

Stefanie Rynders, Yevgeny Aksenov, Daniel L. Feltham, A. J. George Nurser, and Gurvan Madec

Abstract Sea ice retreat and opening of large, previously ice-covered areas of the Arctic Ocean to wind and ocean waves is leading to large changes in the sea ice state. The Arctic sea ice cover is becoming more fragmented and mobile, with large regions of ice cover projected to evolve into a marginal ice zone (MIZ). Fragmented sea ice has different dynamics, necessitating changes in sea ice model rheology. The objective of this study is to improve sea ice dynamics in models for forecasting and climate projections. We introduce granular behaviour in the ice dynamics and assess the impact on sea ice behaviour. For this purpose we have implemented a seamless rheology for MIZ and pack ice in an idealised sea ice and ocean model. The study compares the effect of the combined rheology with that of the standard elastic-viscous-plastic (EVP) rheology. The main effect of granular behaviour in ice rheology is on internal ice pressure. The jostling of the floes causes divergence of the sea ice cover. Sea ice viscosities are only weakly impacted. In idealised simulations the new sea ice rheology results in widening of the MIZ and a more diffuse ice edge in a stand-alone set-up. Oceanic feedbacks counteract and can undo this effect. The resolution of the simulation modifies the effect of the rheology: a rheology

S. Rynders (✉) · Y. Aksenov · A. J. G. Nurser
Marine Systems Modelling Group, National Oceanography Centre, Southampton, UK
e-mail: s.rynders@noc.ac.uk

Y. Aksenov
e-mail: yka@noc.ac.uk

A. J. G. Nurser
e-mail: g.nurser@noc.ac.uk

D. L. Feltham
Centre for Polar Observation and Modelling, Department of Meteorology, University of Reading, Reading, UK
e-mail: d.l.feltham@reading.ac.uk

G. Madec
Laboratoire d'Océanographie et du Climat: Expérimentations et Approches Numériques (LOCEAN), IPSL, Sorbonne Université, Paris 75005, France
e-mail: gurvan.madec@locean.upmc.fr

J. Tuhkuri and A. Polojärvi (eds.), *IUTAM Symposium on Physics and Mechanics of Sea Ice*, IUTAM Bookseries 39, https://doi.org/10.1007/978-3-030-80439-8_13

that accounts for granular effects offers better convergence of the solution than the standard EVP. In conclusion, granular sea ice rheology affects the sea ice in the MIZ in some cases, but the importance of the effect in general is not clear.

13.1 Motivation

The Arctic sea ice state is changing from typical pack ice conditions to becoming more fragmented and mobile and indeed an increase in marginal ice zone (MIZ, region with ice concentration between 15% and 80%) width is already observed (Strong and Rigor 2013). In the future large regions of the ice cover in summer will evolve into a MIZ according to projections from Aksenov et al. (2017) and the Met Office Hadley Centre (Ed Blockley, personal communication). The MIZ is defined as the region that is influenced by waves, which means its dynamics are different from those of pack ice. In addition to wave effects, sea ice floes in the MIZ can have turbulent motion, defined as random fluctuations in the magnitude and direction of ice motion, which differ from the mean velocity over an area. There are also collisions between sea ice floes. Since sea ice constitutive equations currently used in global ocean-sea ice models are designed for central pack ice, the effect of turbulent motion of the sea ice floes is unaccounted for. The floe motion can influence the stresses in the ice cover as well as impact on the ocean. Turbulent motion of floes can cause increased ocean mixing, potentially influencing stratification and consequently currents. An impact on the mixed layer depth also has implications for biogeochemistry and ecosystems, depending on the season. There is also a growing interest in forecasts for applications: shipping, tourism, fisheries and mineral extraction. These activities are all mainly focussed on the marginal ice zones that are easier to access. The objective of this study is therefore to improve sea ice dynamics in models for forecasting and climate projections, focussing on an improved representation of the MIZ. We examine new sea ice rheologies accounting for the turbulent floe motion as this could be of importance for current and future MIZ conditions.

13.2 Marginal Ice Zone Rheology

We introduce granular behaviour in the ice dynamics and assess the impact on the sea ice cover. For this purpose we have implemented a seamless rheology for MIZ and pack ice, that is without switches or other discontinuities, in a coupled sea ice and ocean model. Simulations are performed in an idealised channel model.

Shen first developed a specific rheology for the MIZ, determining the stresses caused by the collisions between floes (Shen et al. 1986, 1987). This approach is known as collisional rheology. The stresses depend on disk diameter, the restitution coefficient (a measure of how inelastic collisions are) and the magnitude of velocity fluctuations. Stresses are then averaged into a continuum expression of the same form as a viscous model. Shen's rheology has rarely been used in large scale realistic numerical sea ice simulations, but e.g. the Nansen Environmental and Remote Sensing Center (NERSC) have used it in Fram Strait simulations. They used a switch to run collisional rheology in the MIZ alongside EVP rheology for pack ice regions. They encountered frequent problems due to discontinuities in the stresses at the MIZ/pack ice boundary causing numerical instabilities and concluded the approach to be not viable (personal communication). There is also a model for the Sea of Okhotsk from the University of Tokyo which uses a collisional rheology (Fujisaki et al. 2010). They combine the collisional rheology of Sagawa (for a semi-lagrangian instead of a continuum model) with viscous-plastic (VP) rheology (Sagawa and Yamaguchi 2006). The model also contains a switch between the rheologies.

Feltham introduced the 'granular temperature' (G_T) of sea ice, an evolving velocity fluctuation magnitude computed from a kinetic energy balance for floes (Feltham 2005). It allows for a smooth transition between MIZ and pack ice type behaviour. He combined Shen's collisional rheology with VP rheology to create a constitutive law that is suitable for both regions, without a need to predefine a boundary between central pack ice and the marginal ice zone. This avoids numerical issues. Since VP and collisional rheology account for stresses due to different mechanisms, it can be assumed that the internal stress at a point is the sum of both contributions. Here the collisional rheology is combined with the elastic-viscous-plastic (EVP) rheology (Hunke and Dukowicz 1997), with again the different contributions summed. This results in a stress tenstor σ as follows:

$$\begin{aligned}\boldsymbol{\sigma}_{ij} &= 2(\eta_{EVP} + \eta_{COL})\boldsymbol{\epsilon}_{ij} + ((\zeta_{EVP} + \zeta_{COL}) - (\eta_{EVP} + \eta_{COL}))\boldsymbol{\epsilon}_{kk}\delta_{ij} \\ &\quad - \frac{1}{2}(P_{EVP} + P_{COL})\delta_{ij},\end{aligned} \tag{13.1}$$

with ϵ the strain rate tensor and δ_{ij} the Kronecker delta. The expressions for the bulk viscosity (for isotropic rate dependent deformation), shear viscosity (non-isotropic part of rate dependent deformation) and internal pressure (for plastic deformation) can be found in Table 13.1, with symbols in Table 13.2. The elastic part of EVP rheology exists only for numerical reasons and does not affect the result provided the internal stress calculation converges. We also use a minimum value for the parameter Δ to prevent spontaneous viscous creep as in Bouillon et al. (2013). For the maximum ice concentration in the parameter γ Feltham choose 1, because if floes have different sizes or shapes this value could be reached. For our simulation we follow the approach of Luepkes et al. for their floe size parameterisation and use A_* instead of A_{max} to avoid a singularity at 1 (Luepkes et al. 2012):

$$A_* = \frac{1}{1 - (D_{min}/D_{max})^{1/\beta}}, \tag{13.2}$$

with $D_{min} = 8$m, $D_{max} = 300$m and $\beta = 0.75$. A_* is always slightly higher than 1. The concentration dependency exhibited by the granular temperature ensures that in practice the collisional stress reaches zero at a "maximum" ice concentration (for

Table 13.1 Combined rheology

	Collisional	EVP
Shear viscosity	$\eta_{COL} = \frac{\gamma(1+e')}{3\pi}\frac{\sqrt{2}G_T^{1/2}}{L_f}$	$\eta_{EVP} = \frac{P}{2\Delta e^2}$
Bulk viscosity	$\zeta_{COL} = \frac{\gamma(1+e')}{\pi}\frac{\sqrt{2}G_T^{1/2}}{L_f}$	$\zeta_{EVP} = \frac{P}{2\Delta}$
Pressure	$P_{COL} = \gamma\frac{\sqrt{2}}{\pi^2}(1+e')\frac{2G_T}{L_f^2}$	$P_{EVP} = g(A)P^*h$
$\gamma = \frac{\rho L_f^2 h}{4}\frac{A^{3/2}}{A_{max}^{1/2} - A^{1/2}}$		
$g(A) = e^{-C(1-A)}$		
$\Delta = \sqrt{(\dot{\epsilon}_{11}^2 + \dot{\epsilon}_{22}^2)(1+e^{-2}) + 4e^{-2}\dot{\epsilon}_{12}^2 + 2\dot{\epsilon}_{11}^2\dot{\epsilon}_{22}^2(1-e^{-2})}$		

Table 13.2 Symbols

G_T	Granular temperature
L_f	Floe size
e'	Restitution coefficient
P^*	Ice strength
ρ	Ice density
h	Ice thickness
A	Ice concentration
A_{max}	Maximum ice concentration
C	Ice strength parameter
e	Eccentricity of the yield curve
$\dot{\epsilon}$	Strain rate

instance the close packed concentration); the exact point depends on the choice of parameters in the evolution equation (Rynders 2017).

The evolution equation for the granular temperature G_T is the following:

$$m\frac{\partial}{\partial t}G_T + m\mathbf{u}\cdot\nabla G_T = -\nabla\cdot\mathbf{q} + F - \boldsymbol{\sigma}^T : \nabla\mathbf{u} - \Gamma; \qquad (13.3)$$

here m is the combined mass of the sea ice and the water in between the floes, down to the bottom of the sea ice; $\mathbf{u}$ is the sea ice velocity and $\boldsymbol{\sigma}^T$ the transposed stress tensor. The first term in the right hand side of the equation is the divergence of the kinetic energy flux $\mathbf{q}$, which restributes kinetic energy through collisions and ice motion; this is a diffusion term. Then there is an energy source due to correlation of the turbulent air and ocean drags with the velocity fluctuation (F). The final terms are internal stress and energy dissipation Γ ($\boldsymbol{\sigma}^T : \nabla\mathbf{u}$ is defined as $\boldsymbol{\sigma}^T : \nabla\mathbf{u} = \sigma_{11}\partial u/\partial x + \sigma_{21}\partial u/\partial y + \sigma_{12}\partial v/\partial x + \sigma_{22}\partial v/\partial y$). The granular temperature is treated as an extra tracer in the sea ice model and the energy sources and sinks added separately. The (non-linear) diffusion term is discretised with a simple central difference scheme. This is not a good solution in general as it exhibits checkerboarding, but presently no high temporal resolution output is used.

13.3 Methods

The configuration is an idealised Southern Ocean channel sea-ice–ocean model. The ocean model is the hydrostatic general circulation model NEMO (Nucleus for European Modelling of the Ocean) (Madec 2014). NEMO is combined with the Los Alamos sea ice model CICE, a dynamic-thermodynamic model with multiple thickness categories (Hunke et al. 2013). A schematic of the channel model is shown in Fig. 13.1. The configuration has been constructed starting from an existing ocean-only Southern Ocean channel model, to which the coupling to CICE has been added for this study. The idealised channel has east–west cyclic boundary conditions with land on the northern and southern edges. It extends 2000 km northward from a latitude of 80 degrees south, and has a flat bottom topography.

To be able to distinguish between ice and ocean effects different degrees of coupling are used: an uncoupled channel, a thermodynamically coupled channel and a fully coupled ice–ocean channel. In the uncoupled channel configuration lateral and bottom melting and freezing of the sea ice is disabled. The ocean current is set to zero

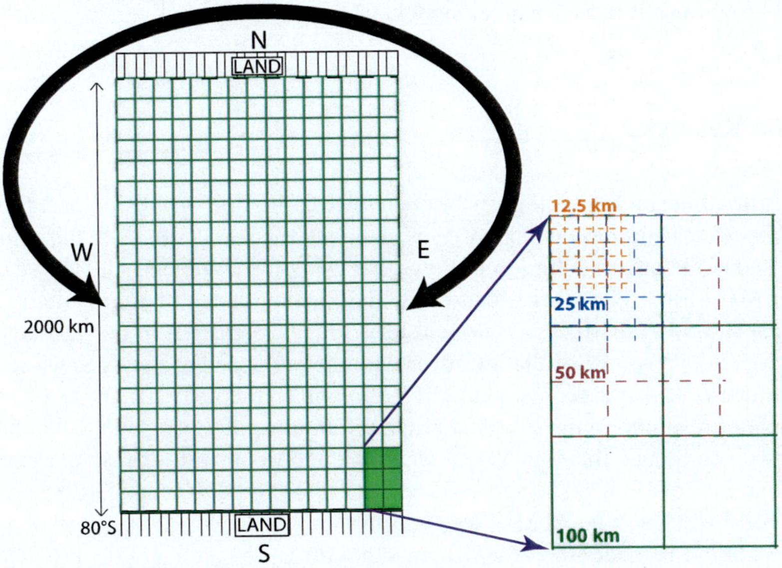

Fig. 13.1 Schematic of the channel model with the different grid resolutions

(in the sea ice model only) and sea surface slope is set to a constant value corresponding to a 3m height difference in the meridional direction over the whole domain. The thermodynamically coupled channel allows all modes of sea ice thermodynamics, but ocean current and sea surface slope are set as in the uncoupled configuration to exclude dynamic interaction with the ocean. A range of configurations is set up at resolutions ranging from 100 km to 12.5 km to examine the convergence of the solution. To limit the resolution dependency of the ocean model solution in all simulations, the lateral diffusion parameters have been adjusted to prevent formation of eddies at high resolution.

All simulations are without daily or seasonal cycles to arrive at a steady state solution. The sea ice is initialised with a 1 m thick ice cover at 100% ice concentration over three quarters of the domain, leaving the northern part ice free; afterwards the ice edge is completely free to move. The sea surface temperature is initialised to a constant value of –1 °C throughout the water column, chosen to prevent large scale melting or freezing of sea ice right at the start of the simulations. For salinity a realistic depth dependent profile (50° S to 80° S average of PHC climatology with surface set to 30.2 psu (Steele et al. 2001)) is used to prevent instability of the water column, which could lead to convection. The atmospheric forcing is a combination of analytical forcing and forcing with realistic latitudinal dependencies. Winds are purely zonal, being westerly at 8.475 m/s over most of the domain and tapering off to zero at the northern and southern edge. Air temperature ranges between –5 and 5 °C, linearly increasing northward. Short and long wave radiation are also increasing northward with a realistic latitudinal dependency. These forcings are chosen to ensure that ice fully melts towards the northern edge of the domain. Liquid and solid precipitation and humidity are held constant over the domain at realistic mean values. There is no run-off in the channel model.

13.4 Results

Firstly the effect of the rheology in the uncoupled configuration is examined. Comparing the ice concentration in the EVP simulation to the simulation with the combined EVP and collisional rheology run shows that using the combined rheology causes a wider MIZ (Fig. 13.2). This is caused by the jostling of the floes creating a higher internal stress in the MIZ, as is typical for a granular material in the gaslike phase. This causes divergence of the ice cover with the ice edge advancing and becoming more diffuse. To test the convergence of the combined rheology (i.e. the result is independent of resolution), the model is run at different resolutions and the width of the MIZ with combined rheology is compared to the EVP model. The simulation is run at resolutions of 12.5, 25, 50 and 100 km (see Fig. 13.3). At 100 and 50 km resolution both rheologies have equal MIZ widths, saturating at 300 km at 50 km resolution. If the resolution is made finer, the MIZ width with EVP rheology gets progressively smaller. Combined rheology gives a constant MIZ width of 300 km for a resolution of 50 km or higher. This means that the effect of the granular behaviour becomes larger

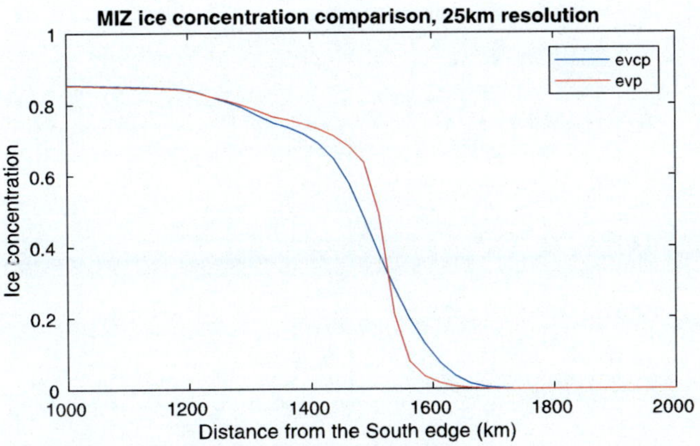

Fig. 13.2 Dependence of the ice concentration on rheology in the uncoupled channel model. The combined rheology simulation is the blue curve marked evcp; the EVP simulation is the red curve marked evp. Granular behaviour causes a wider MIZ and an advancing ice edge in the combined rheology simulation

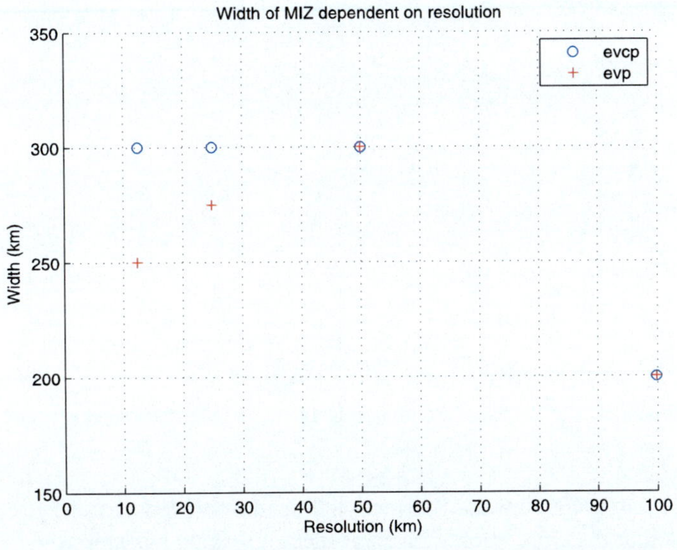

Fig. 13.3 Dependence of the MIZ width on resolution for EVP and combined rheology in the uncoupled configuration. For a resolution of 50 km or higher the combined rheology MIZ width is independent of resolution, while with EVP rheology the MIZ narrows as the resolution becomes finer

compared to the EVP internal stress. It appears the EVP rheology cannot sustain an MIZ at high resolutions due to a missing source of pressure resulting in too little dispersal.

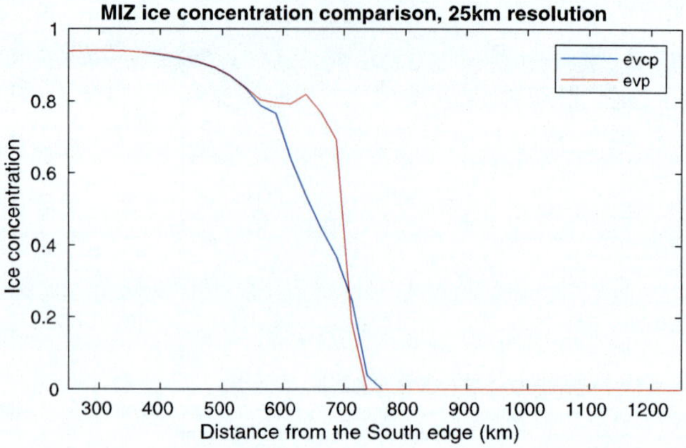

Fig. 13.4 Dependence of the ice concentration on rheology in the thermodynamically coupled configuration. The combined rheology gives a wider MIZ width than the simulation with EVP rheology

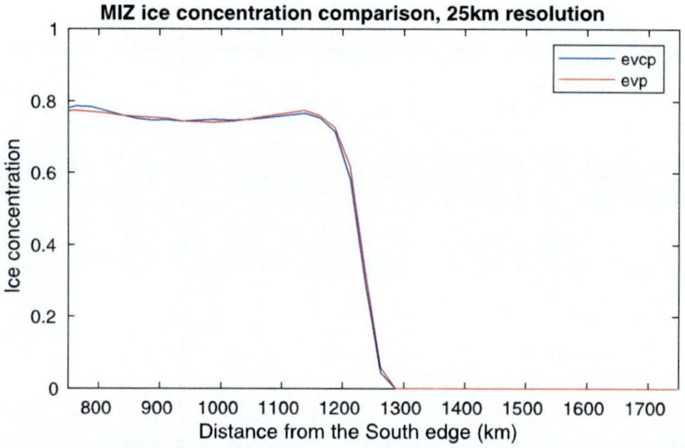

Fig. 13.5 Dependence of the ice concentration on rheology in the fully coupled configuration

In the simulations with the thermodynamically coupled channel it can be seen in Fig. 13.4 that the simulation with EVP rheology again has a narrow MIZ and the simulation with combined rheology a wider MIZ. In this simulation there is a general melting back of the ice cover. Different to the previous simulations the ice edge is in almost the same position in the EVP and combined rheology. This suggests that the position of the ice edge is dictated by the ocean temperature through ice melting.

Lastly the fully coupled configuration is examined, see Fig. 13.5. Remarkably there is no change in MIZ width between the two rheologies. The concentration profiles are almost identical except for perhaps a slight decrease in ice concentration in the MIZ in the combined rheology case. This could be caused by granular behaviour if sea ice is being spread out with the ice edge position dictated by the ocean temperature, but it is unclear whether the change is significant.

The different channel model configurations have shown very different results for the uncoupled and thermodynamically coupled simulations than for the fully coupled simulations. The former configurations indeed showed a clear increase in MIZ width with combined rheology, while the latter showed virtually no influence of the granular effects. To investigate this further we examine the total momentum balance. Figure 13.6 shows the relative size of the different terms of the momentum balance in a meridional cross section of the channel for the uncoupled and the fully coupled simulations, both with combined rheology. There is a striking difference in the relative magnitude of the internal stress and the oceanic drag terms in the two simulations. In the uncoupled simulation the internal stress is the third largest term (order 10^{-2} N/m) and the oceanic drag is negligible except close to the ice edge. In the fully coupled simulation on the other hand the internal stress term is small, but the ocean drag term is comparable to the tilt force and only the Coriolis force is clearly larger. The reason for this changed balance can be found in the motion of the sea ice. In the uncoupled simulation the sea ice is moving very slowly. With a static ocean from the point of view of the sea ice this will result in very little ocean drag. In the fully coupled channel model the sea ice moves much faster (northward), and also moves substantially faster than the ocean below, which results in large ocean drag. In the zonal direction atmospheric drag is balanced by oceanic drag, with internal stress being very small (order 10^{-3} N/m), see Fig. 13.7. Given the unimportance of the internal stress in the fully coupled channel model, it can be expected that the combined rheology has no appreciable effect in this setting.

13.5 Discussion

In the uncoupled channel model it is evident that there is a substantial increase in MIZ width with the granular rheology compared to the EVP rheology as the resolution of the simulation gets higher. This led to the conclusion that EVP rheology cannot sustain an MIZ through internal mechanisms in the sea ice, whereas the combined rheology can. At sufficiently high resolution it can be expected that the EVP rheology will become unable to sustain a certain MIZ width unless this is being imposed by oceanic influence. This is of concern to the sea ice and ocean modelling community as simulations are performed at ever higher resolutions.

However, in the fully ocean coupled simulation this effect due to the rheology alone is not present. It was found that the lack of sensitivity in the fully coupled simulation is due to the relative unimportance of the internal stress term versus the ocean drag term in the momentum balance (Fig. 13.6). When in the future the MIZ forms a major part of the ice cover it is however more likely that low ice concentrations occur in areas of low ocean drag, for instance in places where the ocean current is weak and sea ice velocity low like in the interior Arctic Ocean. This means the effect of the combined rheology is likely to be higher, given the large increase in internal

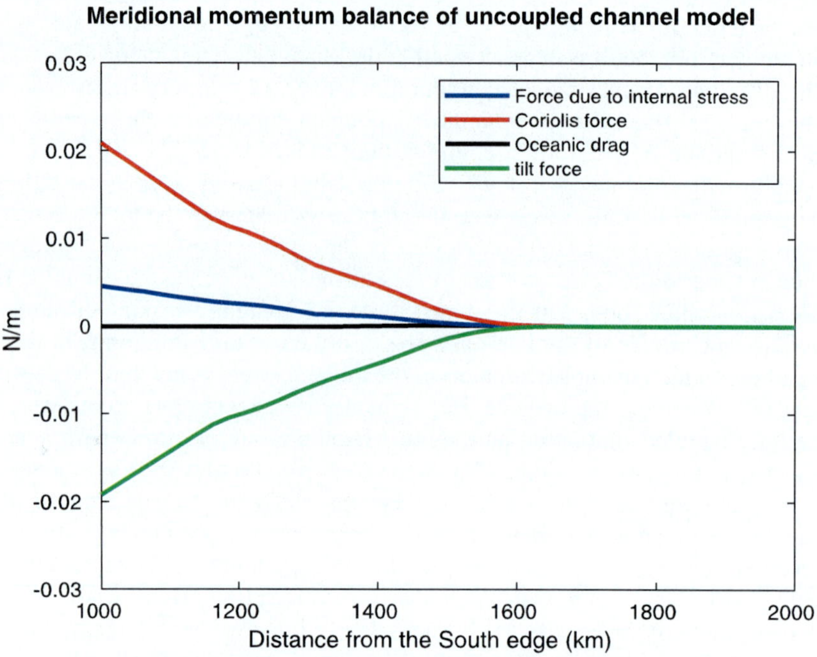

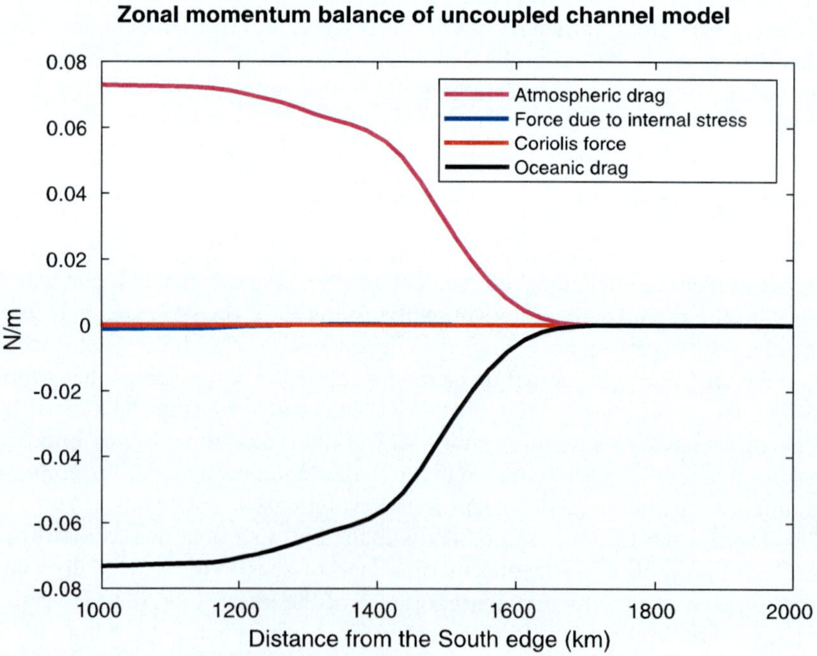

Fig. 13.6 Momentum balance of the uncoupled channel model simulation with combined rheology. **a** meridional direction **b** zonal direction

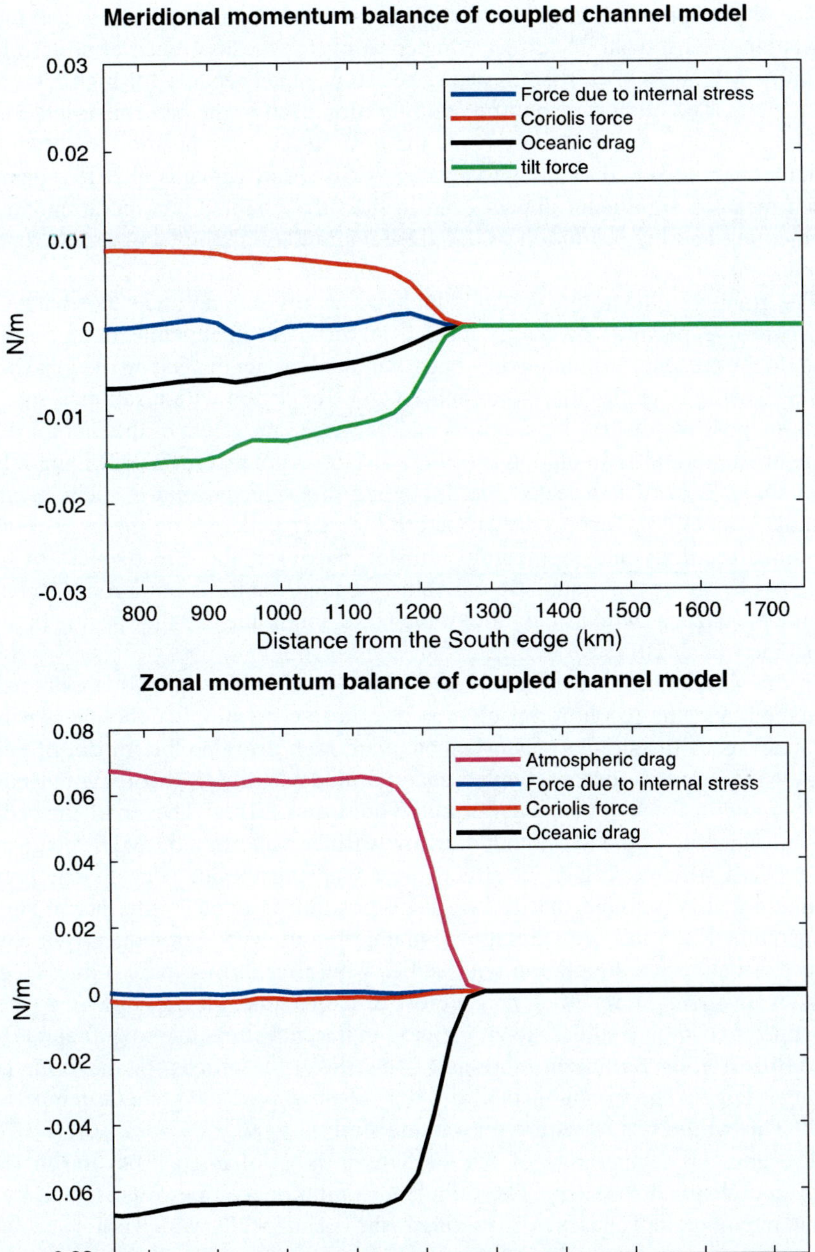

Fig. 13.7 Momentum balance of the coupled channel model simulation with combined rheology. **a** meridional direction **b** zonal direction

sea ice stress compared to the EVP rheology in such circumstances. It also means a possible ocean dynamics effect compensating for the low internal stress of sea ice in the MIZ with EVP rheology may be too weak, especially for high resolution simulations. The highest resolution configuration used in the present simulations is 12.5 km. For operational requirements the desired ice edge position uncertainty is actually lower (10 km for a 5 day forecast, (De Silva and Yamaguchi 2019)) than this which means a significant impact even in the fully coupled configuration can not be ruled out and the resolution of the present simulation is in fact too low for firm conclusions.

The granular temperature equation includes energy sources and sinks that are not well known, especially the energy input from turbulent fluctuations in currents and wind. In the present formulation this energy input does not depend on the local ocean velocity, while in reality more turbulence can be expected with faster currents. It is therefore possible that the simulations underestimate the effect of the rheology. The granular temperature calculation used here also does not include a contribution from wave surge. It can be expected that including this effect would result in a higher granular temperature, especially around the ice edge, increasing the impact of the combined rheology and momentum transfer. However, the effectiveness of wave surge in increasing the granular temperature depends on the ratio of the wavelength and the floe size, as well as the ratio of the wave amplitude and the spacing between floes (Yiew et al. 2017).

In the simulations ocean parameters were set so as to suppress possible eddy formation, and the resolution itself was also coarse, to simplify the ice dynamics and analysis of the simulations. Manucharyan et al. examine the impact of eddies in the MIZ in an ice-ocean coupled channel model with VP rheology, but no sea ice thermodynamics (Manucharyan and Thompson 2017). They find that eddies cause a widening of the MIZ, both drawing warmer water into the MIZ and pushing ice outward where it can melt, effectively advancing the ice edge. In the present simulations it was found that the ice edge position is fixed by the ocean surface temperature when full sea ice thermodynamics is allowed. Consequently it can be expected that such a simulation with eddies would result in a more dilute ice edge without advancing it, resulting in a net loss of ice volume. Ice gets exported when it is trapped in cyclonic eddies, so an increase in internal stress due to granular effects would resist eddy formation more and could delay the process or make the eddy diameter larger. The frontal instabilities may also increase the granular temperature itself. This would be interesting for a future study.

The granular temperature of sea ice floes may be of interest beyond the need for it in collisional rheology. The turbulent motion of ice floes represented by the granular temperature causes extra mixing in the ocean locally, which is not accounted for at the moment. It has also been proposed to use the variance of the ice motion, the granular temperature, to improve oil spill forecasts by estimating the spread in wintertime, when the spill location is inaccessible (Rampal et al. 2016). Mean term long range forecasts are considered accurate enough, but they are missing part of the small-scale motion which disperses the oil. The granular temperature can be used to calculate the cross section of the oil spill within certain probability limits, assuming a Gaussian distribution of the velocity fluctuations. In the MIZ the jostling of ice floes can cause damage to ships if it causes floes to collide with the flanks of ships. Knowledge of the granular temperature could be used to calculate the loads on the ship and hence be taken into account for risk assessment.

Finally, using the granular temperature to distinguish between MIZ type and pack ice type behaviour is not the only option. Some research has been done on phase transitions in granular materials, which can exhibit gaslike, liquidlike and solidlike behaviour. Ji and Shen suggest an index related to shear rate and concentration for shear flow (Ji and Shen 2008). Their simulations also highlight that phase transitions are not always sharp, which means an approach with a smoothly changing constitutive law is appropriate. Ji and Shen's results are not a complete solution as sea ice motion can also have a convergent or divergent component. The phase transition of granular materials in general is an unsolved problem and specific research in phase transitions of sea ice as a granular material is lacking. Sea ice is indeed particularly complicated as it is polydisperse, multishaped and the material properties are temperature dependent.

13.6 Summary

A combined elastic-viscous-plastic and collisional sea ice rheology has been implemented in a numerical model. Simulations have been carried out in uncoupled, thermodynamically coupled and fully coupled ice-ocean channel models. The new rheology increases MIZ width compared to EVP simulations in the uncoupled channel model. There is no change in width when ocean dynamics are present, due to internal stress being unimportant compared to ocean drag for the momentum balance in this configuration. It was also found that in the case of EVP rheology MIZ width is dependent on resolution for high resolution simulations, whereas the new rheology shows convergence of MIZ width in uncoupled simulations. The conclusion of the study is that MIZ rheology may have an impact when ocean current and drag effects are weak and floes relatively large, but otherwise rheology does not affect the MIZ width or edge position in divergent conditions.

Acknowledgements The authors express their gratitude to the International Union for Applied and Theoretical Mechanics for organizing and supporting the Symposium on Physics of Sea Ice in Helsinki in 2019 and providing an excellent forum for discussions and novel ideas. Thanks go to the reviewers for suggestions to extend the discussion and pointing out additional literature sources. This research has been supported by the project 'Ships and waves reaching Polar Regions (SWARP)' supported by the European Union's Seventh Framework Programme (FP7/2007-2013) under grant agreement No 607476 and also by the UK project "Towards a marginal Arctic sea ice cover" (NE/R000654/1). This project has also received funding from the European Union's Horizon 2020 research and innovation programme under grant agreement No 821926 (project IMMERSE—Improving Models for Marine EnviRonment SErvices). This paper, reflects only the authors' view; the European Commission and their executive agency are not responsible for any use that may be made of the information the work contains.

References

Y. Aksenov, E. Popova, A. Yool, A.J.G. Nurser, L. Bertino, T.D. Williams, J. Bergh, On the future navigability of the arctic sea routes: high-resolution projections of the arctic ocean and sea ice decline. Mar. Policy **75**, 300–317 (2017). https://doi.org/10.1016/j.marpol.2015.12.027

S. Bouillon, T. Fichefet, V. Legat, G. Madec, The elastic-viscous-plastic method revisited. Ocean Model. **71**, 2–12 (2013). https://doi.org/10.1016/j.ocemod.2013.05.013

L.W.A. De Silva, H. Grid. Yamaguchi, size dependency of short-term sea ice forecast and its evaluation during extreme Arctic cyclone, in *Polar Science, 21: 204–211* (2019). ISSN 1873-9652 (August 2016). https://doi.org/10.1016/j.polar.2019.08.001. ISAR-5/ Fifth International Symposium on Arctic Research

D.L. Feltham, Granular flow in the marginal ice zone. Phil. Trans. R. Soc. A **363**, 1677–1700 (2005). https://doi.org/10.1098/rsta.2005.1601

A. Fujisaki, H. Yamaguchi, H. Mitsudera, Numerical experiments of air–ice drag coefficient and its impact on ice–ocean coupled system in the Sea of Okhotsk. Ocean Dyn. **60**(2), 377–394 (2010). ISSN 1616-7228. https://doi.org/10.1007/s10236-010-0265-7

E.C. Hunke, J.K. Dukowicz, An elastic–viscous–plastic model for sea ice dynamics. J. Phys. Oceanogr. **27**(9), 1849–1867 (1997). https://doi.org/10.1175/1520-0485

E.C. Hunke, W.H. Lipscomb, A.K. Turner, N. Jeffery, S. Elliott, *CICE: the Los Alamos Sea Ice Model Documentation and Software User's Manual Version 5.0 LA-CC-06-012* (Los Alamos National Laboratory, Los Alamos NM 87545, 2013)

S. Ji, H.H. Shen, Internal parameters and regime map for soft polydispersed granular materials. J. Rheol. **52**(1), 87–103 (2008). https://doi.org/10.1122/1.2807441

C. Luepkes, V.M. Gryanik, J. Hartmann, E.L. Andreas, A parametrization, based on sea ice morphology, of the neutral atmospheric drag coefficients for weather prediction and climate models. J. Geophys. Res. **117**(D13), D13112 (2012). https://doi.org/10.1029/2012JD017630

G. Madec, *NEMO ocean engine (Draft edition r5171)*. Note du Pole de modelisation, Institut Pierre-Simon Laplace (IPSL), France, no 27 issn no 1288-1619 edition (2014)

G.E. Manucharyan, A.F. Thompson, Submesoscale sea ice-ocean interactions in marginal ice zones. J. Geophys. Res.: Oceans **122**(12), 9455–9475 (2017). https://doi.org/10.1002/2017JC012895

P. Rampal, S. Bouillon, J. Bergh, E. Ólason, Arctic sea-ice diffusion from observed and simulated Lagrangian trajectories. Cryosphere **10**(4), 1513–1527 (2016). https://doi.org/10.5194/tc-10-1513-2016

S. Rynders, *Impact of surface waves on sea ice and ocean in the polar regions*. PhD thesis, University of Southampton, Ocean and Earth Sciences (2017). http://eprints.soton.ac.uk/id/eprint/428655

G. Sagawa, H. Yamaguchi, A semi-Lagrangian sea ice model for high resolution simulation, in *Proceedings of the Sixteenth (2006) International Offshore and Polar Engineering Conference* (The International Society of Offshore and Polar Engineers, 2006). pp. 584–590, ISBN 1-880653-66-4

H.H. Shen, W.B. Hibler, M. Lepparanta, On applying granular flow theory to a deforming broken ice field. Acta Mech. **63**(1–4), 143–160 (1986). https://doi.org/10.1007/BF01182545

H.H. Shen, W.B. Hibler, M. Lepparanta, The role of floe collisions in sea ice rheology. J. Geophys. Res. **92**(C7), 7085–7096 (1987). https://doi.org/10.1029/JC092iC07p07085

M. Steele, R. Morley, W. Ermold, PHC: a global ocean hydrography with a high-quality arctic ocean. J. Clim. **14**(9), 2079–2087 (2001). https://doi.org/10.1175/1520-0442(2001)014%3c2079:PAGOHW%3e2.0.CO;2

C. Strong, I.G. Rigor, Arctic marginal ice zone trending wider in summer and narrower in winter. Geophys. Res. Lett. **40**(18), 4864–4868 (2013). https://doi.org/10.1002/grl.50928

L.J. Yiew, L.G. Bennetts, M.H. Meylan, G.A. Thomas, B.J. French, Wave-induced collisions of thin floating disks. Phys. Fluids **29**(12), 127102 (2017). https://doi.org/10.1063/1.5003310

Chapter 14
Physics-Based Modelling of Ice Actions and Action Effects on Marine Structures

Raed Lubbad, Sveinung Løset, Marnix van den Berg, Wenjun Lu, and Shreesha Govinda

Abstract Physics-based time-domain numerical modelling is a powerful approach to study ice-structure interactions and to accurately and effectively calculate ice actions and action effects on marine structures. In this study, we use the classical problem of ship resistance in level ice (continuous ice breaking mode) to demonstrate the advantages of this approach and to discuss current limitations and directions for future research. A case study comprising four different icebreakers transiting with constant speeds in level ice is presented. The Simulator for Arctic Marine Structures (SAMS) is used herein to generate the numerical results. In total 96 simulation runs are completed. In each simulation, the icebreaker under consideration transits at least twice its length in level ice. Beside varying the transit speed, the drag coefficients are also varied, and the simulations are run with and without including the ventilation effects. The numerical experiments reveal that the forces from the ice rubble are the major contributor to the ice resistance and that the increase of the ice resistance with speed comes mainly from the inertia of the ice rubble. Moreover, the ventilation effect is strongly influenced by the ship hull and it generally increases the total ice resistance independent of the transit speed. Further, the numerical experiments show that the rubble transport is largely influenced by the viscous forces of the ambient fluid flow, i.e., larger ice rubble coverage under the hull is achieved at higher transit speeds or when using large drag coefficients. Finally, the findings of the numerical experiments indicate that further improvements to our predictions can be achieved if processes like dynamic ice fracture, backfill effect and gap pressure are accounted for in the model.

R. Lubbad (✉) · S. Løset · M. van den Berg · W. Lu · S. Govinda
Norwegian University of Science and Technology (NTNU), Trondheim, Norway
e-mail: raed.lubbad@ntnu.no

ArcISO aS, Trondheim, Norway

W. Lu
Vista Scholar, the Norwegian Academy of Science and Letters, Oslo, Norway

J. Tuhkuri and A. Polojärvi (eds.), *IUTAM Symposium on Physics and Mechanics of Sea Ice*, IUTAM Bookseries 39, https://doi.org/10.1007/978-3-030-80439-8_14

14.1 Introduction

The design of offshore and coastal structures exposed to sea ice is often governed by sea-ice actions. The latter dominates also the design and operation of ships in icy waters. Therefore, it is important to accurately predict ice actions and the action effects on structures and ships in order to achieve safe and profitable designs and operations.

Simple empirical formulae based on available full- and model-scale data are convenient to use when attempting to estimate ice actions on marine structures and ships. However, one should be cautioned that the applicability of these formulae is limited to the characteristics of the ice and the structure or the ship that were tested and used to derive such formulae. Given that, and the scarcity of available data (especially full-scale data), empirical formulae are generally of less value, especially for the purpose of design optimization where we need to calculate ice actions on variations of an initial design.

Ice-tank testing has for many years been the industry practise for designing Arctic and sub-Arctic structures and ships. Although useful, ice-tank testing is expensive and time consuming which limits the number of parameters one can vary during a test campaign. Moreover, it is usually difficult to reproduce the right ice failure modes during the tests and it is challenging to reliably scale up the measurements to their corresponding full-scale values. Beside this, different ice-tanks have different boundary effects and different testing techniques and procedures, which add to the uncertainty in their estimates of ice actions. Even for the same ice tank, with different initial conditions (e.g., ice floe orientation and positions), the measured data can show an appreciable degree of scattering (van den Berg et al. 2019).

For the aforesaid reasons, physics-based numerical modelling stems as a valuable alternative for accurate and effective predictions of ice actions. The interaction between a structure or a ship and sea ice is a complex phenomenon that involves a number of physical processes at different temporal and spatial scales. The challenge for the modeller is to identify the physical processes that are dominant in the scale of interest; thereafter to find a proper model to describe each one of these processes; and finally, to integrate all these models into a comprehensive unified model for calculating ice actions and action effects on marine structures and ships. In this context, field data and observations are very important to attain proper understanding of the different processes. Full- and model-scale data are also of great value for the verification and validation of the developed models.

Several examples of physics-based models are cited in the literature. Before the start of the new millennium, the available computational power was largely limited. This forced modellers to introduce many simplifying assumptions and sometimes corrections of empirical nature in order to derive analytical closed-form solutions. Under this category, one may refer to Nevel (1992), Croasdale and Cammaert (1994), and Lindqvist (1989). Analytical or semi-analytical models may provide good predictions of the averaged ice actions under idealised conditions, e.g., conical structures, simple hull forms and intact level ice. However, such estimates of averaged ice actions are not enough to assess mooring or dynamic positioning (DP) capacity, where calculations of ice action time history and the peak actions are usually needed. Moreover, the complexity in the ice and the structure or the ship conditions, in a more generic case, is often beyond what the analytical models can handle. For example, the governing mechanisms during floe ice (broken ice) and structures interactions can differ considerably from those dominating when the structures interact with level ice depending on the confinement, ice concentration and floe size distribution (Lu et al. 2015a). The interaction processes are highly nonlinear, and the results depend strongly on the initial and boundary conditions. For this reason, also encouraged by the noticeable increase in available computational power, the focus in the past couple of decades has shifted towards physics-based time-domain numerical modelling of ice-structure or ice-ship interactions (Konno 2009; Konno and Mizuki 2006; Lubbad and Løset 2011; Daley et al. 2014; 2012; Metrikin and Løset 2013).

Numerically, the distinct nature of ice (e.g., ice floes, fragments, rubble) has often promoted the use of discrete element methods (DEM) where ice bodies are treated as rigid bodies and the contact forces are modelled either as compliant contacts (spring-dashpot) or simply as rigid contacts. Dry friction forces are typically calculated using the Coulomb friction model. While the inertia, the contact and possibly the hydrodynamic forces are captured reasonably well in the DEM described above, the effects from global failure of ice floes, e.g. bending and splitting, are not included. In other words, the classical discrete element methods are suited for problems governed by limit energy and perhaps limit force mechanisms but not problems where the limit stress is important. A possible remedy to this shortcoming is to apply lattice-type models (spring-network models) where an ice body is treated as a collection of discrete masses, or simply to combine DEM with the finite element method (FEM) where the latter is used to simulate the fracture of ice floes (Hopkins 1992, 1994; Paavilainen et al. 2010, 2009, 2006, 2011; Ranta et al. 2017; Tuhkuri and Polojärvi 2018). It is also possible to use FEM alone to simulate the dynamics of the multi-body system and the ice fracture (e.g., Gürtner et al. 2008; Konuk et al. 2009a, b; Konuk and Yu 2010; Lu et al. 2014). Although these solutions are theoretically feasible, the high computational expenses may hinder their applicability for simulations of large temporal and spatial scales. A practical solution to include ice fracture in DEM was proposed by Lubbad and Løset (2011) where they showed that the DEM can be supplemented with analytical closed form solutions to represent the icebreaking processes. The low computational cost of the analytical solutions makes it possible to significantly expand the calculation domain, both spatially and temporally, and to include the different limiting mechanisms into the calculations.

In this paper, we use the classical problem of calculating ice resistance to a ship that transits in level ice (continuous ice breaking mode) to demonstrate the advantages of physics-based time-domain numerical modelling and to discuss current limitations of this approach and directions for future research. The term *ice resistance* is used in this study to describe the temporal average of the ice forces acting on the ship in its surge direction. The Simulator for Arctic Marine Structures (SAMS) is used herein to generate the numerical results for the case study, see ArcISo (2020). In the following, the physical processes involved in the continuous ice breaking mode are discussed. Thereafter, the theoretical basis of SAMS and the modelled processes are presented. Subsequently, the case study is described before we show and discuss the simulation results. Finally, the major conclusions are drawn.

14.2 The Physical Processes in Ship and Level Ice Interactions

The physical processes involved in continuous ice breaking mode occurring during ships transit in level ice have been investigated by many researchers, e.g., (Enkvist et al. 1979; Kotras et al. 1983; Lindqvist 1989; Keinonen 1996; Valanto 2001b). The significance of each process relative to the other processes depends largely on the ice properties, the ship hull geometry and the relative velocity between the ice and the ship. For most ships, bending failure of ice dominates over other ice failure modes. For certain bow geometries, pure crushing of ice may dominate in an area around the stem of the ship. The force needed to break the ice constitutes only a portion of the ice resistance (from 30% to 60%). The remaining part of the ice resistance includes forces needed to rotate and submerge the broken ice pieces, in addition to friction forces arising from the sliding of the broken ice pieces along the hull.

The total resistance from ice increases as a function of the ship speed. Several processes can be identified that contribute to the speed dependence. The ice breaking phase is influenced by the inertia from the ice and the surrounding water (added mass) leading to smaller broken pieces at higher speeds. During the rotation of the broken ice pieces, the inertia and hydrodynamic forces play an important role. In addition, ventilation often occurs at higher speeds due to the inability of seawater to immediately fill the gap above the rotating ice pieces. Ventilation leads to increase in buoyancy forces due to slow water backfill above the rotating ice pieces which transmit to the ship through friction, leading to an increase in ice resistance. The forces from the sliding of ice pieces along the hull also increase with the speed. This is mainly due to ice inertia (acceleration of the ice pieces) and due to the viscous effects from water on the ice (drag). Other effects may be attributed to the flow of seawater in the gap between the sliding ice pieces and the ship hull (especially in the bow area). At higher speeds, the pressure in the gap reduces, leading to increase in buoyancy forces on the ice and hence an increase of the ice resistance (Kämäräinen 2007).

14.3 Description of SAMS

The Simulator for Arctic Marine Structures (SAMS) is described briefly in this section. More details about SAMS and its validation can be found in Lubbad et al. (2018) and in other references cited herein.

SAMS is a three-dimensional (3D) numerical model for time-domain simulation of ice-structure or ice-ship interaction. In the remaining of this text, we use mostly the term "the model" or "our model" to refer to SAMS. The model uses the non-smooth discrete element modelling method (NDEM); but it applies a novel implicit time stepping scheme and an improved contact model, enabling general visco-elastic contacts, see van den Berg et al. (2018).

The model adopts the approach of Lubbad and Løset (2011), which supplements the discrete element modelling with analytical closed-form solutions to simulate ice fracture. Published solutions for local bending failure, global splitting failure and radial cracking (e.g., Lu et al. 2015a, b, 2016, 2018a, b) are implemented in the model to determine the ice failure criteria. At each time step, the model calculates the contact forces and compares them with the failure criteria estimated from the analytical solutions. Failure modes are imposed whenever their criteria are fulfilled. Multiple failure modes can be achieved in the simulation, which gives physically sound and realistic results. Moreover, the model can handle situations of pure ice crushing (i.e., continuous ice crushing over a sustained period) by removal of intersection volumes between ice and the hull of the ship (i.e., removal of crushed ice volume). This enables direct simulation of ice crushing at the stem and improves the capability of the model to simulate ship turning in level ice where considerable crushing is expected at the parallel midbody.

In addition to the improved NDEM formulation and the comprehensive set of analytical solutions for ice fracture, the model applies innovative numerical solutions to calculate different hydrodynamic force components on the ship and every ice piece in the calculation domain. This includes, e.g., drag forces from wind and current, and propeller flow. These solutions are calibrated and validated against full-scale and lab-scale data, see (Tsarau and Løset 2015, 2014). Furthermore, the additional forces due to the ventilation effects on the rotating ice pieces are recently included in the model. These forces are applied as additional buoyancy to partly submerged ice pieces in contact with the ship hull. The current implementation of the ventilation in the model is slightly conservative, especially at lower interaction speeds, as it assumes full ventilations and neglects the backfill effects (Lu et al. 2012).

Figure 14.1 illustrates the different calculation modules in the model. Figure 14.1a shows a snapshot of the 3D simulation environment where simultaneous contacts (ice-ship and ice-ice) and multiple ice failure modes are being handled by the model. Figure 14.1b shows how the model idealizes the geometry of ice floes before applying the analytical solutions for ice fracture. Figure 14.1c shows the non-rigid contacts due to ice crushing at the contact interface. Figure 14.1d illustrates the currents in the fluid domain with velocity vectors. Figure 14.1e demonstrates the capability of the model to handle situations of pure crushing. Finally, Fig. 14.1f indicates how the ventilation is implemented in the model. The visualization in Fig. 14.1 is generic and it is valid for level-ice condition as well as floe-ice conditions.

The above description shows that the calculations in the model include several physical processes that cause the major part of the ice resistance. The speed dependency in the model results (i.e., increase of the simulated ice resistance with increased speed) stems mostly from ice inertia during ice rotation and sliding, in addition to

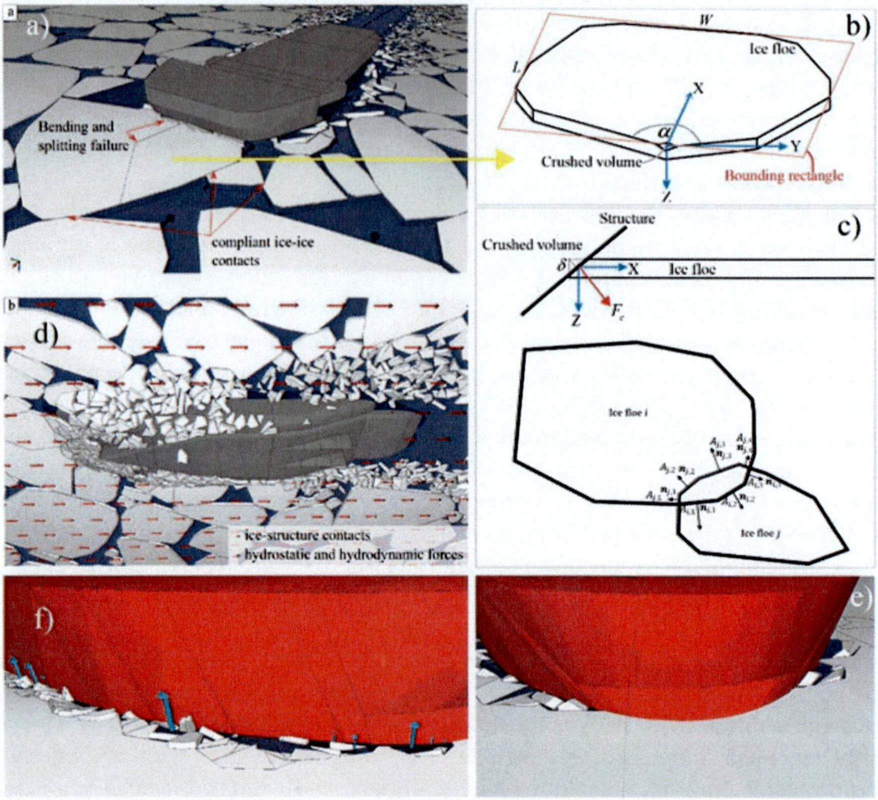

Fig. 14.1 Illustration of different modules within SAMS: **a** snapshot of the 3D simulation environment; **b** idealization of the geometry of ice floes before applying the analytical solutions for ice fracture; **c** illustration of the non-rigid contact due to ice crushing at the contact interface—**A** and **n** denote area and normal vector, respectively; **d** illustration of the fluid domain, e.g., underwater view showing the current velocity vectors; **e** demonstration of a situation of pure crushing at the stem; **f** indication of how the ventilation is implemented in the model

viscous effects in the form of drag forces on the ice. Ventilation may contribute as well to the increase of the simulated resistance with speed. Although the analytical solutions for ice fracture do not consider the interaction velocity, the sizes of the broken ice pieces are adjusted empirically to capture the speed effect on the ice breaking phase (i.e., the breaking length reduces with increasing speed) based on field measurements (Naegle 1980).

All the above-mentioned physical processes shall be further scrutinized in the following case study. We shall focus on the speed dependence of different physical processes and their respective contribution to the total ice resistance. Particularly, the advantage of applying the physics-based time-domain numerical modelling over traditional approaches shall be highlighted.

14.4 Case Study

In this case study, we use the model "SAMS" to simulate the transit of four different icebreakers in level ice. This includes the two Swedish icebreakers Oden and Frej and two hypothetical designs with simplified hull geometries. Table 14.1 presents the main parameters of these icebreakers and Fig. 14.2 shows the geometric representation of these icebreakers in the model. Note that the four icebreakers have relatively

Table 14.1 The main parameters of the simulated Icebreakers

	Oden	Frej	Simple hull #1	Simple hull #2
Ship length at the waterline, L_{WL}, [m]	97	99	100	100
Ship breath, B, [m]	31	24	25	24
Ship Draught, D, [m]	8.5	8.3	8.3	8.3
Stem angle, ϕ, [°]	20	30.6	50.1	22
Waterline entrance angle, α, [°]	90	32.8	24.4	25

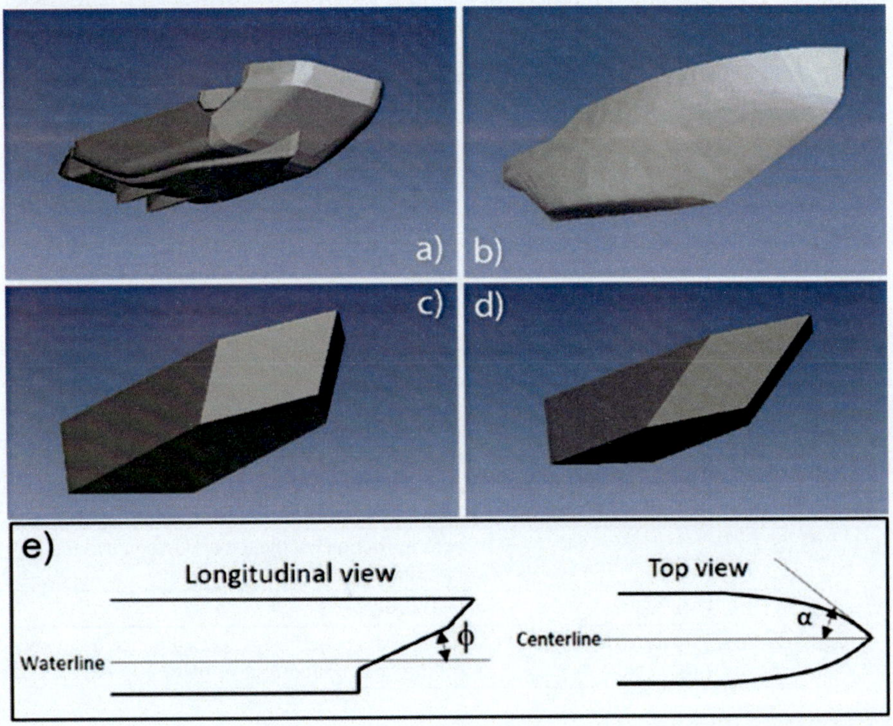

Fig. 14.2 Perspective views of the hull form of the simulated icebreakers in Table 14.1; **a** Oden; **b** Frej; **c** Simple hull #1; **d** Simple hull #2; **e** Definitions of the stem angle and the waterline entrance angle

similar lengths at the waterline, beams, and draughts. The definitions of the stem angle and the waterline entrance angle are shown in Fig. 14.2.

Each icebreaker is simulated with 4 different transit speeds ranging from 1 knot to 6 knots (i.e., from 0.5 m/s to 3 m/s). In each simulation run, the icebreaker is set to transit ahead (only one degree of freedom) in 1 m thick level ice with constant speed for a distance that is twice the icebreaker length. In this way the ice resistance can be estimated as a function of speed for each icebreaker. The simulations are run with and without ventilation to assess the significance of including this physical process in the calculation of ice resistance. The form drag coefficients on the broken ice fragments are varied to study the importance of the viscous effects on the simulated ice resistance. Typically, the ice thickness is smaller than the length or the width of ice floes and most broken ice fragments resulting from level ice and ship interactions. It is expected that the flow separation around any ice fragment will be different if the fragment moves in its in-plane or out-of-plane directions. Therefore, we distinguish between in-plane and out-of-plane form drag coefficients. The earlier is associated with form drag on small projected areas and the latter is associated with form drag on a large projected area.

Table 14.2 summarises the main input parameters for the model. Most of these values are chosen with reference to Timco and Weeks (2010), with a focus on engineering applications.

Table 14.2 Inputs for the simulations

$\Delta t = 0.01\,\text{s}$	Simulation time step;
$h = 1\,\text{m}$	Ice thickness
$\rho_i = 900\,\text{kg/m}^3$	Ice density
$\rho_w = 1025\,\text{kg/m}^3$	Water density
$E = 2\,\text{GPa}$	Young's modulus
$v = 0.3$ [−]	Poisson ratio
$K_{IC} = 150\,\text{kPa}\sqrt{\text{m}}$	Fracture toughness
$\sigma_c = 2\,\text{MJ/m}^3$	Crushing specific energy
$\sigma_f = 500\,\text{kPa}$	Flexural strength
$\mu_{is} = 0.1$ [−]	Ice-structure friction coefficient
$\mu_{ii} = 0.25$ [−]	Ice-ice friction coefficient
$C_f = 0.005$ [−]	Skin friction coefficient applied on ice fragments
$v = 0.5\ ,\ 1\ ,\ 2\ ,\ 3\ \ [m/s]$	Transit speed
$C_d = (0.01/0.05)\ ,\ (0.1/0.5)\ ,\ (0.2/1)$ [−]	Form drag coefficient applied on ice fragments (in-plane motions/out-of-plane motion)

14.5 Results and Discussions

In this section, we present and discuss the simulation results. The model "SAMS" is a time-domain simulator that yields time histories of ice forces. An example of simulated time series of ice forces is displayed in Fig. 14.3. It shows the development of the total ice force acting on the ship in the surge direction. The total ice force starts always from zero before the ship enters into the level ice field and then gradually increases until it stabilizes at around 100 s.

Figure 14.3 shows that among the four icebreakers, the icebreaker "Simple hull #1" experiences a significantly larger total ice force in the surge direction and the force varies in a wide range compared to the other three icebreakers. This can be explained by the waterline processes occurring at such a steep stem angle (50.1°) of Simple hull #1. The steep angle increases the frequency of events where broken ice pieces are being wedged and compressed between the hull and the surrounding level ice. This in turn induces significant crushing of these broken ice pieces at the stem leading to the noticeable increase in the total resistance. In contrast to most empirical formulae, which normally yield an estimate of the average ice force (as shall be discussed), the time history of the ice force gives more information of the ice action and action effects, as demonstrated in Fig. 14.3.

To simplify the comparison, the ice resistance (i.e., the temporal average of the total ice force in the surge direction) is calculated. Note that only the portion of the time series (see Fig. 14.3) that corresponds to the time interval after the ship has transited one ship length into the ice is used to calculate the average ice force. It is ensured that the simulation time is always long enough to achieve a stable ice resistance.

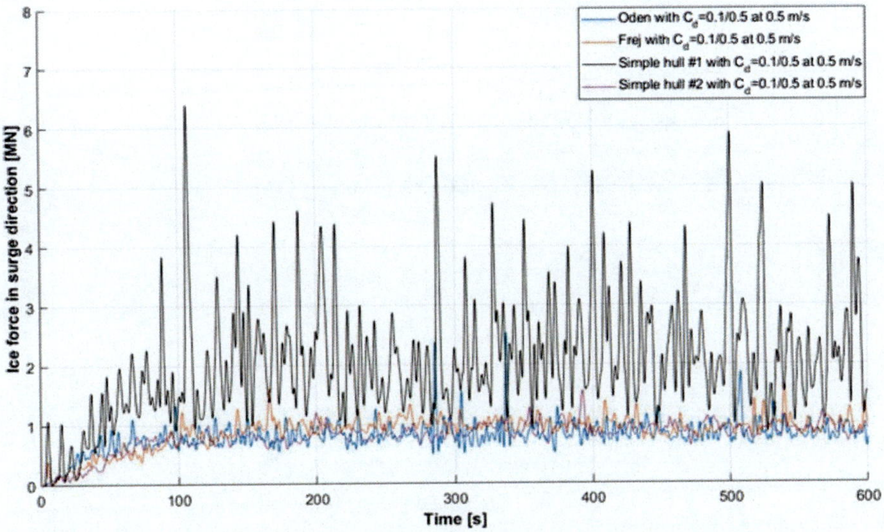

Fig. 14.3 Example of simulated time series of ice force on the different icebreakers

From the calculated ice resistance, we further delineate the ice resistance component that is induced by the ice breaking process (i.e., the breaking force); and the component that is caused by the ice rotation, submerging and sliding phases (i.e., the rubble force). The speed dependence of these two components are investigated.

14.5.1 Ice Resistance Components

Figures 14.4, 14.5, 14.6 and 14.7 illustrate the simulated ice resistance versus the transit speed for the four icebreakers in Table 14.1. In addition, the simulated ice breaking force and rubble force are also shown. Furthermore, a semi-analytical solution of ice resistance based on Lindqvist formula (Eq. (14.1)) is also presented (both the total resistance and the breaking and submergence components). The four figures show that the simulated total ice resistance, illustrated with 'Δ' symbol, matches well with the Lindqvist solution (i.e., solid line). However, for the separate force components, discrepancies are observed between our model and the Lindqvist solution.

The simulated rubble force is generally larger than the rubble force calculated according to the Lindqvist formula. This can partly be explained as discussed earlier by events where broken ice pieces are stuck during their rotation between the level ice and the ship leading to excessive crushing of these pieces especially at the stem and

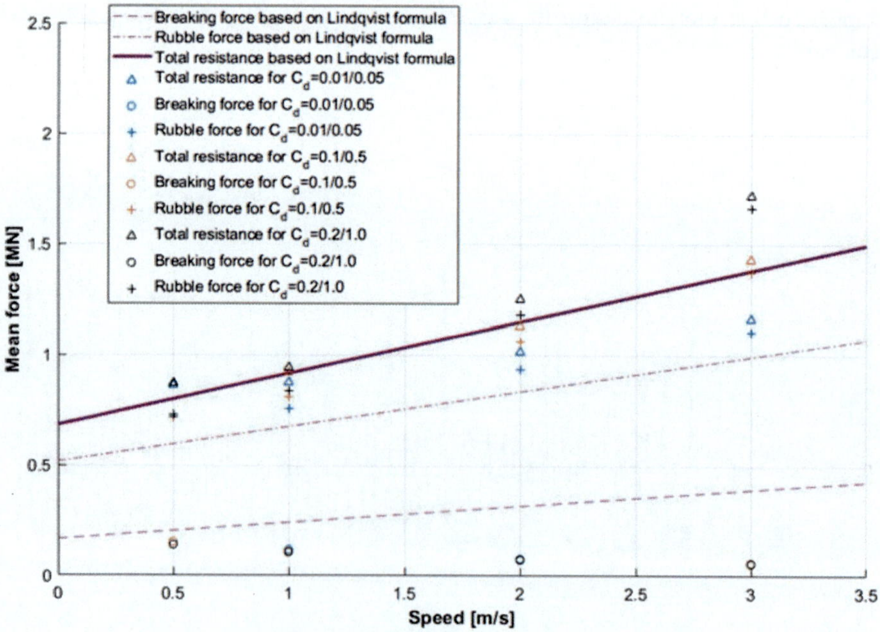

Fig. 14.4 Ice resistance and different force components versus speed for Oden

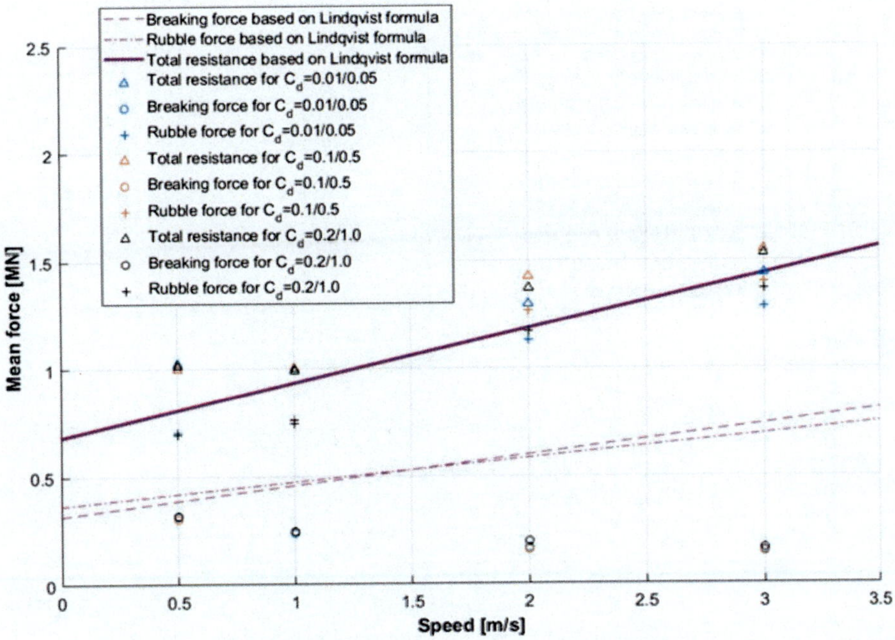

Fig. 14.5 Ice resistance and different force components versus speed for Frej

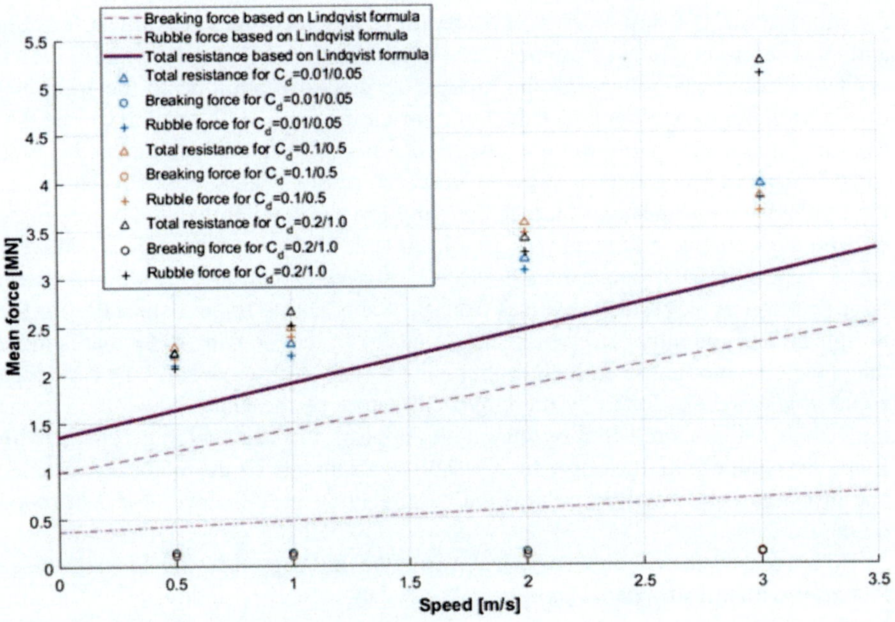

Fig. 14.6 Ice resistance and different force components versus speed for simple hull #1

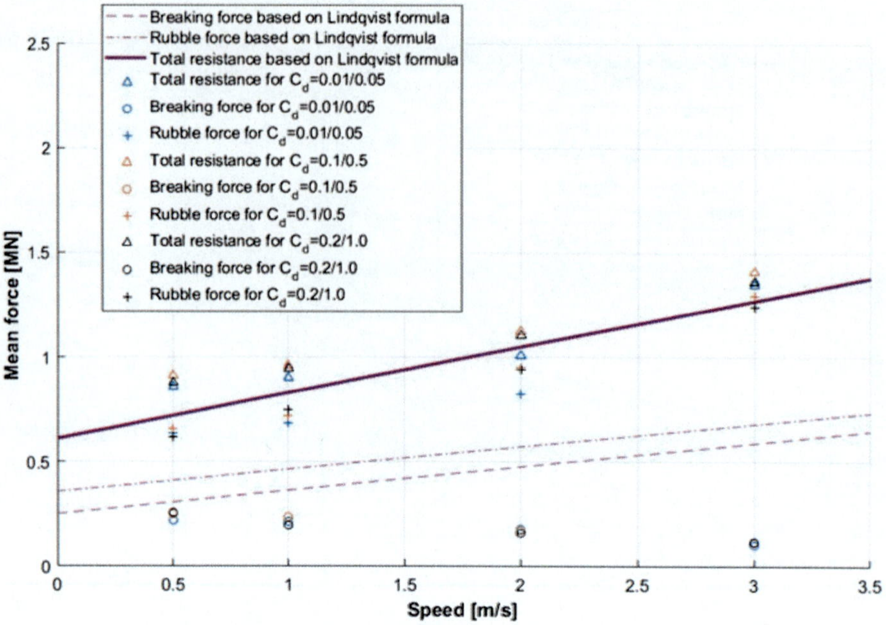

Fig. 14.7 Ice resistance and different force components versus speed for simple hull #2

the shoulders of the ship. Such events are not represented in the Lindqvist formula and most semi-analytical formulae where the rubble force is simplified as the force needed to rotate and submerge the broken ice pieces in addition to the frictional forces from the ice sliding under the hull. For all the simulation results, we see that the simulated ice breaking force is always much smaller than the simulated rubble force. This is in accordance to previous research results (Valanto 2001b). However, the Lindqvist solution shows that ice breaking force is smaller than the rubble force only for the icebreaker Oden and the Simple hull #2, where the stem angle is relatively small. When the stem angle increases, the Lindqvist solution shows that a much larger portion of ice resistance arises from the ice breaking force. Especially for the Simple hull #1, its rather steep hull angle leads to an ice breaking force that is more than twice the rubble force. Such discrepancy between our model and the Lindqvist solution may be explained by the empirical nature of the latter. Nevertheless, the total ice resistance predicted by our model appears in reasonable agreement with Lindqvist solution. In the following, we shall focus mainly on the total ice resistance and investigate the involved influencing parameters, e.g., influence of transit speed on the ice force.

Both measurements and empirical formulae indicate that the ice resistance increases with transit speed (Lindqvist 1989, Valanto 2001b). Lindqvist (1989) states that '*the resistance seems to increase fairly linearly with the speed*' and propose the following empirical formula (Eq. (14.1)) to account for the speed effect on the ice resistance.

$$R_{ice} = (R_c + R_b)(1 + 1.4\frac{v}{\sqrt{gh_{ice}}}) + R_s(1 + 9.4\frac{v}{\sqrt{gL}}) \tag{14.1}$$

where

R_{ice} is the total resistance encountered by the ship.
R_c is the ice resistance connect to crushing of sea ice at ice—hull contact.
R_b relates to the bending failure of sea ice.
R_s accounts for the resistance in relation to ice rubble sliding.
h_{ice} is the thickness of sea ice.
L is the length of the ship.
g is the gravitational acceleration.
v is the ship transit speed.

The formulae for the components R_c and R_b are somewhat lengthy and they are well explained in Lindqvist (1989), so they are not repeated herein. The formula for R_s is shown in Eq. (14.2).

The simulated total ice resistance, illustrated with 'Δ' symbol in Figs. 14.4, 14.5, 14.6 and 14.7 confirms the 'almost linear' trend. In addition, the simulation results highlight that there is much more speed dependence for the case of rubble force ('+' symbol) than for the ice breaking force ('O' symbol).

In the simulation results, it is observed that the simulated ice breaking forces are constant or even decreasing with increasing transit speed. This is in disagreement with the first term on the Right Hand Side (RHS) of Eq. (14.1) which indicates that the ice force due to ice crushing and bending processes linearly scale with transit speed. The causes of this discrepancy are not well understood and need further investigations. One should bear in mind that the decomposition of speed dependence in Eq. (1) between ice breaking and ice submergence phases are purely empirical and prone to uncertainty especially when the contribution of the ice submergence phase are often larger than the contribution of the ice breaking phase. When it comes to our model, the implemented ice fracture algorithms are based on analytical solutions that do not depend on speed. Even though the ice breaking lengths in our model are adjusted empirically to capture the speed effects, the lack of inertia effects (from ice and water) on the analytical solutions used in our model may explain the incapability of the model to capture the speed dependence of the icebreaking forces. Nevertheless, such discrepancy is not jeopardizing the overall capability of our model to capture the speed effects as the ice breaking force covers only a small portion of the total ice resistance. As illustrated by the '+' symbol, the rubble force's speed dependence is well captured in all simulations. For most scenarios, as the rubble force contributes to the major part of the total resistance, the total ice resistance's speed dependence is largely captured.

14.5.2 Physical Processes Behind the Speed Effect

In Eq. (14.1), the observed speed effect was only empirically accounted for by the dimensionless Froude number. However, it is interesting to investigate where the major speed effect stems from during ice-ship interactions. Based on previous section's results and discussions, we observe that most of the speed effect in our simulation results comes from the rubble force.

Within the rubble ice force, we can further investigate the physical source of such speed dependence. Initially, the following parameters are expected to contribute to the observed speed effect:

- Inertia of ice rubble
- Viscous force on ice rubble
- Gap pressure change between sliding ice floes and the ship hull (Kämäräinen 2007)
- Ventilation effects

All the simulations in Figs. 14.4, 14.5, 14.6 and 14.7 are conducted with the full ventilation scenario, which shall be discussed at a later stage. Therefore, we will not discuss the ventilation relationship with the speed effect for the time being. As for the gap pressure between the sliding ice floes and the ship hull, it is a hydrodynamic phenomenon at higher speeds and especially in the bow area that leads to reduction of the pressure in the gap leading to an equivalent increase in buoyancy forces on the ice and hence an increase the ice resistance. This process is not included in the current simulations. Consequently, the increase of the simulated rubble force with the speed (as shown in Figs. 14.4, 14.5, 14.6 and 14.7) comes solely from the inertia of the ice rubble and from the viscous forces (quadratic drag) on the ice rubble. In order to investigate the significance of the viscous forces, each simulation was repeated with three sets of drag coefficients C_d varying over a wide range, see Figs. 14.4, 14.5, 14.6 and 14.7. For all simulated results, even though the viscous forces vary over at least one order of magnitude, the rubble force (see '+' symbols in different colours) vary only slightly over all tested speed ranges. i.e., most '+' symbols with different colours are close to each other in spite of the factor 20 difference in the applied drag coefficients. At a higher interaction speed (i.e., 3 m/s), the drag coefficient change appears to have a larger influence than those at a lower interaction speed (i.e., different coloured '+' symbols are relatively further apart at a higher speed). However, such influence is rather minimal especially considering the factor 20 difference in viscous forces.

Given the above quantification and comparison, it can be concluded that the rubble force's speed dependence arises mainly from the inertia of ice rubble.

14.5.3 Ventilation Effect

Currently, only the full ventilation scenario (Valanto 2001b) is implemented within our model. The formula given by Lindqvist (1989) in Eq. (14.1) does not account for the ventilation effect. Here, by simply activating and deactivating the ventilation in our model, we can quantify the relative importance of the ventilation to the simulated total ice force. Figures 14.8, 14.9, 14.10 and 14.11 show that with the consideration of ventilation, the simulated total ice resistance ('Δ' symbols) is higher than the simulated total ice resistance without the ventilation effect ('∇' symbols). This signifies the importance of considering the ventilation and potential backfill effects in simulating ice-ship interactions.

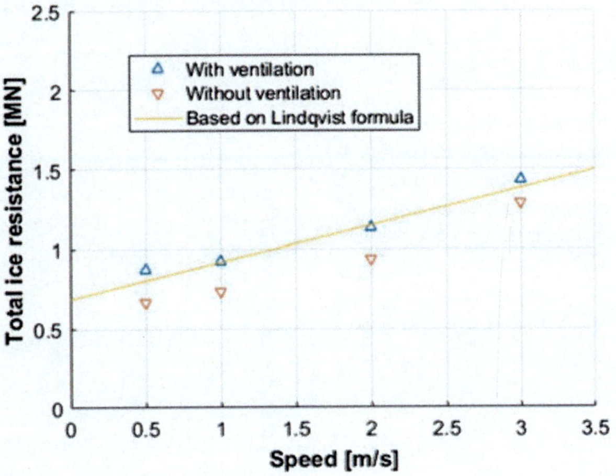

Fig. 14.8 Total ice resistance experienced by Oden with and without ventilation effect at different transit speeds with $C_d = 0.1; 0.5$

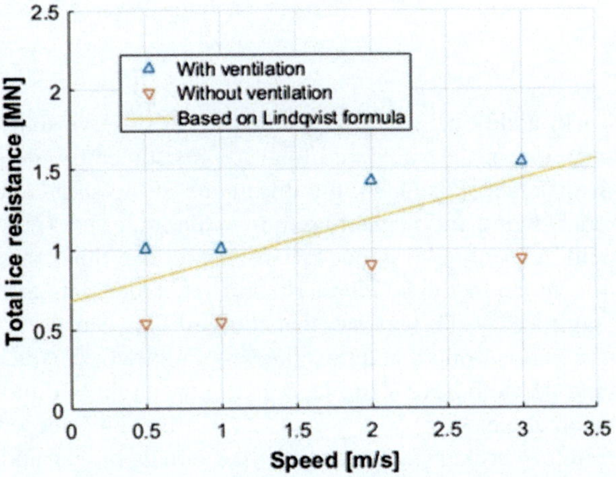

Fig. 14.9 Total ice resistance experienced by Frej with and without ventilation effect at different transit speeds with $C_d = 0.1; 0.5$

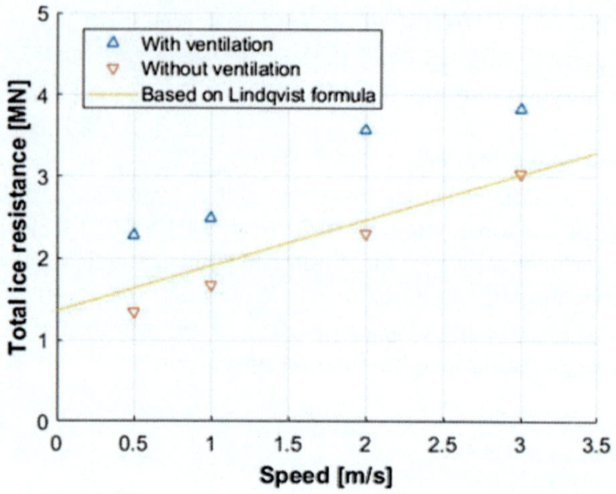

Fig. 14.10 Total ice resistance experienced by simple hull #1 with and without ventilation effect at different transit speeds with $C_d = 0.1; 0.5$

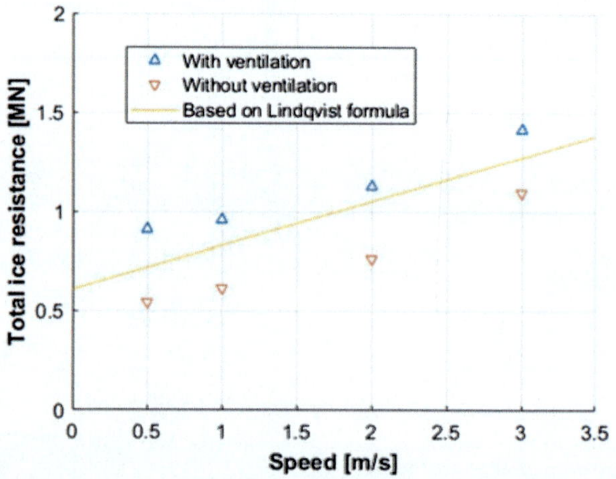

Fig. 14.11 Total ice resistance experienced by simple hull #2 with and without ventilation effect at different transit speeds with $C_d = 0.1; 0.5$

In addition, it is also interesting to observe that for different icebreakers, the increase in the simulated total ice resistance by considering the ventilation effect is different. For Oden, the difference in the simulated total ice resistance between considering and neglecting the ventilation effect (i.e., the distance between the 'Δ' and '∇' symbols) is not as large as for the other three icebreakers. This is largely due to the fact that Oden has a much smaller stem angle (20°) and wide water entry angle (90°). This means that the gap between the rotating broken ice pieces and the hull is limited and thus the ventilation effect is marginal. The simplified hull #1 and #2, both have a similar water entry angle (around 25°), however, with different stem angles (i.e., 50.1° versus 22°). Given the steeper hull #1, the gap between the rotating broken ice pieces and the hull is bigger and thus the ventilation effect is greater compared to the simple hull #2. This is observed by comparing Figs. 14.10

and Fig. 14.11. In Fig. 14.10, the average difference between the cases with and without the ventilation effect is around 0.96 MN (also see Table 14.5); where it is around 0.35 MN in Fig. 14.11 (also see Table 14.6). This is in agreement with our expectation that the simple hull #1 is more susceptible to the ventilation effect increasing the total ice resistance.

In the presented numerical experiments, we show the versatility of our model "SAMS" in evaluating different hull forms and its impact on different ice force components, e.g., the ventilation process. Such detailed comparisons cannot be carried out by simply using an empirical formula such as Eq. (14.1); and cannot be easily investigated through model ice tank tests, as a new ship model would need to be constructed for each design iteration.

In addition to illustrating our model capability in evaluating different hull forms, it is also possible to investigate the speed dependence of the ventilation induced additional force with the above results. This can be done by comparing the difference between the total resistance with and without the ventilation effect (i.e., distance between the 'Δ' and '∇' symbols) over the tested transit speeds.

This is summarised in Tables 14.3, 14.4, 14.5 and 14.6. The results show that the difference in the total resistance is not obviously dependent on the transit speed. For a given icebreaker within Figs. 14.8, 14.9, 14.10 and 14.11, it appears that the total ice resistance increase caused by the ventilation is not varying with the transit speed. This is in accordance to Valanto's (2001b) simulation results, in which the ice rotating force components (including the full ventilation scenario) is not varying much with different transit speeds.

Table 14.3 Total ice resistance of Oden with and without ventilation effect at different transit speeds with $C_d = 0.1; 0.5$

Oden transit speed	Without ventilation [MN]	With ventilation [MN]	Difference [MN]
0.5 m/s	0.661	0.87	0.209
1 m/s	0.729	0.921	0.192
2 m/s	0.933	1.131	0.198
3 m/s	1.27	1.429	0.159

Table 14.4 Total ice resistance of Frej with and without ventilation effect at different transit speeds with $C_d = 0.1; 0.5$

Frej transit speed	Without ventilation [MN]	With ventilation [MN]	Difference [MN]
0.5 m/s	0.5412	1.001	0.4598
1 m/s	0.5425	1.005	0.4625
2 m/s	0.9021	1.423	0.5209
3 m/s	0.9409	1.546	0.6051

Table 14.5 Total ice resistance of simple hull #1 with and without ventilation effect at different transit speeds with $C_d = 0.1; 0.5$

Simple hull #1	Without ventilation [MN]	With ventilation [MN]	Difference [MN]
0.5 m/s	1.348	2.285	0.937
1 m/s	1.671	2.494	0.823
2 m/s	2.299	3.568	1.269
3 m/s	3.039	3.836	0.797

Table 14.6 Total ice resistance of simple hull #2 with and without ventilation effect at different transit speeds with $C_d = 0.1; 0.5$

Simple hull #2	Without ventilation [MN]	With ventilation [MN]	Difference [MN]
0.5 m/s	0.5448	0.9102	0.3654
1 m/s	0.6181	0.959	0.3409
2 m/s	0.7642	1.125	0.3608
3 m/s	1.096	1.411	0.315

14.5.4 Rubble Sliding

The visualization capabilities of our model allow the synchronisation of the ice force history with the view from the simulation environment offering insights into the interaction mechanisms. For most conventional icebreakers, the ice rubble sliding force contributes to the major part of the total ice resistance. However, the understanding of this force component is still rather intuitive and 'empirical'. For example, in Eq. (14.1), the ice sliding force term R_s is formulated in Eq. (14.2) with the assumption that 70% of the ship bottom shall be covered by ice rubble (Lindqvist 1989). This leads to the coefficient 0.7 in Eq. (14.2).

$$
\begin{aligned}
R_s =&(\rho_w - \rho_i) g h_{ice} B [T \frac{B+T}{B+2T} + \mu_{is}(0.7L - \frac{T}{\tan\phi} - 0.25\frac{B}{\tan\alpha} \\
&+ T\cos\psi\cos\phi\sqrt{\frac{1}{\sin^2\phi} + \frac{1}{\tan^2\alpha}})]
\end{aligned} \tag{14.2}
$$

where,

ρ_w and ρ_i	are the water and ice density.
BTL	are the ship beam width, draft, and length.
$\psi = \arctan(\tan\phi / \sin\alpha)$	calculated from the stem angle ϕ and water entry angle α

This formula and the 70% coverage assumption have been extensively utilised and referenced in various research work, e.g., numerical models developed by Valanto (2001a, b, 2006) and Su et al. (2010a, b, c, 2011), Tan et al. (2013), Zhou et al. (2011).

Intuitively, the ice rubble coverage should at least depend on the ship hull form, transit speed and drag force, instead of being a fixed number (70%). With our model, we can investigate the validity of this assumption.

Figures 14.12 and 14.13 illustrate the under-water perspective view of a simulation snapshot showing the ice rubble transport at different transit speeds. The simulation visualizations show that the assumption of 70% hull coverage assumption does not hold for all submerged hull geometries; The hull coverage is strongly dependent on the specific hull geometry.

For the existing icebreakers in Fig. 14.12, we see that the more conventional hull form, Frej, tends to have a higher ice rubble coverage. This may explain how the empirical formula was derived, i.e., mainly through conventional icebreaker experiences. With regard to the more 'unconventional' ship hull form of Oden, the ice rubble coverage is strongly influenced by the presence of the ice knife, which deviates the incoming rubble ice flow sideways, reducing the amount of ice rubble travelling under the ship hull. The ice rubble coverage in this case appears to be is less than 70% based on visual observations.

For both the simplified hull forms in Fig. 14.13, it shows that the ice coverage is by no means a fixed number; it can vary with the interaction speed for the same hull form. The trend appears to be that a higher transit speed leads to a larger coverage. A

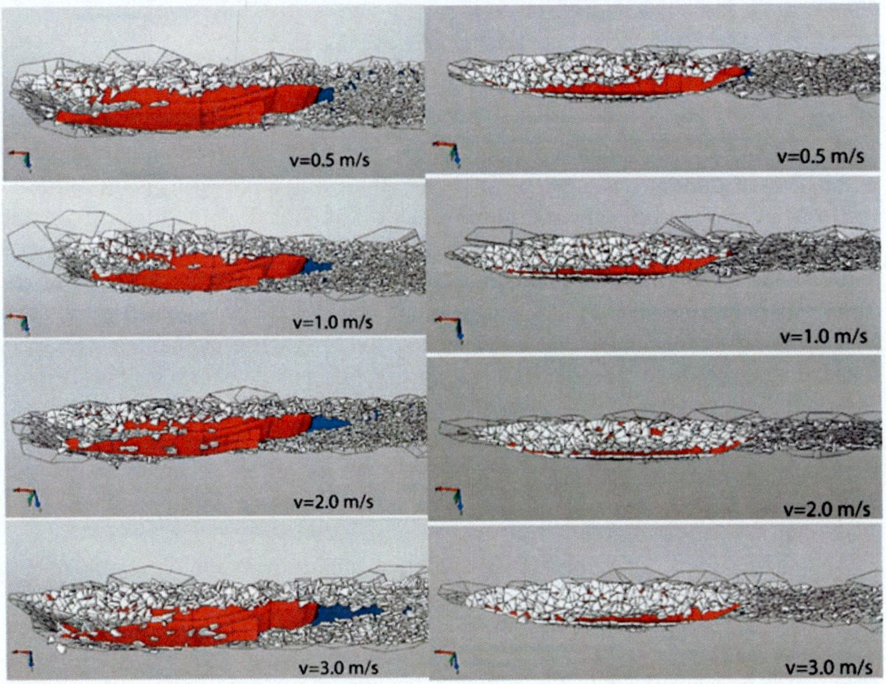

Fig. 14.12 Rubble transport beneath the ship hull, left column: Oden; right column: Frej (C_d = 0.1; 0.5)

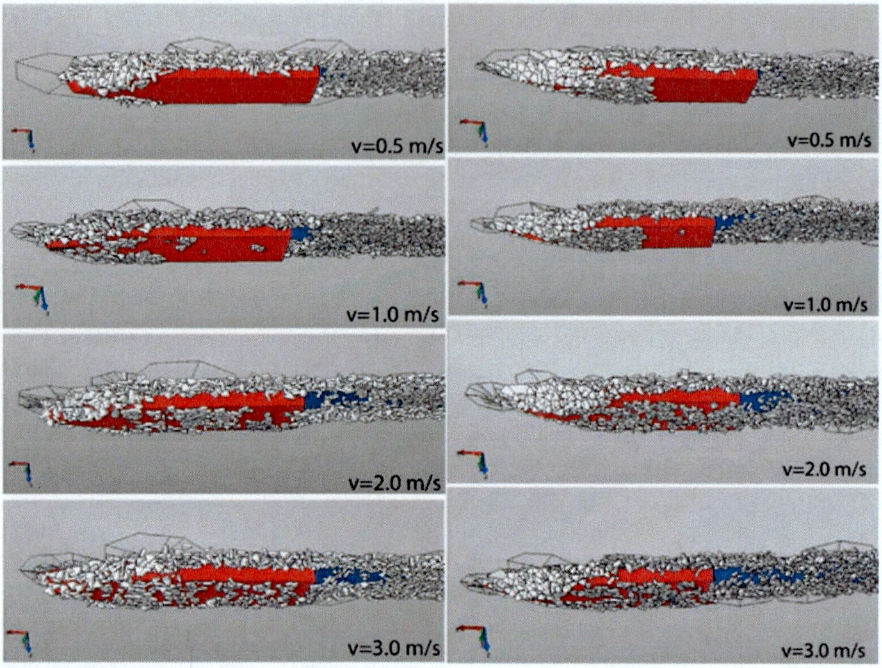

Fig. 14.13 Rubble transport beneath the ship hull, left column: Simple hull #1; right column: Simple hull #2 ($C_d = 0.1; 0.5$)

higher interaction speed means that the submerged rubble has relatively little time to be transported sideways; instead, the increased fluid drag force acting in the opposite-transit direction transports the ice rubble under the hull, leading to a higher coverage. Similar observations were made by Ettema and Huang (1987) in their model tests (with a scale factor of 30). With a barge model and plastic 'ice blocks', Ettema and Huang (1987) found that at a low transit speed, the ice blocks tend to accumulate in front of the barge forming a false bow (see Fig. 14.14a); while at a higher interaction speed, the ice rubble covers a much larger portion beneath the barge (see Fig. 14.14b).

In addition to the speed effect on the rubble sliding phase, the drag coefficient also has an effect on how the ice rubble is transported beneath the icebreakers. Figure 14.15 shows that at a given transit speed (e.g., 2 m/s) the ice rubble accumulation is more packed at a lower drag coefficient but more spread under the hull for a higher drag coefficient.

The observations made for various velocities and drag coefficients' influence on the rubble transport phase indicate the importance of the hydrodynamic forces (e.g., viscous forces). A larger viscous force from the ambient fluid leads to more coverage beneath the ship hull. This is in agreement with the explanations given by Ettema and Huang (1987) to their experimental observations.

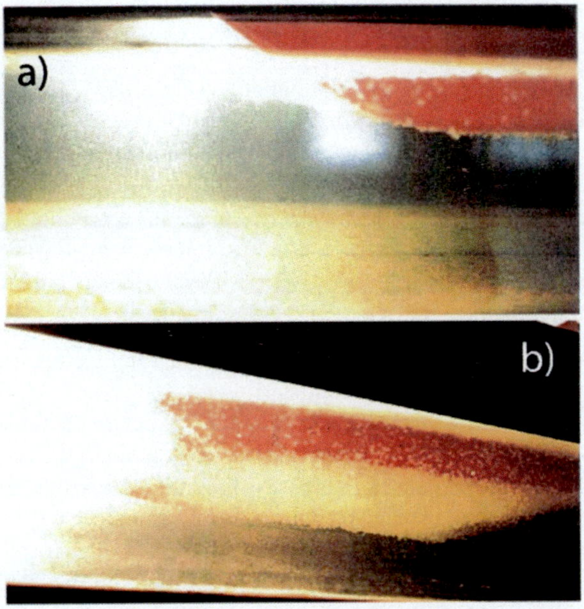

Fig. 14.14 Model tests carried out by Ettema and Huang (1987) illustrating the rubble accumulation beneath a barge at: **a** slow transit speed; and **b** high transit speed

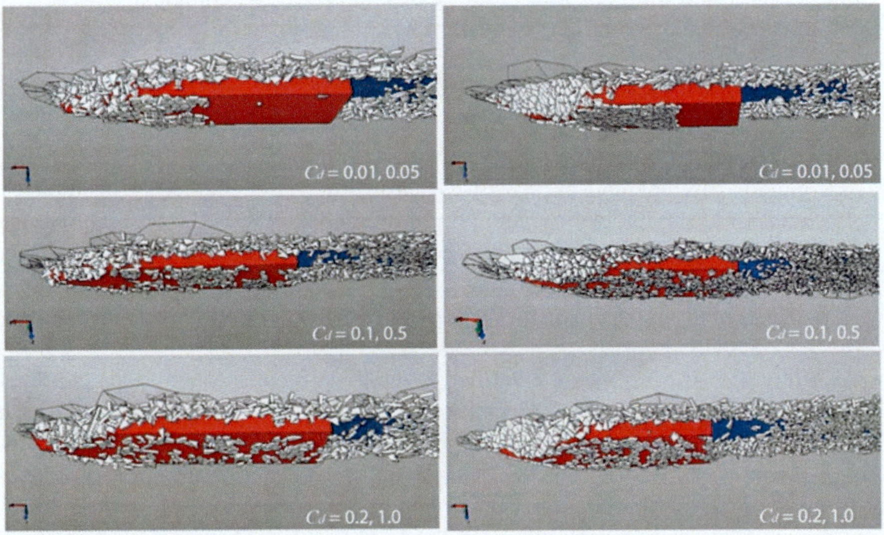

Fig. 14.15 Influences of drag coefficient on ice rubble transport at a transit speed of 2 m/s: left column: simplified hull #1; right column: simplified hull #2

14.6 Conclusions

Physics-based time-domain numerical simulators (e.g., SAMS) allow us to investigate and quantify the importance of different physical process by simply activating and deactivating these processes, or by varying related input parameters. Such versatility is not easily achievable by physical model tests and/or by empirical formulae. In this regard, physics-based numerical models serve as an excellent research tool to explore the physical processes of ice—structure interactions. Based on the numerical experiments conducted using our model "SAMS" in this paper, the following conclusions can be drawn:

- The ice rubble force is the major contributor to the total ice resistance. This is in accordance to previous research results.
- Among the many factors that may lead to a speed effect on the total ice resistance, the inertia of ice rubble is the major contributor.
- The ventilation effect in general increases the total ice resistance; however, it is not sensitive to the transit speed.
- The ventilation effect is strongly influenced by the ship hull form.
- The hull coverage of ice rubble is strongly dependent on the specific hull form. The assumption of 70% hull coverage used in many empirical resistance formulae is not always accurate.
- Rubble transport is influenced by the viscous force of the ambient fluid flow. This means that with a higher transit speed or a larger drag coefficient, a larger ice rubble coverage can be found beneath the ship hull.

From the numerical simulation results, we identify the following simulation capabilities:

- Physics-based time-domain numerical simulation (e.g., SAMS) can realistically account for different physical processes that are not well accounted for in empirical formulae or less effectively accounted for by model tests. For assessing mooring or DP capacity, peak forces are needed that can be derived from the time-force history derived by these simulations.
- The effects of the hull form on the ice resistance can now be effectively investigated by, e.g., SAMS."
- Based on the simulation, we identified several insufficient assumptions that were made in previous empirical approaches. For the often cited Lindqvist formula (1989), these inappropriate assumptions include an oversimplification on the speed effect on the total ice resistance; and the invalid assumption of a fixed amount of ice rubble coverage beneath the hull.

However, the development (both theoretical development and code implementation) of our model is an ongoing process. With the presented numerical experiments, we also identify several tasks for future work. These are:

- Dynamic ice fracture and the ice breaking force's speed dependence have to be implemented.
- The ventilation effect should be improved to account for the 'partial' ventilation effect with the presence of the backfill effect.
- The hydrodynamic force induced by gap pressure change between the sliding ice rubble and a ship hull should be further investigated.
- The validation of the model against full- and model-scale data must be maintained, and it should proceed hand in hand with the model development.

References

ARCISO, *Arctic Integrated Solutions with the Simulator for Arctic Marine Structures (SAMS)* [Online] (2020). Available: https://arciso.com/

K.R. Croasdale, A.B. Cammaert, An improved method for the calculation of ice loads on sloping structures in first-year ice. Power Technol. Eng. (formerly Hydrotechnical Construction) **28**, 174–179 (1994)

C. Daley, S. Alawneh, D. Peters, B. Colbourne, GPU-event-mechanics evaluation of ice impact load statistics. *Society of Petroleum Engineers—Arctic Technology Conference 2014* (2014)

C. Daley, S. Alawneh, D. Peters, B. Quinton, B. Colbourne, GPU modeling of ship operations in pack ice. *International Conference and Exhibition on Performance of Ships and Structures in Ice*, Banff Alberta, Canada September, pp. 20–23 (2012)

E. Enkvist, P. Varsta, K. Riska, The ship-ice interaction. *Proceedings of the 5th International Conference on Port and Ocean Engineering under Arctic Conditions*, Norwegian Institute og Technology, Trondheim, Norway, pp. 977–1002 (1979)

R. Ettema, H.P. Huang, Ice rubble beneath barges in ice-covered waters. 1987 Iowa City, Iowa, USA, pp. 543–555

A. Gürtner, I. Konuk, O. Gudmestad, P. Liferov, Innovative ice protection for shallow water drilling: Part III—finite element modelling of ice rubble accumulation. *Proceedings of the ASME 27th International Conference on Offshore Mechanics and Arctic Engineering*, June 15–20, 2008 Estoril, Portugal. ASME, 1–7 (2008)

M.A. Hopkins, Numerical simulation of systems of multitudinous polygonal blocks, in Technical Report 92–22 Cold Regions Research and Engineering Laboratory, CRREL. 69 (1992)

M. Hopkins, On the ridging of intact lead ice. J. Geophys. Res. **99**, 16351–16360 (1994)

J. Kämäräinen, *Theoretical Investigation on the Effect of Fluid Flow Between the Hull of a Ship and Ice Floes on Ice Resistance in Level Ice*, Helsinki University of Technology, Department of Mechanical Engineering, Laboratory for Mechanics of Materials, Helsinki, Finland (2007)

A. Keinonen, Icebreaker characteristics synthesis-Icebreaker Performance Models, Seakeeping, Icebreaker Escort. Transport Canada (1996)

A. Konno, Resistance evaluation of ship navigation in brash ice channels. *Proceedings of the 20th International Conference on Port and Ocean Engineering under Arctic Conditions.* Luleå, Sweden (2009)

A. Konno, T. Mizuki, Numerical simulation of pre-sawn ice test of model icebreaker using physically based modelling. *Proceedings oh the 18th IAHR International Symposium on Ice.* Sapporo, Japan (2006)

I. Konuk, A. Gürtner, S. Yu, A cohesive element framework for dynamic ice-structure interaction problems—Part I: review and formulation. *Proceedings of the ASME 2009 28th International Conference on Ocean, Offshore and Arctic Engineering*, Honolulu, Hawaii, USA. ASME, pp. 1–9 (2009a)

I. Konuk, A. Gürtner, S. Yu, A cohesive element framework for dynamic ice-structure interaction problems—Part II: implementation. *Proceedings of the ASME 2009 28th International Conference on Ocean, Offshore and Arctic Engineering, Honolulu*, Hawaii, USA. ASME, 1–9 (2009b)

I. Konuk, S. Yu, A cohesive element framework for dynamic ice-structure interaction problems-Part III: case studies. *Proceedings of the 29th International Conference on Ocean, Offshore and Arctic Engineering*, pp. 801–809 (2010)

T.V. Kotras, A.V. Baird, J.N. Naegle, Predicting ship performance in level ice. New York, NY, USA. SNAME, pp. 329–349 (1983)

G. Lindqvist, A straightforward method for calculation of ice resistance of ships. *Proceedings of POAC 1989*, pp. 722–735 (1989)

W. Lu, S. Løset, R. Lubbad, Ventilation and backfill effect during ice-structure interactions, in Li & Lu (eds.) The 21st IAHR International Symposium on Ice, June 11–15, 2012 Dalian, China, pp. 826–841 (2012)

W. Lu, R. Lubbad, S. Løset, Simulating ice-sloping structure interactions with the cohesive element method. J. Offshore Mech. Arctic Eng. **136**, 031501–031501 (2014)

W. Lu, R. Lubbad, S. Løset, In-plane fracture of an ice floe: a theoretical study on the splitting failure mode. Cold Regions Sci. Technol. **110**, 77–101 (2015a)

W. Lu, R. Lubbad, S. Løset, Out-of-plane failure of an ice floe: Radial-crack-initiation-controlled fracture. Cold Regions Sci. Technol. **119**, 183–203 (2015b)

W. Lu, R. Lubbad, S. Løset, Parallel channels' fracturing mechanism during ice management operations. Part II: Experiment. Cold Regions Sci. Technol. (2018a)

W. Lu, R. Lubbad, S. Løset, M. Kashafutdinov, Fracture of an ice floe: local out-of-plane flexural failures versus global in-plane splitting failure. Cold Regions Sci. Technol. **123**, 1–13 (2016)

W. Lu, R. Lubbad, A. Shestov, S. Løset, Parallel channels' fracturing mechanism during ice management operations. Part I: Theory. Cold Regions Sci. Technol. (2018b)

R. Lubbad, S. Løset, A numerical model for real-time simulation of ship-ice interaction. Cold Regions Sci. Technol. **65**, 111–127 (2011)

R. Lubbad, S. Løset, W. Lu, A. Tsarau, M. van den Berg, An overview of the Oden arctic technology research cruise 2015 (OATRC2015) and numerical simulations performed with SAMS driven by data collected during the cruise. Cold Regions Sci. Technol. (2018)

I. Metrikin, S. Løset, Non-smooth 3D discrete element simulation of a drillship in discontinuous ice. *Proceedings of the 22nd International Conference on Port and Ocean Engineering under Arctic Conditions*, 2013 Espoo, Finland

J.N. Naegle, *Ice-Resistance Prediction and Motion Simulation For Ships Operating in The Continuous Mode of Icebreaking.* The University of Michigan (1980)

D.E. Nevel, Ice forces on cones from floes. IAHR Symposium, 11th, 1391–1404 (1992)

J. Paavilainen, J. Tuhkuri, A. Polojarvi, Rubble pile formation against an inclined structure analysis of simulation results. *IAHR International Symposium on Ice*, 1–11 (2010)

J. Paavilainen, J. Tuhkuri, A. Polojärvi, Discrete element simulation of ice pile-up against an inclined structure. *Proceedings of the 18th International Symposium on Ice, IAHR*, 2, 177–184 (2006)

J. Paavilainen, J. Tuhkuri, A. Polojärvi, 2D combined finite-discrete element method to model multi-fracture of beam structures. Eng. Comput. **26**, 578–598 (2009)

J. Paavilainen, J. Tuhkuri, A. Polojärvi, 2D numerical simulations of ice rubble formation process against an inclined structure. Cold Regions Sci. Technol. **68**, 20–34 (2011)

J. Ranta, A. Polojärvi, J. Tuhkuri, The statistical analysis of peak ice loads in a simulated ice-structure interaction process. Cold Regions Sci. Technol. **133**, 46–55 (2017)

B. Su, K. Riska, T. Moan, A numerical method for the prediction of ship performance in level ice. Cold Regions Sci. Technol. **60**, 177–188 (2010a)

B. Su, K. Riska, T. Moan, Numerical simulation of local ice loads in uniform and randomly varying ice conditions. Cold Regions Sci. Technol. (2010b)

B. Su, K. Riska, T. Moan, Numerical simulation of ship turning in level ice. *Proceedings of the asme 29th International Conference on Ocean, Offshore and Arctic Engineering*, 2010, vol 4, Amer. Soc. Mech. Eng., 751–758 (2010c)

B. Su, K. Riska, T. Moan, Numerical simulation of local ice loads in uniform and randomly varying ice conditions. Cold Regions Sci. Technol. **65**, 145–159 (2011)

X. Tan, B. Su, K. Riska, T. Moan, A six-degrees-of-freedom numerical model for level ice–ship interaction. Cold Regions Sci. Technol. **92**, 1–16 (2013)

G.W. Timco, W.F. Weeks, A review of the engineering properties of sea ice. Cold Regions Sci. Technol. **60,** 107–129 (2010)

A. Tsarau, S. Løset, Modelling the hydrodynamic effects associated with station-keeping in broken ice. Cold Regions Sci. Technol. **118,** 76–90 (2015)

A. Tsarau, R. Lubbad, S. Løset, A numerical model for simulation of the hydrodynamic interactions between a marine floater and fragmented sea ice. Cold Regions Sci. Technol. **103**, 1–14 (2014)

J. Tuhkuri, A. Polojärvi, A review of discrete element simulation of ice–structure interaction. *Philosophical Transactions of the Royal Society A: Mathematical, Physical and Engineering Sciences,* 376, 20170335 (2018)

P. Valanto, On the cause and distribution of resistance forces on ship hulls moving in level ice. *Proceedings of the 18th International Conference on Port and Ocean Engineering under Arctic Conditions,* August 12–17, 2001 Ottawa, Ontario, Canada. 803-816 (2001a)

P. Valanto, The resistance of ships in level ice. SNAME **109,** 53–83 (2001b)

P. Valanto, On the ice load distribution on ship hulls in level ice. ICETECH06–110-RF, ICETECH06, Canada (2006)

M. van den Berg, R. Lubbad, S. Løset, An implicit time-stepping scheme and an improved contact model for ice-structure interaction simulations. Cold Regions Sci. Technol. **155**, 193–213 (2018)

M. van den Berg, R. Lubbad, S. Løset, The effect of ice floe shape on the load experienced by vertical-sided structures interacting with a broken ice field. Marine Struct. **65**, 229–248 (2019)

L. Zhou, B. Su, K. Riska, T. Moan, Numerical simulation of moored structure station keeping in level ice. Cold Regions Sci. Technol. (2011)

Chapter 15
Statistics of Ice Loads on Inclined Marine Structures Based on Numerical Experiments

Arttu Polojärvi, Janne Ranta, and Jukka Tuhkuri

Abstract This article focuses on the statistics of ice loads on wide, inclined, offshore structures. The results are based on hundreds of two-dimensional combined finite-discrete element method simulations on ice-structure interaction process. In the simulations, an intact ice sheet collides with, and fails against, a structure. The simulations are sensitive to initial conditions and by varying them, different ice loading processes can be generated even if the simulation parameters are not changed. The statistics of the maximum global horizontal ice load values and the trends in the loading process are studied using the data from the simulations. The analysis shows that the distributions for these maximum load values are non-normal and that numerous repeated measurements are required to reach reasonable estimates for ice loads under given conditions. The results further show that the interaction process evolves and that the observations based on short interaction processes may lead to inaccurate ice load estimates. The sources for the stochasticity of the loading process are also discussed; scatter in the ice load data is inherent to the ice-structure interaction process.

15.1 Introduction

Insight on the mechanical behavior of sea ice and ice loads is important as it leads to safer and more sustainable Arctic operations, including marine transportation, offshore wind energy, and offshore drilling. The ice loads on structures arise from a complex and stochastic ice-structure interaction process, in which an intact ice sheet fails against an offshore structure (Sanderson 1988; Daley et al. 1998; Jordaan 2001).

A. Polojärvi (✉) · J. Ranta · J. Tuhkuri
School of Engineering, Department of Mechanical Engineering, Aalto University, P.O. Box 14100, 00076 Aalto, Finland
e-mail: arttu.polojarvi@aalto.fi

J. Ranta
e-mail: janne.ranta@aalto.fi

J. Tuhkuri
e-mail: jukka.tuhkuri@aalto.fi

J. Tuhkuri and A. Polojärvi (eds.), *IUTAM Symposium on Physics and Mechanics of Sea Ice*, IUTAM Bookseries 39, https://doi.org/10.1007/978-3-030-80439-8_15

Full- and laboratory-scale experiments on ice-structure interaction process are very valuable, yet the stochasticity leads to wide scatter in the results. The scatter is due to several reasons, but in brief, it is challenging to repeat any measurement on ice loading process with the same parameterization and initial conditions. In addition, as the results of ice load measurements are often proprietary, ice load data available are scarce and performing proper statistical analysis on ice loads is difficult.

This article reviews and summarizes the results from our recent statistical studies on virtual ice load data (Ranta et al. 2017a, b, 2018a, b, c; Ranta 2018; Ranta and Polojärvi 2019). The study is based on hundreds of two-dimensional combined finite-discrete element method (FEM-DEM) (Cundall and Strack 1979; Munjiza et al. 1995) simulations on ice-structure interaction process (Fig. 15.1). The simulations are deterministic, but sensitive to initial conditions (Ranta et al. 2017a): Two simulations with identical parameterization, but different initial conditions, yield two ice loading processes that are different in details. This enables performing a statistical study based on repeated numerical experiments with identical, fully controlled, parameterization. This is contrary to the more common statistical ice load studies, which often rely on full-scale ice load data from ships and fixed structures (Jordaan 2001; Kheisin and Popov 1973; Kujala 1994; Brown et al. 2001; Timco and Johnston 2004; Brown and Elseify 2006; Fransson and Lunqvist 2006; Kärnä et al. 2006; Qu et al. 2006; Kujala et al. 2009; Suyuthi et al. 2013) and lack this level of control on parameters. The interest here is on the magnitude of the horizontal global ice load values and the corresponding load records solely; local ice loads are not discussed.

The results presented show that in the case of horizontal global ice loads on a structure, the maximum load data are non-normal distributed and best described by an extreme value distribution. This is the case even when the ice sheet is homogeneous. The results, thus, indicate that it is the complex interaction process itself that leads to the stochastic peak ice load values. The number of observations required for reasonable estimates on peak ice load values is discussed based on the data. The results further show that the ice load values become more predictable once an ice rubble pile has formed in front of the structure. The role of buckling as a load limiting mechanism, which potentially leads to some scatter, is also discussed briefly.

The article is organized as follows. Section 15.2 describes the simulations, the parameters used in them, and the tools used in the analysis of the simulation results. Section 15.3 presents results on peak ice load observations, on their distributions, and on the trends in the ice load records. Section 15.4 addresses the reasons for the stochasticity and discusses the error estimates for the peak ice load observations from the aspect of experimentation. Conclusions are presented in Sect. 15.5.

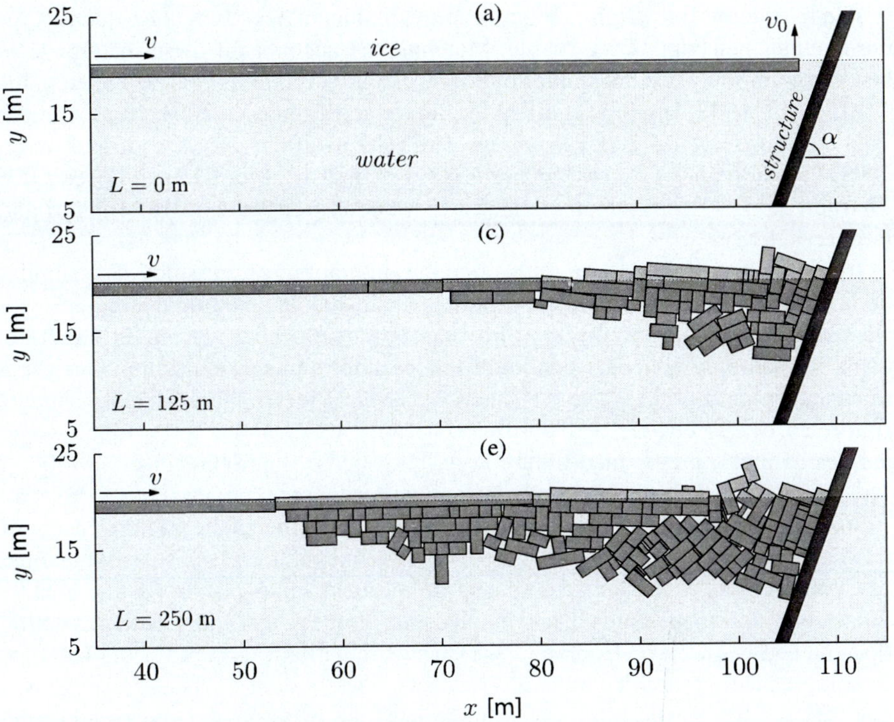

Fig. 15.1 Snapshots of a 2D FEM-DEM-simulated ice-structure interaction process at six stages, described by the length, L, of the ice pushed against the structure. The ice sheet, moving at a constant velocity, v, breaks into ice blocks, which form an ice rubble pile in front of the structure. The load from the still-intact ice sheet is transmitted through the rubble pile to the structure. The first figure shows the initial vertical velocity perturbation, v_0, which was used to vary initial conditions. Here the ice sheet thickness was $h = 1.25$ m. Figure reproduced from Ranta (2018)

15.2 Methods

15.2.1 Simulations

The simulations were performed using the 2D FEM-DEM code developed by the Aalto University Ice Mechanics Group (Paavilainen et al. 2009, 2011; Paavilainen and Tuhkuri 2012, 2013). Paavilainen et al. (2009) successfully compared the model results to those from laboratory scale experiments on ice-structure interaction (Saarinen 2000). Further, Paavilainen et al. (2011) and Ranta et al. (2018a) found the maximum load values yielded by the model to be in the range of the full-scale data collected and reported by Timco and Johnston (2004). Paavilainen et al. (2009; 2011) also discuss the rubble pile profiles and ice rubble behavior during the simulated process. It is advantageous to use FEM-DEM analyses in ice mechanics as this method is able to account for a high number of individual ice blocks and, consequently, for the granular behavior of ice rubble. Methods that account for these features have become commonly used in studies on ice mechanics (Williams et al. 1986; Hopkins 1991, 1992, 1998; Hopkins et al. 1999; Barker and Croasdale 2004; Liferov 2005; Tuhkuri and Polojärvi 2005; Polojärvi and Tuhkuri 2009, 2013a, b; Konuk et al. 2009; Polojärvi and Tuhkuri 2012; Polojärvi et al. 2015; Metrikin and Løset 2013; Metrikin et al. 2015; Shunying et al. 2015; van den Berg et al. 2018; Tuhkuri and Polojärvi 2018).

The snapshots in Fig. 15.1a–f illustrate the simulations of this study. The simulations had an ice sheet of thickness h pushed against an inclined rigid structure with a constant horizontal velocity, $v = 0.05$ m/s. About 100 m away from the structure, a viscous damping boundary condition was used to mimic a semi-infinite ice sheet (Paavilainen et al. 2011). The ice sheet consisted of rectangular discrete elements connected by viscous-elastic Timoshenko beams, which failed at locations where the beams met a pre-defined failure criterion proposed by Schreyer et al. (2006). The beams went through a cohesive softening process upon failure (Hillerborg et al. 1976), with the energy dissipated by softening matching that of the fracture energy of ice (Dempsey et al. 1999). Table 15.1 gives the main parameters of the simulations.

Contact forces were solved by using an elastic-viscous-plastic normal contact force model and an incremental tangential contact force model with Coulomb friction (Paavilainen et al. 2009; Hopkins 1992). The simulations account for the local ice crushing at ice-ice and ice-structure contacts. The amount of local crushing was governed by the plastic limit, σ_p, which relates the maximum contact load to the contact geometry (Hopkins 1992). No new ice features were created nor did the block geometries change during the local crushing and, thus, the simulation tool cannot be used to model continuous crushing of an intact ice sheet on structure; the contact force model estimates the contact area, resulting from the shapes and deformation of the contacting particles, as is standard for the so called soft-contact approaches used in discrete element method simulations (Pöschel and Schwager 2005). Water was accounted for by applying a buoyant force and simplified drag model. The drag coefficient was chosen after that of a rectangular prism, which may underestimate drag on ice blocks and floes with high aspect ratios. The time step was chosen so that the stability of the simulations was ensured. The results from the simulations apply for wide structures in deep water only, since the model is 2D and does not allow for clearing of the ice around the structure. The effect of seabed and rubble pile grounding on maximum peak ice load values and rubble pile profiles is discussed in detail by Lemström et al. (2020).

Table 15.2 summarizes the seven sets of simulations, S1...S7, used as the basis of this study. Each of the sets contained 50 simulations, where all parameters were constant, but which differed by their initial conditions. The initial vertical velocity, v_0, at the free edge of the ice sheet had a unique value of the order of 10^{-12} m/s in each simulation of a given set (Fig. 15.1a). Even if the simulations are deterministic, they are very sensitive to initial conditions (Ranta et al. 2017b): Even a small perturbation in the initial conditions of two simulations with the same parameterization is enough to lead to two ice loading processes that differ in detail. This is demonstrated by

Table 15.1 Main simulation parameters. Ice thickness h and the plastic limit σ_p were varied between simulation sets S1…S6 (Table 15.2). Most parameter values were based on Timco and Weeks (2010). Fracture energy was chosen after Dempsey et al. (1999)

	Description and symbol		Unit	Value
General	Gravitational acceleration	g	m/s^2	9.81
	Ice sheet velocity	v	m/s	0.05
	Drag coefficient			2.0
	Time step		s	$2 \cdot 10^{-5}$
	Duration of the simulation		s	5000
	Total length of pushed ice		m	250
Ice	Thickness	h	m	0.5, 0.875, 1.25
	Elastic modulus	E	GPa	4
	Poisson's ratio			0.3
	Density		kg/m^3	900
	Tensile strength		MPa	0.6
	Shear strength		MPa	0.6
	Fracture energy		J/m^2	12
Contact	Plastic limit	σ_p	MPa	1.0, 2.0, 8.0
	Ice-ice friction coefficient			0.1
	Ice-structure friction coefficient			0.1
Water	Density	ρ_w	kg/m^3	1010
Structure	Slope angle		deg	70

Fig. 15.2a, which shows two ice load records from the simulations of set S5. In the figure, the ice load, F, is plotted against the length of the ice sheet pushed against the structure, L. While both F-records show very similar features, they diverge at $L \approx 10$ m.

As Table 15.2 shows, simulation sets S1…S6 differed by the values of h and σ_p. The ice sheets in the simulations of these sets were homogeneous. The simulations of set S7, on the other hand, were ran with a non-homogeneous ice sheet, with the value of the plastic limit, σ_p, varying after the uniform distribution throughout the ice sheet (the value of σ_p for each discrete element in the ice sheet was random but on the range 1…2 MPa). Simulations in set S7 did not have an initial velocity perturbation; The scatter in the results of simulations in set S7 arises from the non-homogeneity of ice.

Table 15.2 Simulation sets S1. . .S7. Each simulation set included 50 simulations having indices (ID) given in the table. Simulations in each set varied by the values of the initial velocity perturbation (Fig. 15.1a), except the simulations in *S*7, which each had a unique, non-homogeneous, ice sheet (Sect. 15.2.1). Simulation parameters are given in Table 15.1

Set	ID	h [m]	σ_p [MPa]
S1	1–50	0.5	1
S2	51–100	0.5	2
S3	101–150	0.875	1
S4	151–200	0.875	2
S5	201–250	1.25	1
S6	251–300	1.25	2
S7	301–350	1.25	1. . .2

The paper focuses on the statistics of the global maximum load values on the structure, since these are often of first interest in the structural design, and even more commonly, the only available data related to ice loads on a static offshore structure. Fig. 15.2a and b show the maximum peak ice load events in two simulations belonging to the same set. Since the load records were different for all processes, the peak ice load events in them occurred at a different time and gave a different value for the maximum peak ice load, F^p. As the close-ups in Fig. 15.2b show, the ice sheet typically advanced a few meters (up to tens of meters) during a peak ice load event. Further, the data was filtered by using a standard median filter with three data point window. Hence, the F^p values presented and used in the analysis are not due to sudden impacts by individual ice blocks.

15.2.2 Statistical Tools

The distribution types for the peak ice loads observations were not predetermined, but instead, most suitable distribution was chosen based on the histograms of the observations and goodness of fit tests. For these tests, a computer simulation version of an empirical distribution function based technique was used by implementing a suitable procedure outlined in the literature (Lilliefors 1967; Parsons 1982; Keutelian 1991). To improve the reliability of the goodness of fit tests, two different test measures, known as the Kolmogorov-Smirnov and the Anderson-Darling statistics, were used Stephens (1974). All of the tools and their application to ice load data are described in detail by Ranta et al. (2018a, b) and Ranta (2018).

It was also tested how the error estimates for the ice load data develop when the number of observations increases. The error estimates for the different numbers of observations were calculated using sub-sampling and non-parametric bootstrapping procedure known as bootstrap-t method (Efron and Tibshirani 1998). In bootstrapping, confidence intervals are obtained by drawing random samples with a replace-

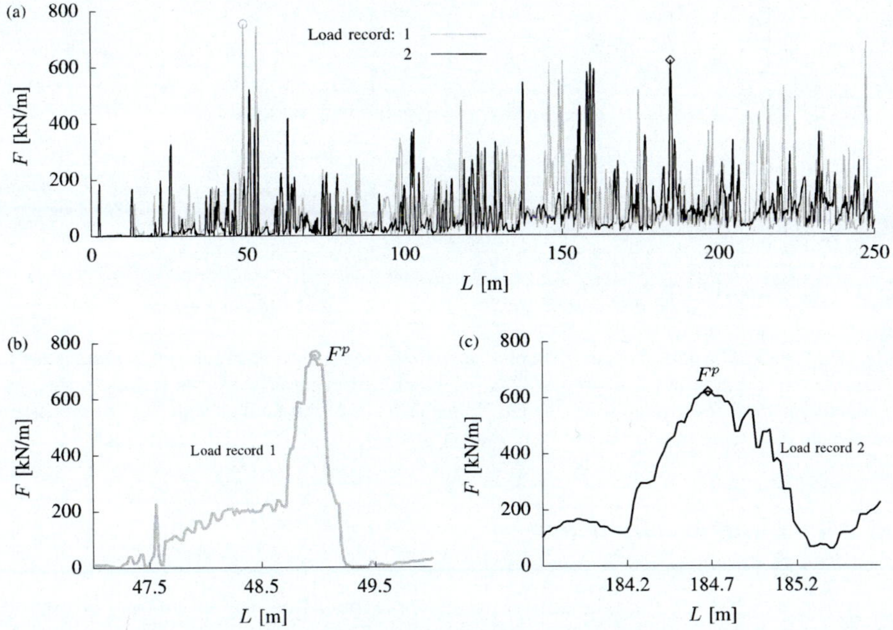

Fig. 15.2 Ice load records: **a** shows the global horizontal ice load, F, from two simulations plotted against length, L, of pushed ice and **b** close-ups of the events leading to the maximum horizontal peak ice load value, F^p, on the structure. The F-records consist of consecutive peak load events, of which one corresponds to F^p. These simulations had the ice thickness $h = 1.25$ m and the plastic limit $\sigma_p = 1$ MPa and they only differed by the value of the initial velocity perturbation v_0 (Fig. 15.1a). Figure reproduced from Ranta and Polojärvi (2019)

ment from the original data. The bootstrap method suited the analysis well as it is distribution independent.

To study the general evolution of the ice loading process, trends for the load records from each of the seven sets of 50 simulations were derived. Derivation of the trends was performed using a simple two-step procedure: (1) calculate the concurrent mean, standard deviation, and maximum statistics from multiple observations on interaction processes (Fig. 15.3 and b) and (2) fit a representative function over the smoothed mean, standard deviation and maximum statistics from step (1). The fitted function for the mean, for the standard deviation, and for the maximum load statistics was chosen to have a form $\beta_1 L^{\beta_2}$, where a non-linear least squares procedure (Marquardt 1963) was used to estimate the case-dependent constants β_1 and β_2.

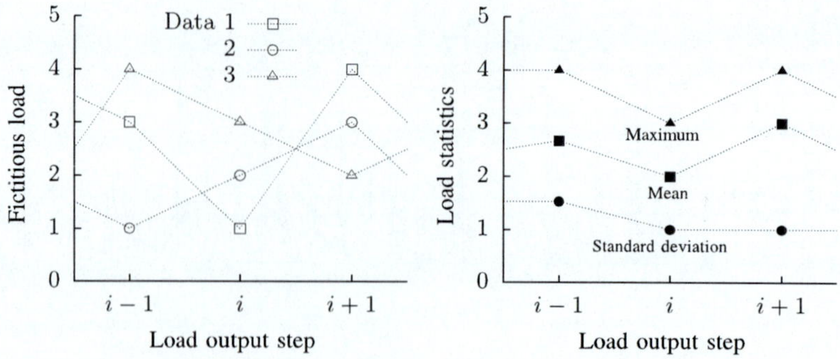

Fig. 15.3 Derivation of the concurrent mean, standard deviation, and maximum ice load statistics in a case of three fictitious ice load processes: **a** load records from output steps $i - 1$, i, and $i + 1$, and **b** concurrent statistics derived from the data. In the analysis of this study, the number of processes in **a** would be 50. Figure reproduced from Ranta (2018)

15.2.3 Limit Mechanisms

Ranta et al. (2018c) introduced a fairly simple buckling model for studying the physical phenomena behind the peak ice load events (Appendix A). The buckling model assumes that the load on the structure is limited by the buckling of a force chain, a chain-like sequence of ice blocks under high compressive load. The model bears resemblance to some earlier buckling-related ice load models (Coon 1974; McKenna et al. 1997; Carter et al. 1998). In essence, the model gives a maximum peak ice load value

$$F^p = \Lambda(\chi)\sqrt{kEI}, \tag{15.1}$$

where $\Lambda(\chi)$ is a buckling-mode-dependent dimensionless multiplier (Appendix A), $k = \rho_w g$ is the specific weight of water (here, ρ_w is the mass density of water and g the gravitational acceleration), E is the elastic modulus of ice, and $I = h^3/12$ is the second moment of area of a beam having a rectangular cross-section with thickness h and unit width.

As will be shown below, h had a strong effect on the values of peak ice loads, F^p. The buckling model could be used in the analysis to eliminate the effect of h. This was done by solving Eq. 15.1 for $\Lambda(\chi)$,

$$\Lambda(\chi) = \frac{F^p}{\sqrt{kEI}}, \tag{15.2}$$

to achieve $\Lambda(\chi)$ values, which did not depend on h. $\Lambda(\chi)$ data from the simulations with different values of h could be, thus, combined in the analysis. The buckling model could also be used to justify on the physical phenomena behind the peak ice load events (Ranta 2018; Ranta and Polojärvi 2019).

15.3 Results

15.3.1 Maximum Peak Load Observations

Figure 15.4a shows the magnitudes of maximum peak ice loads, F^p, given by the simulations. The data were originally introduced and used by Ranta et al. (2017a, b, 2018a, b, c); Ranta and Polojärvi (2019). The error bars in the figure describe the mean values and the standard deviations for the data of each simulation set S1...S7. The figure shows that the mean value of F^p depends strongly on the ice thickness, h. Also the relative scatter in the F^p data grows with h or, in other words, with the values of F^p. This observation aligns with a study by Suominen and Kujala (2014).

The plastic limit, σ_p, also had an effect on the values of F^p as Fig. 15.4a further shows. However, this effect was weaker than that of h. Depending on the case, the

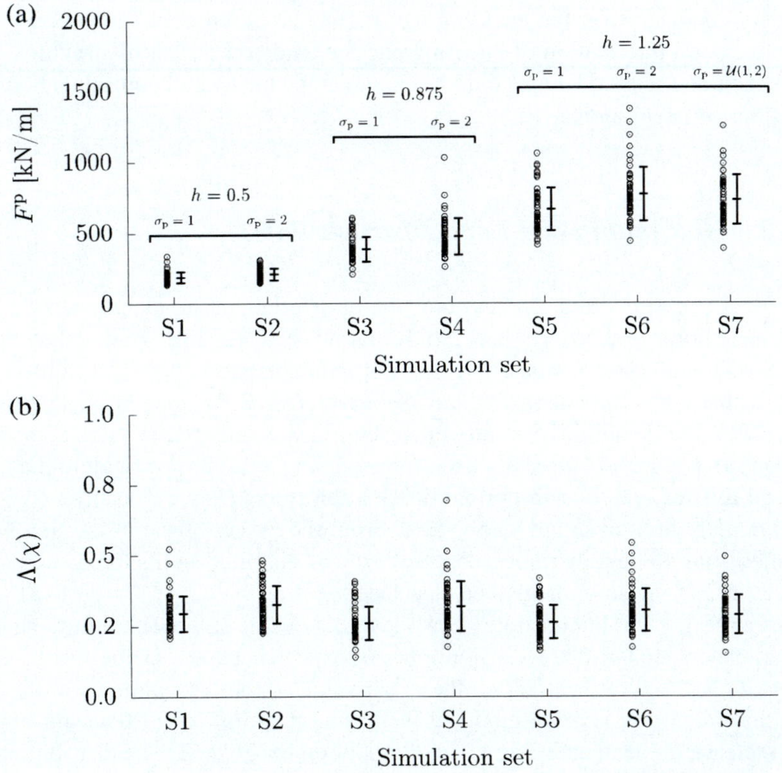

Fig. 15.4 Peak ice load data for simulation sets S1...S7 (Table 15.2): **a** values of the peak ice loads, F^p, and **b** dimensionless maximum peak loads, $\Lambda(\chi) = F^p/\sqrt{kEI}$ (Eq. 15.2). The values for ice thickness h [m] and the plastic limit σ_p [MPa] are shown in **a**. Error bars describe the mean values and the standard deviations of the data. Data in **b** shows that dividing F^p data by buckling related factor $\sqrt{kEI}$ eliminates the effect of h on the values of F^p. Figure reproduced from Ranta (2018)

mean value of F^p increased by 10–20 % when σ_p was increased from 1 MPa to 2 MPa. The amount of increase did not depend on h, but was statistically significant for all h used in the simulations. The simulations of set S7 had a non-homogeneous ice sheet of thickness $h = 1.25$ m and σ_p varying randomly from 1 to 2 MPa throughout the sheet (Sect. 15.2.1). The mean value of F^p for the simulations of set S7 was between the mean values for the simulations in sets S6 and S8. The latter two sets had a homogeneous ice sheet of same h as set S7 with σ_p, respectively, having values 1 MPa and 2 MPa.

Figure 15.4b presents the values for dimensionless maximum peak loads $\Lambda(\chi) = F^p/\sqrt{kEI}$ (Sect. 15.2.3 and Appendix A). As was demonstrated in Ranta et al. (2018c), the buckling model quantifies the F^p data and that buckling is one of the governing mechanisms that limit the values of ice loads: The values of F^p are remarkably well normalized by the buckling-related factor $\sqrt{kEI}$. The mean values of $\Lambda(\chi)$ are on the same range (the difference is about 15 % at maximum) and the effect of h is eliminated. It is likely that the scatter in the F^p data partly owes to this mechanism, as the buckling force would be expected to vary due to the varying boundary conditions for the buckling ice fragments. Sodhi et al. (1983) performed experiments on the relation of buckling and ice loads on a vertical structure. Their results included scatter, which could be explained by the lack of control on boundary conditions of buckling ice.

15.3.2 Maximum Peak Load Distributions

Figure 15.5 presents fitted distributions for the F^p data of simulation sets S1…S6, which had a homogeneous ice sheet. Study in Ranta et al. (2018a) showed that the data for all sets were best described by a non-normal distribution having a right-skewed shape with longer right than left tail (see Sect. 15.2.2 for details on distribution fitting). In more detail, the best fitted distribution was of extreme value distribution (EVD) type 1, which is typically used for modeling extremes of random data; here the need for this type of non-normal distribution stems from an occurrence of rare and extremely high peak ice load values. From the aspect of the discussion below, it is interesting to already notice, that this type of distribution fits the data from the simulations having an homogeneous ice sheet.

Further, Fig. 15.5 demonstrates how ice thickness, h, and plastic limit, σ_p, affect the parameters of the fitted distribution. As expected based on the results above (Fig. 15.4), h has a clear effect on the mean value of the distribution. On the other hand, the effect of σ_p is more subtle, as increasing σ_p from 1 to 2 MPa leads to about 15 % increase in the mean value of the distribution for given h. The distributions for the simulation sets with $\sigma_p = 2$ MPa also indicate higher scatter than the distributions for those with $\sigma_p = 1$ MPa.

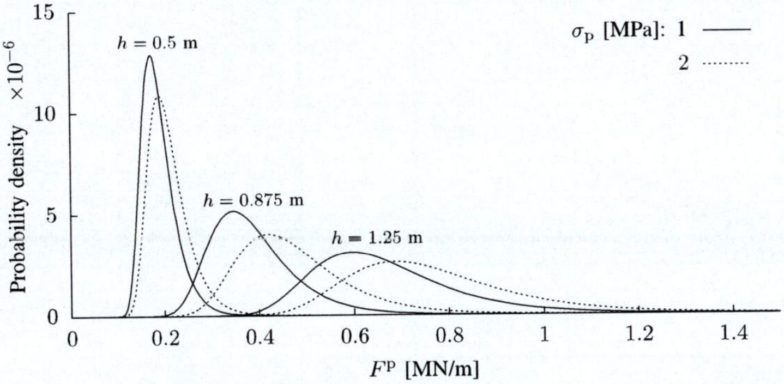

Fig. 15.5 EVD type 1 (Gumbel) probability distributions for the magnitudes of peak ice load, F^p, from the simulations in sets S1...S6. Solid and dashed lines describe the cases with the plastic limit values of σ_p = 1 MPa and σ_p = 2 MPa, respectively. Ice thickness h is shown for all distributions. Figure reproduced from Ranta (2018)

15.3.3 Evolution of the Process

Figure 15.6, originally presented in Ranta et al. (2018b), shows typical trends for the mean, the standard deviation, and the maximum load statistics on concurrent ice load observations (Fig. 15.3). In addition to the load statistics, each graph shows the exponential fit on the data. The data of the figure is from the 50 simulations of simulation set S5, where $h = 1.25$ m and $\sigma_p = 1$ MPa, but the concurrent ice load data from the other sets showed very similar general features. Figure 15.6a shows a high load peak in the early process stage, on interval $L = 10 \ldots 15$ m, for the mean statistics data. This peak is due to all processes in a given set following each other closely until the solutions diverge after the first bending failure producing a peak load of small value. Thus, the exponential fits only account for the data with $L > 15$ m. The simulation data in Fig. 15.6c shows, that the actual magnitudes of the peak ice loads were significantly higher than the fit indicates. The fit is, anyhow, suitable for the discussion here, as it best describes the trends in the maximum load data.

Exponential fits describing the trends for the load statistics from simulation sets S1...S6 are shown in Fig. 15.7. These graphs were used in Ranta et al. (2018b) to show that the ice-structure interaction process did not reach a stationary state: None of the fits reach a constant value. The rate of increase in all trends is strong during the initial stages of the process, but weakens towards the end; reliable limit for a stationary interaction process could only be established through a computationally infeasible numerical test campaign with hundreds of very long simulations to be performed. The fit for the maximum load has a significantly higher magnitude than that for the mean load. The effect of ice thickness, h, is strong on all load statistics.

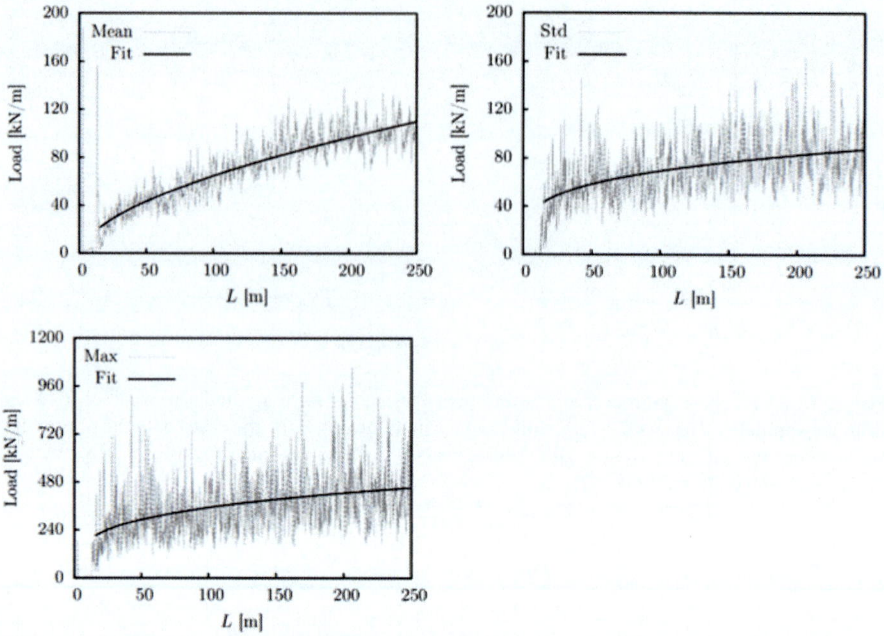

Fig. 15.6 Concurrent ice load data and exponential fits: **a** mean ice load, **b** standard deviation, and **c** maximum ice load. The data is from the fifty simulations of set S5, where the ice thickness $h = 1.25$ m and plastic limit $\sigma_p = 1$ MPa (Table 15.2). L is the amount of the ice pushed against the structure. The data from the simulations is shown in gray. Figure reproduced from Ranta (2018b)

Interestingly, σ_p has only a very weak effect on the mean load, but a stronger effect on the standard deviation and the maximum load. This implies that the effect of σ_p is related to the mechanisms, which limit the maximum ice load value in ice-structure interaction process, whereas it does not affect the mean ice load levels.

15.4 Discussion

15.4.1 Sources of Stochasticity

While non-homogeneity of ice is likely to lead to variability in ice load measurement results, it is still reasonable to ask, if non-homogeneity is a necessary condition for the scatter. To see if the non-homogeneity is a necessary condition for the scatter, Ranta et al. (2017b) compared the F^p observations from the simulations with homogeneous ice sheets to those from the simulations with non-homogeneous ice sheets. Figures 15.8a and b, respectively, show the F^p observations and their cumulative probability distributions for the simulation sets S5, S6 and S7. Set S7 had

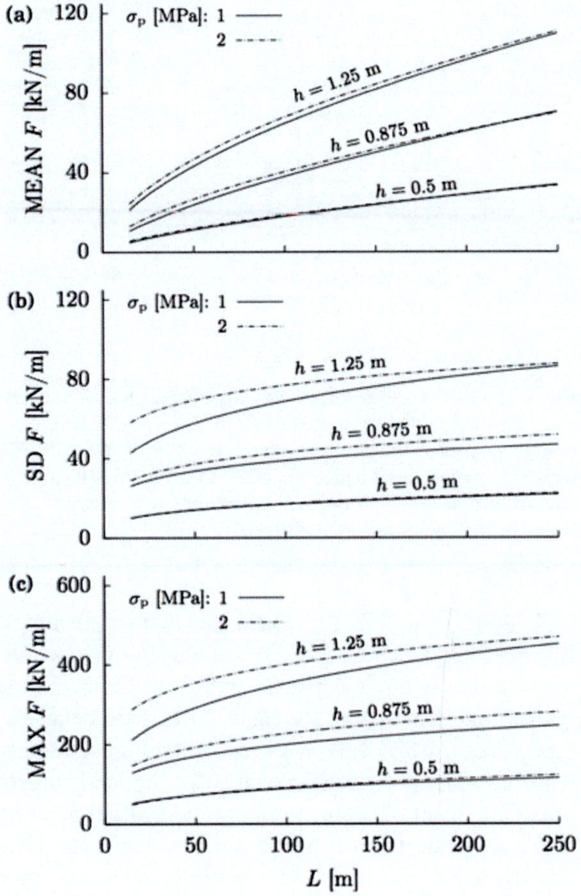

Fig. 15.7 Exponential fits on the concurrent the ice load data from simulation sets S1 …S6 (Table 15.2): **a** mean ice load, **b** standard deviation, and **c** maximum ice load statistics. L m is the amount of pushed ice, h the ice thickness, and σ_p the plastic limit. Figure reproduced from Ranta (2018)

simulations with non-homogeneous ice sheet with σ_p varying between 1 and 2 MPa throughout the sheet (Sect. 15.2). Sets S5 and S6 had a homogeneous ice sheet, with $\sigma_p = 1$ MPa and $\sigma_p = 2$ MPa, respectively. All simulations in these three sets had ice thickness $h = 1.25$ m.

Figure 15.8a shows the values of F^p from the simulations of sets S5…S7 as a function σ_p together with the mean values and standard deviations for the F^p data. For set S7, F^p data points are plotted at $\sigma_p = 1.5$ MPa, which is the expected value for this parameter in this case. Dotted lines of the figure connect the error bars of simulation sets S5 and S6. These lines practically intersect with the extremes of the error bars of set S7. The figure also shows that the mean value of F^p for simulations in set S7 is between that for simulations in sets S5 and S6. The data from the interaction processes with non-homogeneous ice sheets, thus, appear to be well in line with the data from the processes with homogeneous ice sheets.

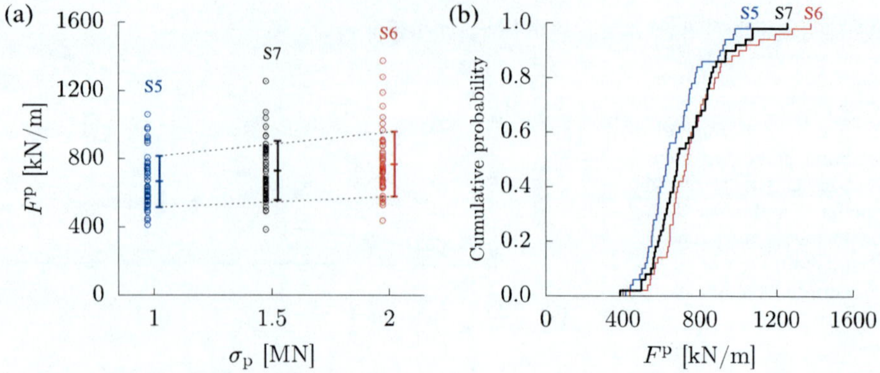

Fig. 15.8 Effect of inhomogeneity: **a** peak ice load, F^P, observations and **b** cumulative distributions for them in the simulations of sets $S5 - S7$ (Table 15.2). In sets S5 and S6, ice sheets were homogeneous and the plastic limit σ_p was 1 MPa and 2 MPa, respectively. In set S7, ice sheets were non-homogeneous and had an expected value of 1.5 MPa for σ_p. The dotted lines in **a** connect error bars of data sets S5 and S6. All sets had the ice thickness $h = 1.25$ m. Figures reproduced from Ranta (2018) and Ranta et al. (2017b)

In addition to this, Fig. 15.8b shows that the cumulative probability distributions for the three data sets are very similar. In other words, the non-homogeneity of ice does not cause any clearly observable difference in the peak ice load distributions: Best fitting distribution for each of the three data sets is a non-normal and of EVD type 1, regardless of the ice sheet being homogeneous or non-homogeneous. (Even if not shown here, also the trends for the load statistics derived for the concurrent load data of set S7 were similar to those of all other sets.)

Thus, while the stochastic nature of ice load observations can be due to non-homogeneity of ice, the results here show that this stochasticity may just as well arise from the ice-structure interaction process itself; the ice-structure interaction process is an inherently a stochastic process. This observation aligns with that presented by Daley et al. (1998). Further, Suyuthi et al. (2013) presented a similar notion related to the randomness of local ice loads on ship hull during an ice breaking process. It is important to emphasize that here this finding applies on the global load on a wide, fixed, structure and it is an outcome from observations on tens of interaction processes, where all parameters related to each process are fully controlled.

15.4.2 Scatter and Error

From the aspect of experimentation, it is interesting to discuss how much error the measurements on peak ice loads, F^P, may yield if the number of observations, m, is limited. This topic was addressed in Ranta et al. (2018a) by studying how the relative error margins for the peak ice load data change as a function of m. Figure 15.9 shows

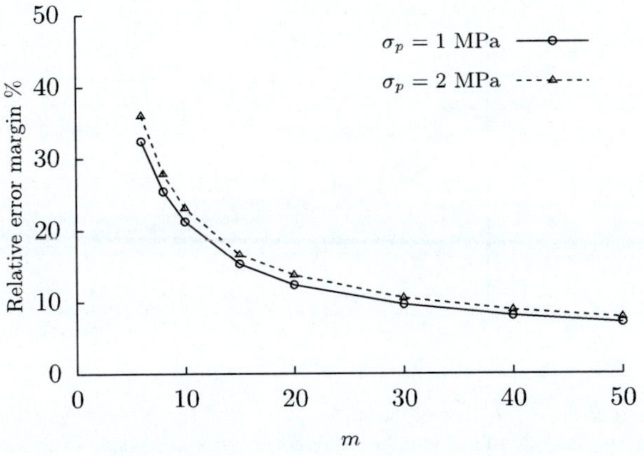

Fig. 15.9 The error estimates for maximum peak load, F^p, data as a function of sample size m. Data were derived by using bootstrap analysis as described in detail in Ranta et al. (2018a). The analysis was performed using scaled and combined data from sets S1...S6: Two sets, each containing data from 150 simulations with constant plastic limit σ_p ($\sigma_p = 1$ MPa or 2 MPa) and all ice thicknesses h included (see text for details). Figure reproduced from Ranta et al. (2018a)

these error margins, derived by applying bootstrapping procedure and random subsampling of F^p data (Sect. 15.2.2). To include more data in the analysis, the data from the simulations in sets S1...S6 was scaled and combined to form two combined sets of 150 simulations. Each combined set included F^p data from the simulations having constant plastic limit, σ_p, and all three ice thicknesses, h. The values of F^p were scaled by dividing them by the buckling-related factor $\sqrt{kEI}$ to eliminate the effect of h (Fig. 15.4b). Similar technique for scaling and combining data was used by Ranta et al. (2018a).

Three distinct features of Fig. 15.9 are: (1) Small sample sizes lead to significant error margins, but (2) the error margins initially decrease fast with increasing m, while (3) high value of m is required for high accuracy. Feature (1) is demonstrated by the fact that with $m < 10$, the data leads to error margins of about 25 % at minimum. Feature (2) suggests that aiming for repeated measurements with the same parameterization is beneficial, as especially with low m, adding even some few observations is enough to improve accuracy of the results significantly. Feature (3) is manifested by the requirement to have sample sizes $m > 30$ for reaching a 10 % error margin. It should noticed here, that it is rare for well-controlled ice load experiments to reach sample sizes of even $m > 5$.

Another issue related to the accuracy of measured data is the required length of the observed ice-structure interaction process. This issue was discussed in Ranta et al. (2018b) by using the trends in the load data to derive the coefficient of variation, CV, for the load statistics of Fig. 15.7. The trend for CV can be calculated for a given set of simulations by dividing the exponential fit for the standard deviation (Fig. 15.7a) by the exponential fit for the mean (Fig. 15.7b). The result of this procedure is presented by Fig. 15.10, which shows CV plotted against the length of pushed ice, L.

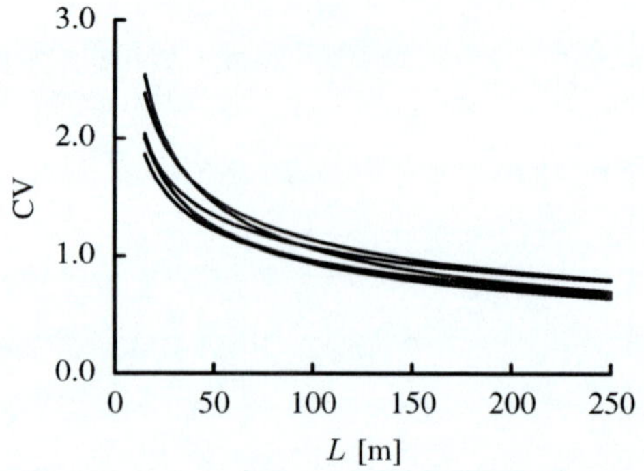

Fig. 15.10 Trends for the coefficient of variation CV = SD/MEAN, derived using the exponential fits for the concurrent mean load and standard deviation statistics, MEAN and SD, respectively, of sets S1...S6 (Fig. 15.7). CV is here plotted against the length of ice pushed against the structure, L. Figure reproduced from Ranta (2018)

Figure 15.10 shows that the value of CV is always high during the initial stages of the interaction process. This means that predicting the ice load values during an early process is especially challenging: When CV > 1, the standard deviation is greater than the mean of the underlying data. With increasing L, CV continuously decreases. This indicates that the loading process becomes more predictable when it evolves, that is, when the volume of ice rubble in front of the structure increases. This implies that it is favorable to perform experiments, or to observe interaction processes, that are fairly long. Further this implies that the distributions assuming a constant value for CV should not be used to predict the ice loads at different stages of the process. Further insight on this could be also gained by performing a numerical simulation campaign, where hundreds of longer interaction processes, reaching a potential stationary state, are modeled.

Related to real-life observations, it should be further noticed that the numerical data is likely to lead smaller error margins than the full-scale observations. This is due to the control on the parameters: There is practically always some variation between the parameters and the initial conditions of any two real-life ice loading processes. This is not the case with the sets of numerical experiments, which can be performed by using identical parameters and randomly perturbed initial conditions.

15.5 Conclusions

The focus of this article was on the statistics of a process, where an ice sheet is failing against an inclined, rigid, structure. The article reviewed and summarized the work presented in Ranta et al. (2017a, b, 2018a, b, c); Ranta and Polojärvi (2019) and Ranta (2018). The analyzed data were produced by using 2D combined finite-discrete element method (2D FEM-DEM) simulations. 350 simulated interaction processes were used in total. These included six sets of 50 slightly perturbed simulations with homogeneous ice sheets and one set of 50 simulations with non-homogeneous ice sheet. The focus was on the magnitudes of the maximum horizontal global ice loads and on the trends in the corresponding load histories. The ice thickness and the plastic limit of ice were varied between the sets. The three main findings described above are:

- Distributions for the peak ice load observations were non-normal (Fig. 15.5). The best fitted distribution was of EVD type 1 regardless of the ice sheet being homogeneous or non-homogeneous; the scatter in the ice load data stems from the complex ice failure process itself.
- High scatter in the observations leads to large error margins (Fig. 15.9). When the number of repeated observations is low, adding even few more of them will decrease the potential error significantly. Tens of repeated processes are required to reach error margins of less than 10%.
- The trend in the coefficient of variation was initially of high value and then continuously decreased during the ongoing interaction process (Fig. 15.10). This indicates that the process becomes more predictable with the increasing volume of ice rubble in front of the structure.

The findings here call for more field- and laboratory-scale experimental campaigns, which are comprehensive from the aspect of ice load statistics. High scatter, owing to the inherently stochastic ice loading process, clearly leads to large error margins. All estimates on peak ice loads and parameter effects should be based on well-controlled and parameterized experiments with sufficient sample sizes.

Acknowledgements The authors are grateful for the financial support from the Academy of Finland research projects (309830) Ice Block Breakage: Experiments and Simulations (ICEBES) and (268829) Discrete Numerical Simulation and Statistical Analysis of the Failure Process of a Non-Homogenous Ice Sheet Against an Offshore Structure (DICE). The authors also wish to acknowledge the support from the Research Council of Norway through the Centre for Research-based Innovation SAMCoT and the support from all SAMCoT partners.

Appendix A: Buckling Model

Figure 15.11 shows the buckling model introduced by Ranta et al. (2018c). The model considers a rigid system consisting of an ice floe, or a few ice floes compressed together, having total length L_f resting on an elastic foundation having a modulus of k and discrete springs having spring constants K_1 and K_2. The modulus k of the elastic foundation was $k = \rho_w g$, where ρ_w is the mass density of the water and g is the gravitational acceleration. The axial compressive load P results from an interaction with an adjacent ice floe or from an interaction with an ice rubble pile. The critical buckling load of the system is

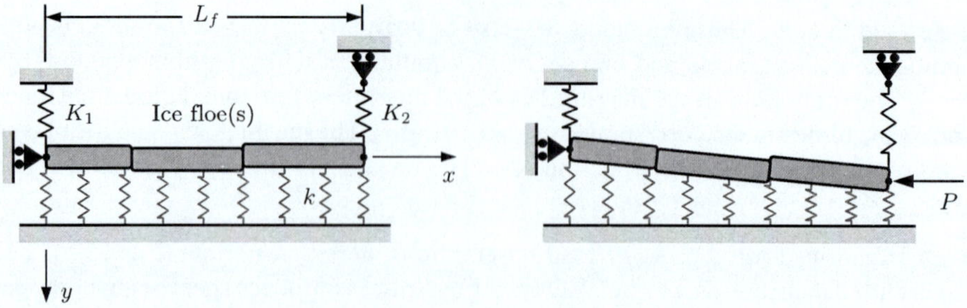

Fig. 15.11 The buckling model in its initial (left) and buckled (right) states. The model consists of an ice floe or several ice floes having total length of L_f resting on an elastic foundation with modulus k presenting water. Springs with spring constants K_1 and K_2 modeled the boundary conditions for the buckling modes shown in Table 15.3. Compressive force P is due to the other floes or the structure. Figure reproduced from Ranta (2018)

Table 15.3 Four buckling modes used in the analysis with the corresponding spring constants K_1 and K_2 (Fig. 15.11). The buckling load is $P = \Lambda\sqrt{kEI}$, where $\Lambda = \Lambda(\chi)$ a mode-dependent factor representing the dimensionless peak load. Factor χ relates the buckling length, L_f, to the characteristic length through $L_f = \chi L_c$ as described in the text. Table reproduced from Ranta (2018)

Mode		K_1	K_2	$\Lambda(\chi)$
1	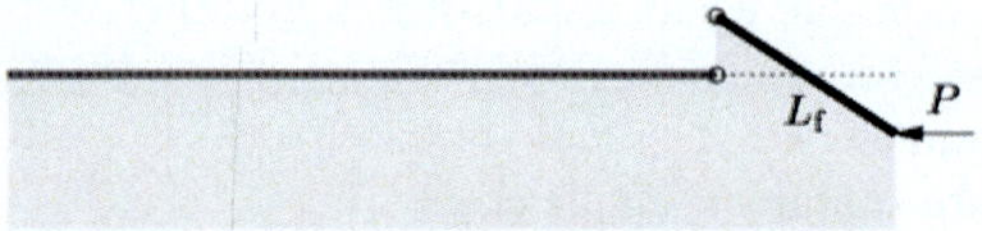	0	0	$\frac{\chi^2}{6}$
2		∞	0	$\frac{2\chi^2}{3}$
3	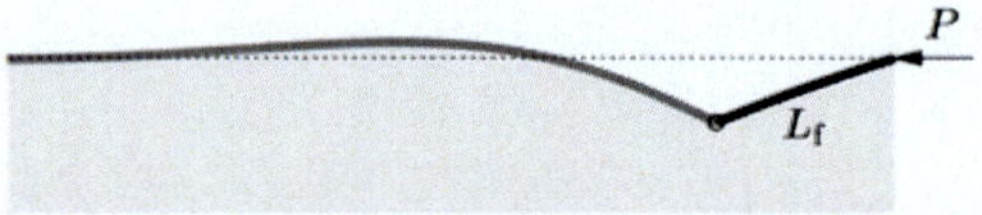	$\tilde{K}_3$	∞	$\frac{12\chi + 8\chi^2}{9\chi + 12}$
4	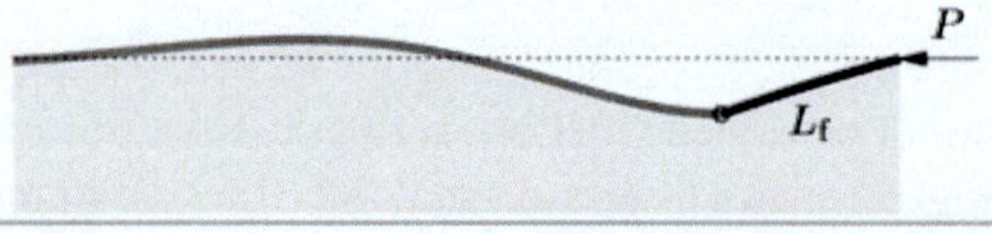	$\tilde{K}_4$	∞	$\frac{12\chi + 4\chi^2}{3\chi + 6}$

$$P_{cr} = \frac{k^2 L_f^3 + 4k(K_1 + K_2)L_f^2 + 12K_1K_2L_f}{12(kL_f + K_1 + K_2)} \tag{15.3}$$

Spring constants $K_1 = C_1 k L_c - C_2 P/L_c$ and $K_2 = C_3 k L_c - C_4 P/L_c$, where L_c is the characteristic length (Hetényi et al. 1979), account for different boundary conditions depending on the choice of constants $C_1 \dots C_4$ (Table 15.3). Characteristic length $L_c = \sqrt[4]{4EI/k}$, where E is the elastic modulus and $I = h^3/12$ is the second moment of area of the floe having a unit width. By plugging the spring constants into Eq. 15.3, and further by substitutions $L_f = \chi L_c$ and $P = P_{cr}$, the critical load may be written concisely as

$$P_{cr} = \Lambda(\chi)\sqrt{kEI}, \tag{15.4}$$

where $\Lambda(\chi)$ is dimensionless buckling load and χ dimensionless buckling length factor. Table 15.3 shows four different buckling modes considered here with the corresponding mode-dependent expressions for $\Lambda(\chi)$. In modes 1 and 2, the system of length L_f buckles alone, whereas in modes 3 and 4 the system of length L_f is affected on the left-hand side by an intact semi-infinite ice sheet. Buckling in the ice load models by Coon (1974) and McKenna et al. (1997) occurred in mode 1. Mode 2 was used by Carter (1998) to describe level ice failure against a vertical structure. The applicability of different buckling modes in the analysis of peak ice loads on inclined structures was discussed in detail by Ranta et al. (2018c).

References

T. Sanderson, *Ice Mechanics, Risks to Offshore Structures* (Graham & Trotman Inc., Kluwer Academic Publishers Group, 1988)

C. Daley, J. Tuhkuri, K. Riska, Cold Reg. Sci. Technol. **27**(3), 197 (1998)

I. Jordaan, Eng. Fract. Mech. **68**, 1923 (2001)

J. Ranta, J. Tuhkuri, A. Polojärvi, Cold Reg. Sci. Technol. **133**, 46 (2017a)

J. Ranta, A. Polojärvi, J. Tuhkuri, in *Proceedings of the 24th International Conference on Port and Ocean Engineering under Arctic Conditions, POAC'17* (Busan, Korea (Electronic Publication), 2017b)

J. Ranta, A. Polojärvi, J. Tuhkuri, Cold Reg. Sci. Technol. **148**, 1 (2018a)

J. Ranta, A. Polojärvi, J. Tuhkuri, Mar. Struct. **57**, 72 (2018b)

J. Ranta, A. Polojärvi, J. Tuhkuri, Cold Reg. Sci. Technol. **147**, 34 (2018c)

J. Ranta, Discrete element modeling of ice failure against an inclined structure - statistical analyses of peak loads and the failure process. Ph.D. thesis, Aalto University (2018)

J. Ranta, A. Polojärvi, Cold Regions Sci. Technol. **67**(102633) (2019)

P. Cundall, O. Strack, Géotechnique **29**(1), 47 (1979)

A. Munjiza, D. Owen, N. Bićanić **12**, 145 (1995)

D. Kheisin, Y. Popov, *Ice Navigation Qualities of Ships* (Technical report, Cold Regions Research and Engineering Laboratory (CRREL), 1973)

P. Kujala, Acta Polytechnica Scandinavica. Mechanical Engineering Series, vol. 116 (Helsinki, 1994)

T. Brown, I. Jordaan, K. Croasdale, Canadian J. Civil Eng. **28**(562–573) (2001)

G. Timco, M. Johnston, Cold Reg. Sci. Technol. **38**(2–3), 185 (2004)

T. Brown, M. Elseify, in *Proceedings of the 18th International Symposium on Ice, IAHR* (2006)

L. Fransson, J.E. Lunqvist, in *Proceedings of OMAE06, 25th International Conference On Offshore Mechanics and Artcic Engineering* (Hamburg, Germany, 2006)

T. Kärnä, Y. Qu, Q. Yue, in *ICETECH-06 Proceedings of the International Conference and Exhibition on Performance of Ships and Structures in Ice* (Alberta, Canada, 2006)
Y. Qu, Q. Yue, X. Bi, T. Kärnä, Cold Reg. Sci. Technol. **45**, 148 (2006)
P. Kujala, M. Suominen, K. Riska, in *Proceedings of the 20th International Conference on Port and Ocean Engineering under Arctic Conditions* (Luleå, Canada, 2009)
A. Suyuthi, B. Leira, K. Riska, Struct. Saf. **40**, 1 (2013)
J. Paavilainen, J. Tuhkuri, A. Polojärvi, Eng. Comput. **26**(6), 578 (2009)
J. Paavilainen, J. Tuhkuri, A. Polojärvi, Cold Reg. Sci. Technol. **68**(1–2), 20 (2011)
J. Paavilainen, J. Tuhkuri, Cold Reg. Sci. Technol. **81**, 1 (2012)
J. Paavilainen, J. Tuhkuri, Cold Reg. Sci. Technol. **85**, 157 (2013)
S. Saarinen, Description of the pile-up process of an ice sheet against an inclined plate. Master's thesis, Helsinki University of Technology (2000)
R. Williams, G. Mustoe, K. Worgan, in *Proceedings of the 8th IAHR International Symposium on Ice* (Iowa City, IA, USA, 1986)
M. Hopkins, W.D. Hibler III., Cold Reg. Sci. Technol. **19**(2), 201 (1991)
M. Hopkins, Numerical simulation of systems of multitudinous polygonal blocks. Technical report, 92-22, Cold Regions Research and Engineering Laboratory, CRREL (1992). 69 p
M. Hopkins, J. Geophys. Res. **103**(C10), 21,883 (1998)
M. Hopkins, J. Tuhkuri, M. Lensu, J. Geophys. Res. **104**(C6), 13605 (1999)
A. Barker, K. Croasdale, in *Proceedings of the 17th IAHR International Symposium on Ice, Saint Petersburg, Russia* (2004), pp. 257–264
P. Liferov, Cold Reg. Sci. Technol. **41**(2), 153 (2005)
J. Tuhkuri, A. Polojärvi, in *Proceedings of the 18th International Conference on Port and Ocean Engineering Under Arctic Conditions, POAC'05*, vol. 2 (Potsdam, New York, 2005), pp. 939–948
A. Polojärvi, J. Tuhkuri, Cold Reg. Sci. Technol. **56**(1), 18 (2009)
A. Polojärvi, J. Tuhkuri, in *DEM6, 6th International Conference on Discrete Element Methods* (August 5 - 6, Golden, Colorado, USA (Electronic Publication), 2013a)
A. Polojärvi, J. Tuhkuri, Cold Reg. Sci. Technol. **85**, 191 (2013b)
I. Konuk, A. Gürtner, S. Yu, in *Proceedings of the 20th International Conference on Port and Ocean Engineering under Arctic Conditions, POAC'09 (Electronic Publication)* (2009)
A. Polojärvi, J. Tuhkuri, O. Korkalo, Cold Reg. Sci. Technol. **81**, 11 (2012)
A. Polojärvi, A. Pustogvar, J. Tuhkuri, Cold Regions Sci. Technol. (116), 12 (2015)
I. Metrikin, S. Løset, in *Proceedings of the 22nd International Conference on Port and Ocean Engineering under Arctic Conditions, POAC'13* (Espoo, Finland (Electronic Publication), 2013)
I. Metrikin, A. Gürtner, B. Bonnemaire, X. Tan, A. Fredriksen, D. Sapelnikov, in *Proceedings of the 23rd International Conference on Port and Ocean Engineering under Arctic Conditions, POAC'15* (in Trondheim, Norway (Electronic Publication), 2015)
J. Shunying, D. Shaocheng, L. Shewen, Eng. Comput. **32**(4), 1121 (2015)
M. van den Berg, R. Lubbad, S. Løset, Cold Reg. Sci. Technol. **155**, 193 (2018)
J. Tuhkuri, A. Polojärvi, Philos. Trans. R. Soc. A: Math., Phys. Eng. Sci. **376**(2129) (2018)
H. Schreyer, D. Sulsky, L. Munday, M. Coon, R. Kwok, J. Geophys. Res.: Oceans **111**(11) (2006)
A. Hillerborg, M. Modéer, P.E. Petersson, Cem. Concr. Res. **6**, 773 (1976)
J. Dempsey, R. Adamson, S. Mulmule, Int. J. Fract. **95**, 347 (1999)
G. Timco, W. Weeks, Cold Reg. Sci. Technol. **60**(2), 107 (2010)
T. Pöschel, T. Schwager, *Computational Granular Dynamics* (Springer Science & Business Media, Berlin, 2005)
I. Lemström, A. Polojärvi, J. Tuhkuri, Cold Regions Sci. Technol. **176**(103088) (2020)
H. Lilliefors, J. Am. Stat. Assoc. **62**(318), 399 (1967)
F. Parsons, Microelectron. Reliab. **22**(2), 163 (1982)
H. Keutelian, The kolmogorov-smirnov test when parameters are estimated from data. Technical report, Fermilab note, Version 1.0 (1991)
M. Stephens, J. Am. Stat. Assoc. **69**(347), 730 (1974)
B. Efron, R. Tibshirani, *An Introduction to the Bootstrap* (CRC Press LLC., 1998)

D.W. Marquardt, J. Soc. Ind. Appl. Math. **11**(2), 431 (1963)
M. Coon, J. Petrol. Technol. **26**, 466 (1974)
R. McKenna, D. Spencer, M. Lau, D. Walker, G. Crocker, in *Proceedings of OMAE'97, 16th International Conference on Offshore Mechanics and Arctic Engineering*, vol. IV (Yokohama, Japan, 1997), pp. 329–338
D. Carter, D. Sodhi, E. Stander, O. Caron, T. Quach, J. Cold Regions Eng. **12**(4), 169 (1998)
M. Suominen, P. Kujala, Cold Reg. Sci. Technol. **106–107**, 131 (2014)
D.S. Sodhi, F.D. Haynes, K. Kato, K. Hirayama, Ann. Glaciol. **4**, 260 (1983)
M. Hetényi, *Beams on Elastic Foundation: Theory with Applications in the Fields of Civil and Mechanical Engineering* (The University of Michigan Press, Ann Arbor, 1979)

Chapter 16
Numerical Study of Oil Spill Behavior Under Ice Cover

Hajime Yamaguchi and Liyanarachchi Waruna Arampath De Silva

Abstract The recent demands of commercial shipping through the Arctic region and exploitation of its natural resources increase the risk of an oil spill in the Arctic Ocean. State-of-the-art modeling techniques are necessary for prevention measures and responses and to determine the fate and effect of oil in the environment. Thus, we developed a numerical simulation model for spilled oil in ice-covered areas. The model performs high-resolution computations at a low computational cost. The model results for oil spreading over open water, under full ice cover and under a broken ice field agreed reasonably well with previous experimental results. The results of the behavior of spilled oil under the ice-covered area in the East Siberian Sea showed the possibility for establishing an accurate and practical forecasting system for oil spread under marginal sea ice.

Keywords Oil spill · Ice cover · Numerical modeling of oil

16.1 Introduction

The retreat of summer sea ice in the Arctic Ocean continues to attract interest for natural resource exploration and exploitation and for the establishment of shorter commercial shipping routes compared with the traditional Suez Canal Route. An increase in those activities has increased the risk of oil spills in ice-covered areas. If an oil spill occurs in this region, it could adversely impact the marine ecosystem as well as the economy of coastal areas and beyond (Kelly et al., 2018).

The behavior of oil after a spill depends on two main processes. The first is weathering where the physical and chemical properties of the oil change because of environmental conditions (e.g., evaporation, emulsification, bio-degradation) (Díez et al., 2007; Boehm et al., 2008). The second process is oil spread through the ocean by currents, waves and sea ice (French-McCay, 2004). The weathering and movement processes can also overlap to produce complex interactions in oil spreading.

H. Yamaguchi (✉) · L. W. A. De Silva
Graduate School of Frontier Sciences, University of Tokyo, 5-1-5 Kashiwanoha, Kashiwa 277-8561, Japan
e-mail: h-yama@edu.k.u-tokyo.ac.jp

J. Tuhkuri and A. Polojärvi (eds.), *IUTAM Symposium on Physics and Mechanics of Sea Ice*, IUTAM Bookseries 39, https://doi.org/10.1007/978-3-030-80439-8_16

Many researchers have studied the spreading of oil in open water after a spill and many numerical models have been introduced to simulate the processes (Fay, 1971; Fannelop and Waldman, 1972) and also a few models introduced to simulate oil under sea ice (Sebastião and Soares, 1995; Yapa and Chowdhury, 1990). In general, predictions of oil behavior in water under ice cover is more complex than in open water. Due to the lack of information about sea ice conditions (ice cracks, leads and ice floe distribution) and uncertainties and unknowns about ice and oil interaction. If spilled oil density lower than sea ice density, oil can penetrate on the ice surface, and migrate go through the ice from ice bottom to the ice surface. Therefore, research on the spread of oil under ice cover has been very limited.

However, when an oil spill happens in an ice-covered area, the ability to predict oil spreading in daily to weekly timescale is important to assist cleanup operations, and monthly to seasonal predictions to evaluate the environmental impact and prevention planning (Afenyo et al., 2015). Therefore, highly accurate modeling is necessary to predict oil spill behavior for spill prevention and response. Main obstacle of the clean-up process under sea ice is the difficulty to locate oil under the ice cover. Oil also can be trapped in the leads where it may become fully encapsulated into growing ice, transported with it and release back to the ocean at a different location when sea ice melts. In terms of oil recovery, it is impossible to use conventional open-ocean methods for oil mechanical containment and recovery; oil degrading by dispersants and of in-situ burning are often impossible in the ice-covered seas (e.g. Wilkinson et al., 2017).

Yapa and Chowdhury (1990) introduced theoretical equations based on Navier–Stokes equations for oil spread under ice. Izumiyama et al. (1998) improved on this theory by introducing net interfacial tension. Izumiyama et al. (2002) validated their previous results by conducting a series of laboratory experiments and those results are very important for validating and parameter adjustment of subsequent numerical models.

In 2006, the Engineering Advancement Association of Japan began a three-year research program to develop a prototype model to predict spilled oil diffusion and advection in the Sea of Okhotsk (Terashima, 2003; Rheem and Yamaguchi, 2004; Hara et al., 2008). In this study, we used the same oil spill simulation model to predict short-term (15-day) oil trajectories and simulate a variety of scenarios for an oil spill in the East Siberian Sea to create a hazards map. We also compared the results of laboratory experiments with the numerical model in an open water area, a fully ice-covered area and a partially ice-covered area.

16.2 Model Description

The oil spill model for predicting the short-term oil distribution used in this study is based on the model developed by Terashima (2003) and Rheem and Yamaguchi (2004). This is a 2-D model with a treatment for vertical motion. The vertical diffusion of oil in water column under an ice-covered area was defined in four stages (Fig. 16.1).

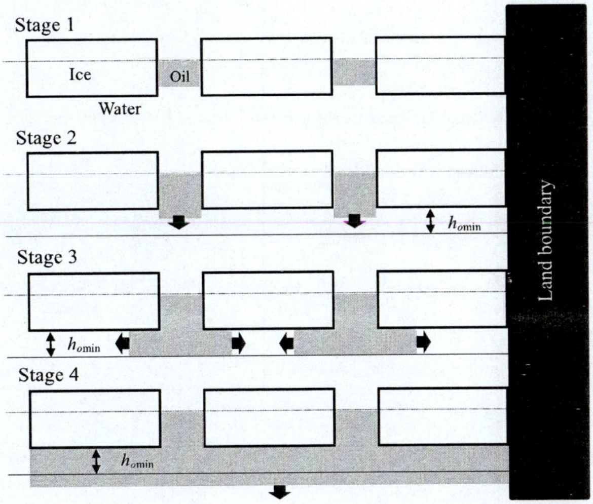

Fig. 16.1 Schematic diagram of the vertical spread of oil in water under ice, where h_{omin} is the minimum oil thickness for spreading under the ice (see Eq. 16.1)

In stage 1, spilled oil spreads horizontally through gaps between the ice floes. After stage 1, if the oil contacts land or becomes surrounded by a dense ice area, it thickens below the bottom of the ice until the minimum oil thickness (Mackay et al., 1976) for spreading under ice (h_{omin}) is obtained (i.e. stage 2). This minimum oil thickness is defined as:

$$h_{o\,\min} = \sqrt{\frac{2\sigma_{o-i-w}}{g(\rho_w - \rho_o)}} \tag{16.1}$$

where σ_{o-i-w} denotes surface tension between ice, water and oil, g (9.81 m/s^2) is gravitational acceleration and ρ_w(1025.9 kg/m^3) and ρ_o denote the density of water and oil, respectively. If oil thickens further, it starts spreading horizontally under the ice (stage 3) while maintaining a constant thickness (h_{omin}). In stage 3, surface tension forces between ice, water and oil dominate the buoyancy force. In the final stage (stage 4), the oil is spread along the entire base of sea ice and then becomes even thicker due to the ongoing supply of oil.

To improve the ability of the oil spill model to accurately simulate the boundary of the oil spill we used the Eulerian–Lagrangian method for oil spill advection (i.e., subgrid-scale oil motion; Rheem et al., 1997). First, we solved the momentum equation in a Eulerian grid and then followed oil conservation on a Lagrangian grid. Spilled oil was rendered as rectangular oil patch in each model grid cell (Fig. 16.2) as described by Rheem et al. (1997).

The ice field was represented by a set of square floe particles (side length D) with a given ice thickness and in-between oil patch as shown in Fig. 16.3. We assumed ice floes were homogenously distributed in the model cells with the shortest distance between any pairs of the neighbouring ice floes defined as l and number of ice floes in

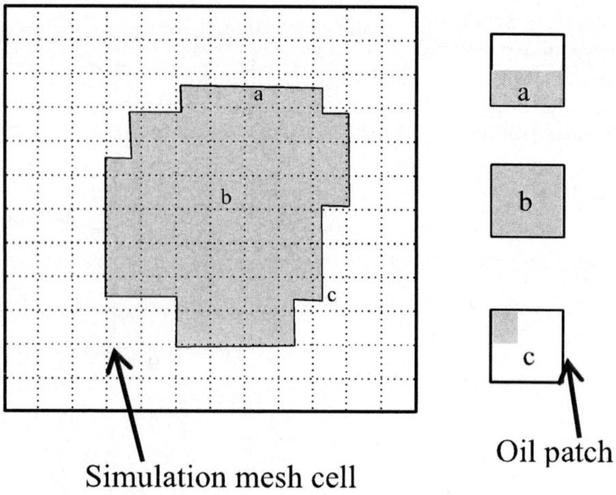

Fig. 16.2 Model grid cell and oil patch locations in the computational grid **a** the bottom half is filled with oil, **b** oil patch fully covered the grid cell and **c** oil patch only covered the top left corner of grid cell

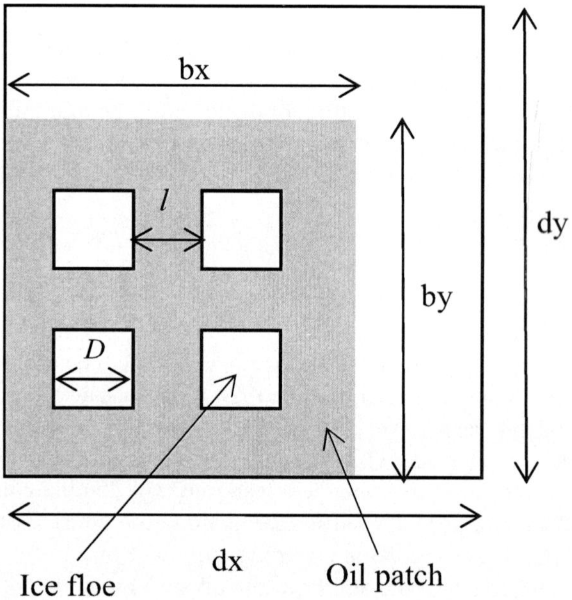

Fig. 16.3 Simulation mesh, ice floe distribution and oil patch location, where dx and dy are cell size and bx and by are size of the oil spill in x and y direction. D is the length of a square ice floe and l is the shortest distance between the two neighboring ice floes

the x and y directions represented by nx and ny, respectively. The sea ice concentration (A) can be expressed as:

$$A = \frac{(Dnx)(Dny)}{dx \cdot dy} \tag{16.2}$$

where dx and dy represent the cell size. Equations 16.3 and 16.4 show the number of ice floes in the x and y directions.

$$nx = \frac{\sqrt{A}dx}{D} \tag{16.3}$$

$$ny = \frac{\sqrt{A}dy}{D} \tag{16.4}$$

Equation 16.5 shows the distance between two ice floes.

$$l = \frac{\left(1 - \sqrt{A}\right)D}{\sqrt{A}} \tag{16.5}$$

In each Eulerian grid cell, the momentum equation (Eq. 16.6) was solved for the velocity of oil diffusion:

$$m_o\left[\frac{\partial \overline{u}_o}{\partial t} + f\overline{k} \times \overline{u}_o\right] = \overline{F}_p + \overline{F}_f + \overline{F}_m + \overline{F}_{st} \tag{16.6}$$

where m_o is oil mass, u_o is oil velocity, f is the Coriolis parameter, k is the unit vector in the vertical direction and F_p, F_f, F_m and F_{st} are the forces of pressure, friction, added mass and surface tension, respectively (the over bar denotes the vectors). In the open water, surface tension force is applied into the oil with the direction of oil diffusion and under ice counter direction of oil diffusion. Surface tension force in open water is defined in Eqs. 16.7 and 16.8 and under ice in Eq. 16.9.

$$F_{st} = R_{st}\sigma_{o-w-a}L \tag{16.7}$$

$$R_{st} = \begin{matrix} 1 & h_o \geq 10h_{o\min} \\ \frac{h_o - h_{o\min}}{9h_{o\min}} & h_{o\min} \leq h_o < 10h_{o\min} \\ 0 & h_o < h_{o\min} \end{matrix} \tag{16.8}$$

$$F_{st} = \sigma_{o-w-i}L \tag{16.9}$$

where, σ_{o-w-a} surface tension between oil, water and air σ_{o-w-i} surface tension between oil, water and ice, L is width of oil leading edge and h_o is oil thickness.

External forces change under different sea surface conditions. When oil spills on a sea surface free of ice, the directional diffusion forces acting on the oil are surface tension and pressure force from gravity in calm sea. The forces counteracting diffusive spread of oil spill are: (i) friction due to the relative motion between the oil spill and the ambient water and (ii) added mass of water around the spill (Fig. 16.4a). However, when oil spills under sea ice, only the pressure force from

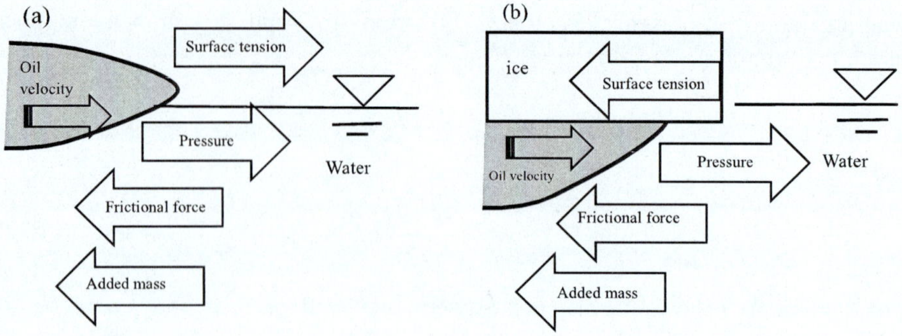

Fig. 16.4 The external forces acting on oil **a** on the water surface and **b** under sea ice

gravity is an active directional diffusion force. The directional counter-acting forces include (i) surface tension, (ii) added mass and (iii) frictional forces between oil spill and sea ice and between oil spill and water (Fig. 16.4b).

After calculating the cell face velocities, each Eulerian cell oil patch was advected in space, creating new configurations. Finally, a new oil state was obtained for the Eulerian grid by summing and redistributing the advected configurations using the linear area remapping. More details about redistribution can be found in Rheem et al. (1997).

The atmospheric boundary conditions of the oil diffusion model were provided by 6-hourly atmospheric reanalysis data of air-temperature at 2 m height and wind at 10 m height from the European Centre for Medium-Range Weather Forecast Re-Analysis Interim (ERA-Interim). The sea ice conditions (velocity, concentration and thickness) and ocean velocity were provided by the high-resolution ice–ocean coupled model IcePOM (De Silva et al., 2015) with 1 h frequency outputs. The weathering process and bio-degrading of the oil was neglected for these computations, which implies that oil density and viscosity is constant.

16.3 Results

Past oil spill experiments conducted with different sea ice concentrations were compared with the oil spill numerical model. Three patterns of the spilled oil diffusion radius were computed: (1) spilled oil in open water by Yamaguchi; (2) spilled oil with full ice cover (100% sea ice concentration) by Izumiyama et al. (2002); (3) spilled oil under partial ice cover (65% sea ice concentration) with the effects of an ice floe motion experiment by Gjosteen and Loset (2002). The physical properties of the oil and experimental conditions are shown in Table 16.1.

Table 16.1 Experimental conditions and calculations of oil spill accuracy

Experiment	Open water	Full ice cover	Partial ice cover
Oil (name)	Idemitsu Dafny #32	Lubrication oil	IFO 30 bunker oil
Density [kg/m^3]	863	890	938.63
Water temperature [°C]	12	−2	−2
Viscosity [Pa/s]	1.1×10^{-1}	1.23×10^{-1}	6.87×10^{-1}
Oil–water surface tension [N/m]	1×10^{-2}	2.56×10^{-2}	2.32×10^{-2}
Spilled oil volume [m^3]	1.0725×10^{-3}	16.32×10^{-3}	9.8×10^{-3}
Oil spilled time [s]	205.6	600	57
Sea ice concentration	–	100%	65%
Sea ice thickness [m]	–	0.01	0.038
Ice bottom	–	Flat	Flat
Sea ice rotational velocity [rad/s]	–	–	0.003
Time step Δt [s]	0.1	0.1	0.1
Grid resolution [m]	0.03	0.03	0.04
Number of grids	100 × 100	100 × 100	100 × 100

16.3.1 *Oil Diffusion Model in Open Water*

The experiment regarding an oil spill in open water (Yamaguchi, 2007) was carried out at the Institute of Industrial Science (University of Tokyo) in a tsunami and storm surge water tank. Calm water, square shape of 1.4 m length and 0.25 m depth tank is used for the experiment. And high-speed camera installed directly above the water tank is used to apographs the oil patch. We have setup the numerical model with same laboratory experiment conditions. And numerical model is integrated for 400 s. Figure 16.5 shows the comparison of the oil diffusion radius obtained in the experiment (blue dots) with the numerical model (black line). Our numerical model showed good agreement with experimental results.

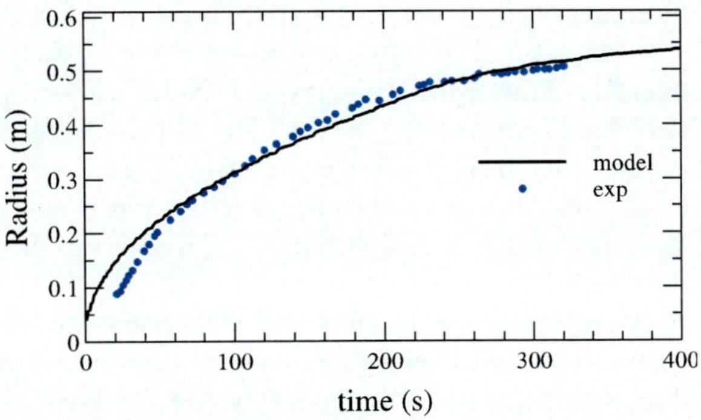

Fig. 16.5 Accuracy of the oil diffusion radius results in open water between the experiment (Yamaguchi, 2007) and model (this study). Blue dots denote the experimental results and the black line indicates the modeling results

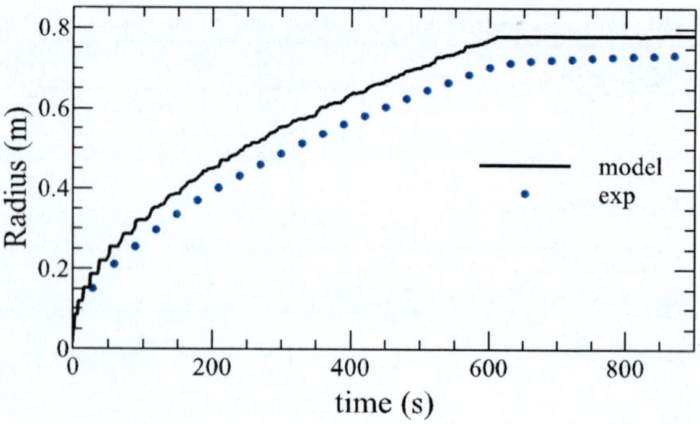

Fig. 16.6 Accuracy of the oil diffusion radius results from the experiment (Izumiyama et al., 2002) and model (this study) under fully ice-covered conditions. Blue dots denote the experimental results and the black line indicates the numerical results

16.3.2 Oil Diffusion Model Under Full Ice Cover

The oil spill experiment under full ice cover conditions (Izumiyama et al., 2002) was carried out in the ice tank of the National Maritime Research Institute (Japan). The test was performed for level ice sheets (100% concentration) with flat bottoms and without ice sheet movement. Simulation is integrated for 1000 s.

Figure 16.6 shows the comparison of experimental and modeling results. The distribution of the oil radius with respect to time showed a similar pattern in both the experiment and simulations. However, the model overpredicted the radius by about 0.05 m.

16.3.3 Oil Diffusion Model Under Partial Ice Cover

Gjosteen and Loset (2002) conducted the laboratory experiment for an oil spill under a broken ice field in the ice tank at the Hamburg Ship Model Basin (Germany). The oil spread and ice floe motion were monitored by video cameras. The ice floes were convex polygons with diameter of about 0.5–1 m were used in the tank experiment. The results were very useful to validate our model under the influence of ice flow dynamics.

Figure 16.7 shows the comparison of the oil diffusion radius between the experimental results (Gjosteen and Loset, 2002) and our numerical model results. Simulation is integrated for 1000 s. The numerical model reasonably reproduced the experimental results. In the experiment of Gjosteen and Loset (2002), ice floes had a clockwise, circular motion; they exerted drag on the oil and enhanced its spread. In the experiment, oil was placed near the center of the domain and constantly spilled for 57 s.

Shortly after the injection was completed, the ice cover began to move. At first the oil spread horizontally along the ice floes with very slow spreading in the vertical direction. After the oil thickness reached the sea ice, it began to spread under the ice floes while experiencing friction from the bottom of the ice. Our model reproduced

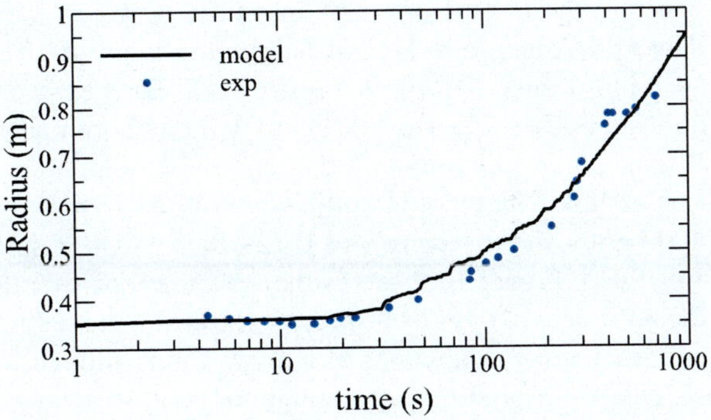

Fig. 16.7 Accuracy of the oil diffusion radius results from the experiment (Gjosteen and Loset, 2002) and model (this study) under partial ice cover. Blue dots denote the experimental results and the black line indicates the numerical results

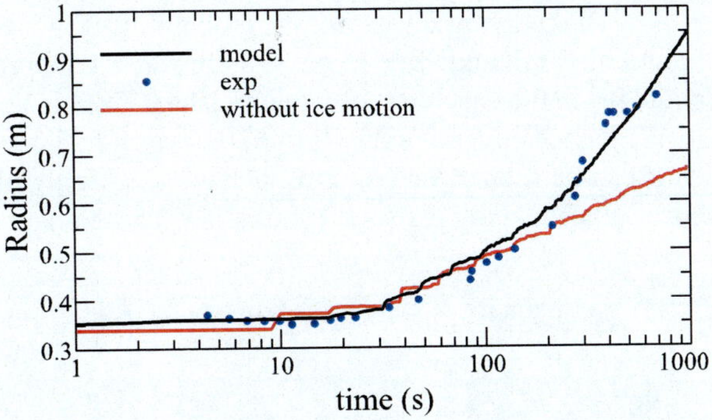

Fig. 16.8 Verification of oil diffusion radius results from the experiment (Gjosteen and Loset, 2002) and model (this study) under partial ice cover with and without ice motion. Blue dots denote the experimental results; the numerical results are shown by black (with ice motion) and red (without ice motion) lines

those treads accurately. We also tried to simulate the effect of ice motion on an oil spill; Fig. 16.8 shows the numerical results with and without ice rotational motion and blue dots are experimental results of oil distribution with ice floe rotational motion. Before 200 s elapsed, the oil spread in a similar radius both with and without ice rotation. However, after 200 s, the oil spread accelerated exponentially with ice rotation, which was not the case when there was no ice rotation. There results suggest that the frictional force between oil and ice is important in oil spreading under the ice and requirement of accurate prediction of ice conditions.

16.4 Spilled Oil Behavior in the East Siberian Sea

We carried out simulations of sea ice and spilled oil behavior in the East Siberian Sea. The initial and boundary sea ice distribution (sea ice concentration, thickness and ice velocity) and oceanic conditions (surface ocean current and ocean temperature) were provided by a ice-ocean coupled computation from IcePOM. Ice floe distribution is assumed to be uniformly distributed 60 m squares. IcePOM model is high-resolution 2.5 km regional model along the Northern Sea route (De Silva et al., 2015). To generate the hazard map, 30 ensemble members were generated using reanalysis data (air-temperature at 2 m height and wind at 10 m height) from ERA-Interim spanning 1 July 2015 to 9 August 2015. Each ensemble member contained 10-day atmospheric data for the first set of computations and 15-day data for the second set. It took 3 min to compute 1-day ensemble computation with 3.7 GHz processor. It was assumed that the oil spilled at a rate of 4 m^3/s for 3000 s. The oil density was 890 kg/m^3 and viscosity was 0.123 Pa/s. Oil was spilled south of Novaya Sibir in July 2015. Figure 16.9 shows the ice and oil distribution at the time of the spill and then 3, 7 and 10 days later. It demonstrated that oil spilled near the East Siberian Sea can reach the coastal areas of Novaya Sibir within a few days.

The 10-day and 15-day simulations (Fig. 16.10) were made with the same oil spill conditions in July 2015; 30 ensemble computations were made by changing the date of the initial oil spill. The results showed the probable distribution of oil in the area; within 10 days, there was a 23% probability (Fig. 16.10a) that oil could reach the Sannikov Strait and after 15 days, there was a 43% probability (Fig. 16.10b) that oil could reach the east coast of Kotelny Island. We also calculate the amount of oil trap inside the ice cover, it's 29% after 10 days and 13.8% after 15 days.

It can be seen clearly that oil behavior is highly affected by sea ice distribution because ice usually resists oil spreading. These results suggest that the complicated

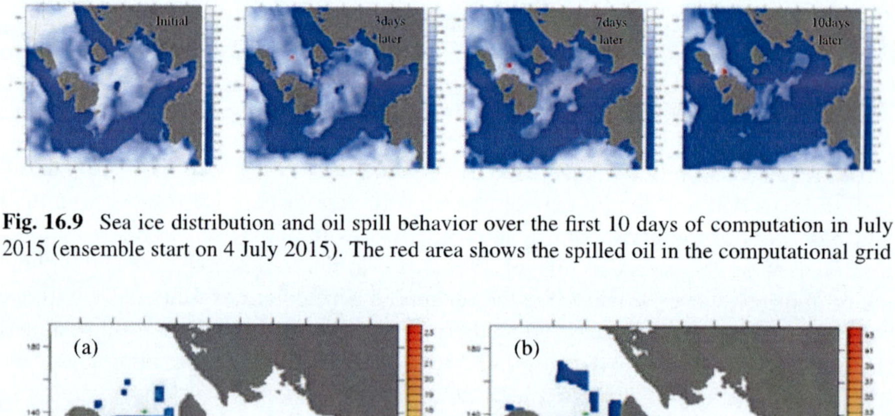

Fig. 16.9 Sea ice distribution and oil spill behavior over the first 10 days of computation in July 2015 (ensemble start on 4 July 2015). The red area shows the spilled oil in the computational grid

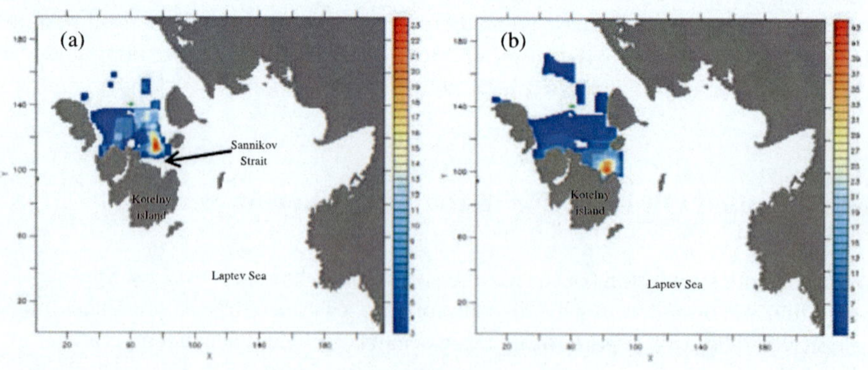

Fig. 16.10 Probability distribution of oil (in %) after **a** 10 days and **b** 15 days. The simulation was calculated with 30 ensembles from 1–30 July 2015. Note that the color scales differ

ocean structure in the Laptev Sea may lead to significantly different results with even slight differences in the initial conditions of oil spill events. Finally, repeated computations with various ice and weather conditions in different areas can produce a hazard map. This map indicates areas where ships should exercise particular attention with reduced speed or avoid altogether.

16.5 Conclusions

Data from published oil spill experiments using different sea ice concentrations were compared with our oil spill numerical model. Overall, our model showed good agreement with the experimental results. We also performed numerical simulations of sea ice and spilled oil behavior in the ice-covered East Siberian Sea. We showed that if oil is spilled in the East Siberian Sea, it can reach the coastal area of Kotelny Island in a few days. Thus, to accomplish high accuracy forecasting of spilled oil behavior, we have to use highly accurate sea ice simulations from IcePOM.

Acknowledgements The authors acknowledge support from the Green Network of Excellence Program-Arctic Climate Change Research Project (GRENE-Arctic) and Arctic Challenge for Sustainability Research Project (ArCS) by the Japanese Ministry of Education, Culture, Sports, Science and Technology, as well as a Kakenhi grant (no. 26249133) from the Japan Society for the Promotion of Science. The authors acknowledge the Arctic Data archive System (ADS) for providing the gridded AMSR2 data. The authors thank Yihao Hong for his useful comments, and Kara Bogus, PhD, from Edanz Group (https://en-author-services.edanzgroup.com/) for editing a draft of this manuscript. The authors' gratitude is extended to anonymous reviewers for their helpful comments and suggestions.

References

M. Afenyo, B. Veitch, F. Khan, A state-of-the-art review of fate and transport of oil spills in open and ice-covered water. Ocean Eng. **119**, 233–248 (2015)

P.D. Boehm, D.S. Page, J.S. Brown, J.M. Neff, J.R. Bragg, R.M. Atlas, Distribution and weathering of crude oil residues on shorelines 18 years after the Exxon Valdez spill. Environ. Sci. Technol. **42**(24), 9210–9216 (2008). http://www.ncbi.nlm.nih.gov/pubmed/19174894. Accessed 13 Sept 2016

L.W.A. De Silva, H. Yamaguchi, J. Ono, Ice–ocean coupled computations for sea-ice prediction to support ice navigation in Arctic sea routes. Polar Res. **34**(1), 25008 (2015). https://doi.org/10.3402/polar.v34.25008

S. Díez, E. Jover, J.M. Bayona, J. Albaigés, Prestige oil spill. III. Fate of a heavy oil in the marine environment. Environ. Sci. Technol. **41**(9), 3075–3082 (2007). https://doi.org/10.1021/es0629559

T.K. Fannelop, G.D. Waldman, Dynamics of oil slicks. AIAA J. **10**(4), 506–510 (1972). https://doi.org/10.2514/3.50127

J.A. Fay, (1971) Physical processes in the spread of oil on a water surface. Int. Oil Spill Conf. Proc. **1**, 463–467 (1971). https://doi.org/10.7901/2169-3358-1971-1-463

D.P. French-McCay, Oil spill impact modeling: development and validation, *Environmental Toxicology and Chemistry*, **23**(10), 2441–2456 (2004). http://www.ncbi.nlm.nih.gov/pubmed/15511105. Accessed: 13 Sept 2016

J.K.O. Gjosteen, S. Loset, Laboratory studies of oil spreading in broken ice, in *Proceedings of the 16th IAHR International Symposium on Ice* (Dunedin, New Zealand, 2002), pp. 110–119

K. Hara, H. Yamaguchi, G. Sagawa, A numerical simulation of spilled oil behaviors of ice-covered sea in the Sea of Okhotsk, in *Proceedings of the 23rd International Symposium on Okhotsk Sea and Sea Ice* (Mombetsu, Japan, 2008), pp. 108–113

K. Izumiyama, A. Konno, S. Sakai, Experimental study on the spreading of oil under ice covers, in *Proceedings of the Twelfth International Offshore and Polar Engineering Conference* (Kitakyushu, Japan, 2002), pp. 821–826

K. Izumiyama, S. Uto, S. Narita, R. Tasaki, Effect of interfacial tension on the spreading of oil under an ice cover, in *Ice in Surface Waters*, ed. by Shen (1998), pp. 419–426

S. Kelly, E. Popova, Y. Aksenov, R. Marsh, A. Yool, Lagrangian modeling of Arctic Ocean circulation pathways: impact of advection on spread of pollutants. J. Geophys. Res. Ocean. **123**, 2882–2902 (2018). https://doi.org/10.1002/2017JC013460

D. Mackay, M. Medir, D.E. Thornton, Interfacial behaviour of oil under ice. Can. J. Chem. Eng. **54**(1–2), 72–74 (1976). https://doi.org/10.1002/cjce.5450540110

C.-K. Rheem, H. Yamaguchi, A numerical simulation model for oil spills in an ice-covered sea (in Japanese). SEISAN KENKYU **56**(6), 483–486 (2004). https://doi.org/10.11188/SEISANKENKYU.56.483

C.K. Rheem, H. Yamaguchi, H. Kato, Distributed mass/discrete floe model for pack ice rheology computation. J. Mar. Sci. Technol. **2**(2), 101–121 (1997). https://doi.org/10.1007/BF02491524

P. Sebastião, C.G. Soares, Modelling the fate of oil spills at sea. Spill Sci. Technol. Bull. **2**(2/3), 121–131 (1995). https://doi.org/10.1016/S1353-2561(96)00009-6

K. Terashima,*A study of numerical models for oil diffusion in ice-covered seas* (in Japanese). (University of Tokyo, Tokyo, Japan (Graduation Thesis), 2003)

J. Wilkinson, C.J. Beegle-Krause, K.U. Evers, N. Hughes, A. Lewis, M. Reed, P. Wadhams, Oil spill response capabilities and technologies for ice-covered Arctic marine waters: A review of recent developments and established practices. Ambio **46**(3), 423–441 (2017). https://doi.org/10.1007/s13280-017-0958-y

D. Yamaguchi, *Research on a numerical simulation model of spilled oil behavior in an ice-covered sea* (in Japanese). Nihon University, Tokyo, Japan (Graduation Thesis) (2007)

P.D. Yapa, T. Chowdhury, Spreading of oil spilled under ice. J. Hydraul. Eng. **116**(12), 1468–1483 (1990). https://doi.org/10.1061/(ASCE)0733-9429(1990)116:12(1468)